CZECHOSLOVAK ACADEMY OF SCIENCES

TRANSACTIONS

of the

NINTH PRAGUE CONFERENCE

on

INFORMATION THEORY, STATISTICAL DECISION FUNCTIONS, RANDOM PROCESSES

held at

Prague, from June 28 to July 2, 1982

VOLUME A

1983

D. REIDEL PUBLISHING COMPANY

DORDRECHT / BOSTON / LANCASTER

The Library of Congress Cataloged the First Issue of this Title as Follows:

Conference on Information Theory, Statistical Decision Functions, Random Processes.

Transactions. 1 st- conference; 1956-
Prague, Publishing House of the Czechoslovak Academy of Sciences.
v. diagrs. 25 cm. (Československá akademie věd. Sekce technická.
Studie a prameny, sv. 16
English, Russian, French, and German.
1. Information theory-Congresses. 2. Statistical decision. 3. Stochastic processes.
QA273.C743 519 58-42106

ISBN 978-94-009-7015-1 ISBN 978-94-009-7013-7 (eBook)
DOI 10.1007/ 978-94-009-7013-7

TRANSACTIONS include contributions of authors reprinted directly by a photographic method.
For this reason the authors are fully responsible for the correctness of their text.

Sold and distributed in the U.S.A. and Canada by Kluwer Boston Inc., 190 Old Derby Street, Hingham, MA 02043, U.S.A.

Published by D. Reidel Publishing Company, P. O. Box 17, 3300 AA Dordrecht, Holland, in co-edition with ACADEMIA, Publishing House of the Czechoslovak Academy of Sciences, Prague

Softcover reprint of the hardcover 1st edition 1983

CONTENTS

Preface 9

Invited Papers

Helstrom C. W. : Performance of Exotic Quantum Signals in Free-Space Optical Communications 13

Jurečková J.: Robust Estimators of Location and Regression Parameters and Their Second Order Asymptotic Relations ... 19

Jurek Z. J.: Generators of Some Classes of Probability Measures on Banach Spaces 33

Kramosil I.: Monte-Carlo Methods from the Point of View of Algorithmic Complexity 39

Perez A.: Discrimination Rate Least Favorable Pairs of Distributions for ε-Contaminated Statistical Hypotheses or with f-Divergence Like Neighborhoods 53

Révész P.: On the Local Time of Brownian Bridge 67

Rieder H.: Robust Estimation of One Real Parameter when Nuisance Parameters Are Present 77

Rosenmüller J.: Discrete Methods in Cooperative Game Theory 91

Vajda I.: A New General Approach to Minimum Distance Estimation 103

Žilinskas A.: On Models of Complicated Functions under Uncertainty 113

Communications

Ahlbehrendt N., Draeger U.: An Attempt to Solve Approximately the Optimal Estimation Problem for Markov Processes by Expansion of the A-Posteriori Density in an Edgeworth Series 121

Anděl J.: Marginal Distributions of Autoregressive Processes 127

Антамошкин А.: Об оптимальных алгоритмах оптимизации функционалов с булевыми переменными 137

Basar T.: An Equilibrium Theory for Multi-Person Multi--Criteria Stochastic Decision Problems with Multiple Subjective Probability Measures 143

Benassi A.: Processus D´Ornstein-Uhlenbeck généralisé. Mesures stationnaires dans le cas gaussien........................ 151

Bretthauer G.: Application of the Statistical Decision Theory to System Identification 157

Butnariu D.: Computing Fixed Points for Fuzzy Mappings 165

Campbell L. L.: Information Submartingales 171

Хашимов Ш. А.: О квадратической мере отклонения оценки линии регрессии 177

Demongeot J.: Coupling of Markov Processes and Holley´s Inequalities for Gibbs Measures. Statistical Application of Gibbs Measures 183

Domański C., Tomaszewicz A.: An Empirical Power of Some Tests for Linearity 191

Drăguţ M.: The Optimal Control of Partially Observable Semi--Markov Processes over the Infinite Horizon: Discounted Costs 199

Durgaryan I. S., Pashchenko F. F.: Information Methods in Identification 207

Dvurečenskij A., Pulmannová S.: Quantum Stochastic Processes 215

Elia M.: Symbol Error Rate of Binary Block Codes 223

Ericson T.: Binary Communication over a Channel Subject to Active Interference 229

Fischer T.:Generalizations of the Maximum Entropy Principle and Their Applications 235

Gerencsér L.: A Modification of the Extended Kalman Filter Algorithm with Application in Hydrology 241

Gerstenkorn T.: Relations between the Crude, Factorial and Inverse Factorial Moments 247

Gillert H.: The Bernstein-von Mises Theorem for Non-Stationary Markov Processes 253

Girlich H.-J., Küenle H.-U.: On Dynamic Min-Max Decision Models 257

Grandell J.: Estimation of Precipitation Characteristics from Time-Integrated Data 263

Halilov A., Mirzahmedov M. A.: The Central Limit Theorem for Statistics of a Spectral Density with Time Shift 269

Harman B., Riečan B.: On the Martingale Convergence Theorem in Quantum Theory 275

Havránek T.: Some Complexity Considerations Concerning Hypotheses in Multidimensional Contingency Tables 281

Idzik A.: Theorems on Selectors in Topological Spaces I ... 287

Idzik A., Simonsen P. B.: A Game-Theoretic Arrow-Debreu Model 293

Ignatov Z.: Asymptotic Results for an Epidemic Process on Random Graphs 301

Ihara S.: Maximum Entropy Spectral Analysis and Arma Processes II 307

Janžura M.: Discrete Finite State Random Fields and Their Reduced Versions as Information Sources 313

Jiroušek R.: Strategical Test - a Generalization of the Wald´s Sequential Test 319

Kaňková V.: Sequences of Stochastic Programming Problems with Incomplete Information 327

PREFACE

The Ninth Prague Conference on Information Theory, Statistical Decision Functions, and Random Processes was organized by the Institute of Information Theory and Automation of the Czechoslovak Academy of Sciences from June 28 to July 2, 1982. Similarly as the preceding Prague Conferences, during their twenty six years history, it provided a space for the presentation and discussion of recent scientific results, as well as for personal contacts of many scientists both from abroad and from Czechoslovakia. Nearly 150 specialists from 17 countries participated in the Conference and they read more than 100 papers (including 18 invited ones), 88 of which have been published in the present two volumes of the Transactions of the Conference. Namely invited papers, having been read by outstanding specialists, have brought invaluable offer for participants to create themselves an orientation in the modern trends of the above mentioned scientific branches.

Allow us to use this opportunity to express our sincere thanks to all who have contributed to the success of the Conference, especially to those who prepared and presented papers.

Our gratitude is also due to Academician Jaroslav Kožešník, the scientific editor of the Transactions, and to the editorial board for reviewing all papers and fulfilling many printing management duties.

We would like to appreciate a great work of all colleagues from the Institute of Information Theory and Automation, who participated in the preparation and in the organization of the Conference.

As many of regular participants of the Prague Conferences remember, the Transactions of the Eighth Prague Conference were published before the Conference to be available for the participants. Nevertheless, it has required to stipulate a deadline of papers rather long before the Conference and so considerable differences between the content of submitted papers and lectures read on the Conference occured. Therefore, taking into consideration all the advantages and disadvantages of such publication, the Organizing Committee decided to publish these Transactions after the Conference.

To ensure the publication of them in the shortest possible time, they were printed by offset. It implied the necessity of a little greater care in preparation of the manuscripts delivered by participants and the Organizing Committee sincerely thanks to all that contributed to these Transactions for their effort connected with it.

We also thank Academia Publishing House for prompt printing the Transactions.

ORGANIZING COMMITTEE
of the
NINTH PRAGUE CONFERENCE
on
INFORMATION THEORY,
STATISTICAL DECISION FUNCTIONS,
RANDOM PROCESSES

INVITED PAPERS

PERFORMANCE OF EXOTIC QUANTUM SIGNALS IN FREE-SPACE OPTICAL COMMUNICATIONS

Carl W. Helstrom
La Jolla

Key words: optical communications, quantum communication theory, coherent states, binary channel, error performance.

ABSTRACT

The performance of integral-quantum and two-photon-coherent-state signals in free-space optical communications is compared with that of ordinary coherent signals. The farther apart the transmitter and receiver, the smaller the advantage of the former over the latter.

A simple classical model of a free-space optical channel contains a transmitter that sends out light energy by exciting the electromagnetic field in a planar aperture A. For the sake of efficiency only a single spatio-temporal mode of that field is excited, and its complex amplitude we designate by α, in such units that the energy in the mode equals $h\nu|\alpha|^2$; h = Planck's constant, ν = the frequency of the light field. At some great distance R is the aperture of a receiver, lying in a plane parallel to the aperture of the transmitter and coaxial with it. We assume that the receiver is so far from the transmitter that a single normal mode of the magnetic field of its aperture is excited. The complex amplitude of the receiving mode is $\beta = \kappa\alpha$, where κ is the amplitude transmittance of the channel. The energy transmittance $|\kappa|^2$ is roughly given by

$$|\kappa|^2 \approx A_t A_r/(\lambda R)^2,$$

where A_t is the area of the transmitter aperture, A_r is the area of the receiver aperture, and $\lambda = c/\nu$ is the wavelength.

Interest has lately arisen in the potentialities of such a channel when the principal mode of the transmitting aperture is placed in one of a class of special quantum states. If that could be done, lower average probability of error could be attained than when the state of the transmitter is an ordinary coherent state, which corresponds to a simple classical oscillation of the transmitting mode and to the emission of a pulse from an ideal laser. Here we shall examine those potentialities; we shall show that when, as usually, the transmitter and receiver are very far apart, exotic quantum states lose their advantage; and we shall try to explain why their performance deteriorates in this way.

The electromagnetic field in the plane of the transmitting aperture, but outside it, can also be decomposed into normal modes in a complementary manner, so that we can consider a particular one of those external modes as the source of the thermal noise that enters the aperture of the receiver. We call this external mode a "parasitic" mode and denote its complex amplitude by γ. The field in the plane of the receiving aperture is decomposed into a correlative set of modes, and a single such external mode can be thought of as taking up the transmitted energy that misses the aperture of the receiver. It too is a parasitic mode, and we denote its complex amplitude by δ. Then conservation of energy and reciprocity dictate the relations

$$\beta = \kappa\alpha + \kappa'\gamma, \quad \delta = -\kappa'\alpha + \kappa\gamma, \tag{1}$$

between these modal amplitudes, with $\kappa' = (1 - |\kappa|^2)^{\frac{1}{2}}$,

At optical frequencies it is necessary to treat the electromagnetic field in terms of quantum mechanics. The complex amplitudes α, γ, β, and δ become nonselfadjoint operators, related by the counterparts of (1),

$$b = \kappa a \otimes \underset{\sim}{1}' + \kappa' \underset{\sim}{1} \otimes c, \quad d = -\kappa' a \otimes \underset{\sim}{1}' + \kappa \underset{\sim}{1} \otimes c, \tag{2}$$

where $\underset{\sim}{1}$ and $\underset{\sim}{1}'$ are identity operators in the Hilbert spaces $\mathcal{H}_0$ and $\mathcal{H}_0'$ associated with the transmitting mode and its parasite. Here a and c are photon annihilation operators for the transmitting mode and its parasite, their adjoints a^+ and c^+ are photon creation operators, and the operators a^+a and c^+c determine the numbers of photons in those modes. These operators and their adjoints obey commutation relations of the form

$$\begin{aligned} aa^+ - a^+a = [a, a^+] = \underset{\sim}{1}, \quad [c, c^+] = \underset{\sim}{1}', \\ [b, b^+] = \underset{\sim}{1} \otimes \underset{\sim}{1}', \quad [d, d^+] = \underset{\sim}{1} \otimes \underset{\sim}{1}', \end{aligned} \tag{3}$$

and they otherwise commute among themselves (Takahasi, 1965).

The operators b and d acting on the vacuum state of the system generate Hilbert spaces $\mathcal{H}$ and $\mathcal{H}'$ associated with the receiving mode and its parasite; b and d annihilate photons in those modes, and b^+ and d^+ create them. The state of the receiving mode, represented by the density operator ρ_r, is determined from the states ρ_0 of the transmitting mode and ρ_0' of its parasite by

$$\rho_r = \mathrm{Tr}_{\mathcal{H}'}(\rho_0 \otimes \rho_0'), \tag{4}$$

where $\mathrm{Tr}_{\mathcal{H}'}$ indicates a trace with respect to the states in $\mathcal{H}'$. This trace operation expresses the unobservability of the field outside the receiving aperture, and it has important consequences: even though ρ_0 and ρ_0' may represent pure quantum states, the received state ρ_r is in general mixed; that is, it suffers a statistical uncertainty corresponding to a kind of random noise. How to use (4) to calculate the received state ρ_r has been shown by Takahasi (1965), Helstrom (1967), Yuen and Shapiro (1978), and Helstrom (1979b).

If, for instance, the transmitted state ρ_0 contains an integral number M of photons, $\rho_0 = |M\rangle\langle M|$, where $a^+a|M\rangle = M|M\rangle$, and if its parasite is in the ground state, $\rho_0' = |0\rangle\langle 0|$, thermal noise being absent, then the received state ρ_r contains a random number of photons having a binomial distribution with mean $|\kappa|^2 M$; photons randomly enter the receiving aperture with probability $|\kappa|^2$ and miss it with probability $1 - |\kappa|^2$ (Takahasi, 1965).

The only kind of pure transmitted state that does not suffer from this "partition" phenomenon is the coherent state $\rho_0 = |\alpha_0\rangle\langle\alpha_0|$, where $|\alpha_0\rangle$ is a right eigenvector of the photon annihilation operator a with arbitrary complex eigenvalue α_0; putting the transmitter into the state $\rho_0 = |\alpha_0\rangle\langle\alpha_0|$ is equivalent to transmitting a pulse from an ideal laser (Glauber, 1963). The state it produces in the receiving aperture is the coherent state $\rho_r = |\kappa\alpha_0\rangle\langle\kappa\alpha_0|$ when thermal noise is absent. That these are the only states not affected by the partition phenomenon has been shown by Helstrom (1981).

It is instructive to compare communication systems utilizing exotic quantum states with those transmitting ordinary coherent states such as $|\alpha_0\rangle\langle\alpha_0|$. Two kinds of transmitted states have been extensively analyzed, the integral-quantum states such as $|M\rangle\langle M|$ and the generalized coherent states, also known as two-photon coherent states (TCS) (Yuen and Shapiro, 1978). These are the right eigenvectors of the operator $\mu a + \nu a^+$, $\mu^2 - \nu^2 = 1$, and they reduce to ordinary coherent states when $\mu = 1$. In a binary on-off communication system sending 0's and 1's, each '0' is transmitted by leaving the transmitting aperture in the ground state $|0\rangle\langle 0|$; each '1' is transmitted by putting the aperture mode into a TCS.

Preliminary calculations of the average error probability P_e in such a TCS system in the absence of thermal noise showed it to be minimum for a certain choice of the parameter $\mu > 1$, and that minimum error probability is less than that incurred by a system transmitting ordinary coherent states ($\mu = 1$) (Helstrom, 1979a).

As the transmitter recedes from the receiver, however, and $|\kappa|^2 \to 0$, with the average number of received photons kept fixed, the optimum value of μ decreases toward 1, and the minimum error probability rises to that for $\mu = 1$. A more extensive comparison has now been carried out, in which the average error probability P_e was kept fixed, and the minimum average number S_r of received photons needed to attain it was calculated as a function of the energy transmittance $|\kappa|^2$ and compared with the average number S_c needed when ordinary coherent signals are transmitted (Helstrom, 1983). In both cases the phase of the received signal was assumed known, the optimum quantum detector was utilized, and various amounts of thermal noise were supposed present at the receiver. It was found that although $S_c > S_r$, the ratio S_c/S_r goes to 1 as $|\kappa|^2 \to 0$ as a result of the effect of the partition phenomenon on the TCS signals. The presence of thermal noise further reduces the relative advantage of the TCS system.

Other systems investigated included the on-off binary TCS system on a channel of random phase arg κ, for which the optimum detector counts photons, a system transmitting antipodal TCS signals, and one transmitting integral-quantum signals $|M\rangle\langle M|$. In all cases their relative advantage over a system transmitting ordinary coherent signals vanishes as the channel transmittance $|\kappa|^2$ decreases to zero, the average transmitted energy increasing at the same time so as to maintain the same average probability P_e of error (Helstrom, 1983).

With this increase in transmitted energy as $|\kappa|^2 \to 0$ goes an increased average number of photons per mode, and quantum fields with a large number of photons per mode behave much like classical fields. The quantum counterpart of a classical field is one in a pure coherent state or in a statistical mixture of such states. Indeed, for any quantum transmitted state ρ_0, when $|\kappa|^2 << 1$, the received state ρ_r in the absence of thermal noise can be shown to lie close to a classical state ρ_r' whose P-representation is

$$\rho_r' = \int P_r'(\alpha)\,|\alpha\rangle\langle\alpha|\,d^2\alpha, \quad P_r'(\alpha) = |\kappa|^{-2}\langle\alpha/\kappa|\rho_0|\alpha/\kappa\rangle/\pi,$$

in the sense that $\mathrm{Tr}\,(\rho_r - \rho_r')^2 = o(|\kappa|^2)$ (Helstrom, 1979b). This state ρ_r, like its approximation ρ_r', is a mixed state as a result of the partition phenomenon, which reduces its detectability as compared with that of the pure coherent received state $|\kappa\alpha_0\rangle\langle\kappa\alpha_0|$, which results from the transmitted state $|\alpha_0\rangle\langle\alpha_0|$. It is for these reasons that the relative advantage of a system transmitting exotic quantum signals vanishes as $|\kappa|^2 \to 0$, and for free-space communication over great distances such signals cannot be expected to be significantly superior to ordinary coherent signals (Helstrom, 1980).

REFERENCES

Glauber, R. J. (1963): Coherent and incoherent states of the radiation field. Phys. Rev. 131, 2766-2788.

Helstrom, C. W. (1967): Quasi-classical analysis of coupled oscillators. J. Math. Physics 8, 37-42.

(1979a): Optimum quantum decision between a pure and a mixed state with application to detection of TCS signals. Trans. IEEE, IT-25, 69-76.

(1979b): Quasiprobability distributions and the analysis of the linear quantum channel with thermal noise. J. Math. Physics 20, 2063-2068.

(1980): Nonclassical states in optical communication to a remote receiver. Trans. IEEE, IT-26, 378-382.

(1981): The conversion of a pure state into a statistical mixture by the linear quantum channel. Optics Comm., 37, 175-177.

(1983): Comparative performance of quantum signals in unimodal and bimodal binary optical communications. Trans. IEEE, vol. IT-29, no. 1.

Takahasi, H. (1965): Information theory of quantum-mechanical channels. In: Advances in Communication Systems, A. Balakrishnan (ed.), 1, 227-310, McGraw-Hill, New York.

Yuen, H. P. (1978): Optical communication with two-photon coherent states -- Shapiro, J. H. Part I: Quantum state propagation and quantum noise reduction. Trans. IEEE, IT-24, 657-668.

Department of Electrical Engineering &
Computer Sciences, C-014
University of California, San Diego
La Jolla, California, 92093, U.S.A.

ROBUST ESTIMATORS OF LOCATION AND REGRESSION PARAMETERS AND THEIR SECOND ORDER ASYMPTOTIC RELATIONS

Jana Jurečková
Prague

Key words : Huber's estimator of location and regression parameters, α - trimmed mean, α - trimmed least-squares estimator, one-step version of M-estimator

---

ABSTRACT

Second order asymptotic relations-namely the orders of asymptotic equivalence - of some pairs of robust estimators are studied. More specifically, we shall study the relations of α-trimmed mean to Huber's estimator of location and of α-trimmed least-squares estimator to Huber's estimator of regression, respectively, and the order of the approximation of an M-estimator by its one-step version.

1. INTRODUCTION

Let $X_{n1},\dots,X_{nn}$ be independent random variables, X_{ni} distributed according to the distribution function $F(x-\sum_{j=1}^{p} c_{ij}\,\theta_j)$, $i = 1,\dots,n$. The problem is that of estimating the parameter $\underset{\sim}{\theta} = (\theta_1,\dots,\theta_p)$ in the situation that F is not fully specified; it is only assumed that F belongs to a system $\mathcal{F}$ of absolutely continuous distribution functions(d.f.) satisfying some general regularity conditions.

Among various robust estimators, less sensitive to deviations from a specific distribution shape, three broad classes play the most important role: M-estimators (estimators of maximum likelihood type, originated by Huber (1964)), R-estimators (estimators based on rank tests, originated by Hodges and Lehmann (1963)) and L-estimators (estimators based on ordered observations). All these estimators follow the same idea: to reduce the influence of the extreme values of observations and yet to estimate well the parameter $\underset{\sim}{\theta}$ even if, for instance, the basic distribution is normal.

Jaeckel (1971) observed the close relations of the above types of estimators in the location case which hold asymptotically if number n of observations increases. He proved that, if the weight function generating the respective estimators are related to each other in a special way, the difference of L- and M-estimators is of the order $O_p(n^{-1})$. However, his Theorem 2 does not cover the asymptotic relation of the famous pair of robust estimators, consisting of Huber's M-estimator and of the trimmed mean (Huber's function ψ does not admit the expansion (1) in Jaeckel (1971)). Anyway, his conjecture of the asymptotic equivalence of the order $O_p(n^{-1})$ of Huber's estimator and of the trimmed mean appears to be right (see Section 2 of the present paper for the proof).

Recently Koenker and Bassett (1978) introduced the concept of regression quantile and proposed the trimmed least-squares estimator of regression parameter as an extension of the trimmed mean to linear model. This estimator was later studied by Ruppert and Caroll (1978), (1981). We shall prove (see Section 3), utilizing the results of Ruppert and Caroll and of Jurečková and Sen (1981b), that the trimned least-squares estimator is asymptotically equivalent to Huber's estimator of regression parameter with the order $O_p(n^{-3/4})$.

Asymptotic relations of other pairs of estimators are studied in Jurečková (1977), (1982) and in Hušková and Jurečková (1981). In all cases we encounter either $O_p(n^{-1})$ or $O_p(n^{-3/4})$ as the order of asymptotic equivalence; the lower order appears if the functions generating the M- or R-estimators have jump-discontinuities or if the L-estimator involves single sample quantiles.

Bickel (1973) suggested a one-step version M_n^* of the M-estimator M_n in the linear model which is asymptotically equivalent to M_n. We shall also provide the order of the approximation of M_n by M_n^* (see Section 4).

2. HUBER'S ESTIMATOR AND THE TRIMMED MEAN IN THE LOCATION CASE

Let $X_1, X_2, \ldots$ be independent random variables, identically distributed according to the d.f. $F(x-\theta)$; assume that F is absolutely continuous and symmetric, i.e. $F(x)+F(-x)=1$, $x \in R^1$.

Let $L_n(\alpha)$ denote the α-trimmed mean,

(2.1) $$L_n(\alpha) = (n-2[n\alpha])^{-1} \sum_{i=[n\alpha]+1}^{n-[n\alpha]} X_{n:i}$$

where $0<\alpha<\frac{1}{2}$ and $[x]$ denotes the largest integer k satisfying $k\leq x$; $X_{n:1}\leq \ldots \leq X_{n:n}$ are the order statistics corresponding to $X_1,\ldots,X_n$. Let M_n be Huber's M-estimator, defined as

(2.2) $$M_n = \tfrac{1}{2}(M_n^+ + M_n^-)$$

with

(2.3) $$M_n^+ = \inf\left\{t : \sum_{i=1}^n \psi(X_i-t) < 0\right\}$$
$$M^- = \sup\left\{t : \sum_{i=1}^n \psi(X_i-t) > 0\right\}$$

and

(2.4) $$\psi(x) = \begin{cases} x & \text{if } |x|\leq c \\ c.\text{sign}x & \text{if } |x|>c; \end{cases}$$

$c>0$ is a constant.

There has been some confusion in the history of the problem of asymptotic relations of $L_n(\alpha)$ and M_n. One might intuitively expect that the Winsorized mean rather than the trimmed mean resembles Huber's estimator (cf. Huber (1964)). Bickel (1965) was apparently the first who recognized the close connection between Huber's estimator and the trimmed mean. Jackel (1971) studied the asymptotic relations of M- and L-estimators but, in fact, his theorem does not cover this special case.

The following theorem shows that, with a proper choice of the constant c in (2.4), the difference of both estimators is of the order $O_p(n^{-1})$ as $n\to\infty$.

THEOREM 2.1. Let $X_1,X_2,\ldots$ be independent random variables, identically distributed according to the d.f. $F(x-\theta)$ such that $F(x)+F(-x)=1$, $x\in R^1$ and which satisfies the following conditions:

(i) F has an absolutely continuous density and finite Fisher's information, i.e.

$$\mathcal{I}(f) = \int (f'/f)^2\, dF < \infty \; .$$

(ii) $f(x)>a>0$ for all x such that
$\alpha-\varepsilon \leq F(x)\leq 1-\alpha+\varepsilon$, $0<\alpha<\frac{1}{2}$, $\varepsilon>0$.

(iii) f' exists in the interval $F^{-1}(\alpha-\varepsilon)$, $F^{-1}(1-\alpha+\varepsilon)$.

Then

(2.5) $$L_n(\alpha) - M_n = O_p(n^{-1}), \quad \text{as} \quad n\to\infty \quad ,$$

where $L_n(\alpha)$ is the α-trimmed mean (2.1) and M_n is the M-estimator defined in (2.2)-(2.4) with $c = F^{-1}(1-\alpha)$.

P r o o f . We may put $\theta = 0$ without a loss of generality. It follows from Theorem 3.2 of Jurečková (1980) that

$$M_n = (1-2\alpha)^{-1}\Big\{n^{-1}\sum_{i=1}^{n} X_{n:i}\, I\big[F^{-1}(\alpha) \le X_{n:i} \le F^{-1}(1-\alpha)\big]$$
$$+ \; n^{-1}F^{-1}(\alpha)\sum_{i=1}^{n} I\big[X_{n:i} < F^{-1}(\alpha)\big]$$
$$+ \; n^{-1}F^{-1}(1-\alpha)\sum_{i=1}^{n} I\big[X_{n:i} > F^{-1}(1-\alpha)\big]\Big\} + O_p(n^{-1})$$

(2.6)
$$= (1-2\alpha)^{-1}\Big\{n^{-1}\sum_{i=i_n}^{j_n} X_{n:i}$$
$$+ F^{-1}(1-\alpha)\big[1-F_n(F^{-1}(1-\alpha)) - F_n(F^{-1}(\alpha)-)\big]\Big\}$$

where

(2.7) $$i_n = n F_n(F^{-1}(\alpha)-)+1, \qquad j_n = n F_n(F^{-1}(1-\alpha)).$$

(2.7) implies that

(2.8) $$X_{n:i} = X_{n:n-[n\alpha]} + O_p(n^{-1/2})$$

for each integer i between $n-[n\alpha]$ and j_n and

(2.9) $$X_{n:i} = X_{n:[n\alpha]+1} + O_p(n^{-1/2})$$

for i being an integer between $[n\alpha]+1$ and i_n. Thus, combining (2.6)-(2.9), we get

$$(1-2\alpha)(M_n - L_n(\alpha))$$
$$= F^{-1}(1-\alpha)\big[1-F_n(F^{-1}(1-\alpha)) - F_n(F^{-1}(\alpha)-)\big]$$

(2.10)
$$-\big[X_{n:[n\alpha]+1} + O_p(n^{-1/2})\big]\big[F_n(F^{-1}(\alpha)-)-\alpha+O_p(n^{-1/2})\big]$$
$$+\big[X_{n:n-[n\alpha]} + O_p(n^{-1/2})\big]\big[F_n(F^{-1}(1-\alpha))-(1-\alpha)+O_p(n^{-1/2})\big] + O_p(n^{-1})$$

and this, by Bahadur (1966), is further equal to

$$F^{-1}(1-\alpha)\big[1-F_n(F^{-1}(1-\alpha)) - F_n(F^{-1}(\alpha)-)\big] -$$

$$-\left\{F^{-1}(\alpha) + [f(F^{-1}(\alpha))]^{-1}\left[\alpha - F_n(F^{-1}(\alpha))\right] + 0_p(n^{-1/2})\right\}\cdot$$
$$\cdot\left[F_n(F^{-1}(\alpha)-) - \alpha + 0_p(n^{-1/2})\right]$$
$$+\left\{F^{-1}(1-\alpha) + [f(F^{-1}(\alpha))]^{-1}\left[1-\alpha - F_n(F^{-1}(1-\alpha))\right] + 0_p(n^{-1/2})\right\}\cdot$$
$$\cdot\left[F_n(F^{-1}(1-\alpha)) - (1-\alpha) + 0_p(n^{-1/2})\right] + 0_p(n^{-1}) = 0_p(n^{-1}).$$

Q.E.D.

3. HUBER'S ESTIMATOR AND THE TRIMMED LEAST SQUARES ESTIMATOR IN THE REGRESSION CASE

We shall now study the asymptotic relation of Huber's estimator of regression parameter to the trimmed least square estimator, suggested by Koenker and Bassett (1978). We shall show that with a proper choice of c in (2.4), both estimators are asymptotically equivalent with the order $0_p(n^{-3/4})$; this relation will be proved for the case of two-parameters regression model, but an analogous conclusion may be apparently done for a more general regression model.

Let $X_{n1},\ldots,X_{nn}$ be independent observations, X_{ni} distributed according to the d.f. $F(x-\beta_1-c_{ni}\beta_2)$, $i=1,\ldots,n$, where F is an abeolutely continuous and symmetric d.f.

Let $\underset{\sim}{M}_n = (M_{n1},M_{n2})'$ be Huber's estimator of $\underset{\sim}{\beta} = (\beta_1,\beta_2)'$, defined as a solution of the system of equations (we write X_i, c_i instead of X_{ni}, c_{ni}, etc.)

$$\sum_{i=1}^{n} \psi(X_i - t_1 - c_i t_2) = 0$$

(3.1)

$$\sum_{i=1}^{n} c_i \psi(X_i - t_1 - c_i t_2) = 0$$

with respect to t_1, t_2, where ψ is the function of (2.4). The question is whether there exists an appropriate L-estimator counterpart of $\underset{\sim}{M}_n$.

The L-estimators are computionally appealing and have further convenient properties in the location model (cf. Bickel and Lehmann (1975)). However, they do not extend to the linear model in a straightforward way. A possible regression analog of L-estimators was suggested and studied by Bickel (1973). His estimators, defined in the two-step way with the aid of an initial estimator, have good efficiency properties but they are computionally complex.

Recently Koenker and Bassett (1978) extended the concept of

quantiles to the linear model in the following way: for $\alpha \in (0,1)$, denote

(3.2) $$\psi_\alpha(x) = \begin{cases} \alpha & \text{if } x \geq 0 \\ \alpha-1 & \text{if } x < 0 \end{cases}$$

and

(3.3) $$\rho_\alpha(x) = x.\ \psi_\alpha(x), \qquad x \in R^1.$$

The α-th regression quantile $\hat{\beta}(\alpha)$ is defined as any value of $t = (t_1, t_2)'$ which solves

(3.4) $$\sum_{i=1}^n \rho_\alpha(X_i - t_1 - c_i t_2) := \min.$$

Notice that $\hat{\beta}(\alpha)$ is scale-equivariant and it is, in fact, an M-estimator. Koenker and Bassett (1978) then proposed the α-trimmed least-squares estimator $L_n(\alpha)$, $0 < \alpha \leq 1/2$, in the following way: Assume that there is a rule which selects a unique solution of (3.4). $L_n(\alpha)$ is defined as the least-squares estimator calculated after removing all observations satisfying

(3.5) $$X_i - \hat{\beta}_1(\alpha) - c_i \hat{\beta}_2(\alpha) < 0 \quad \text{or} \quad X_i - \hat{\beta}_1(1-\alpha) - c_i \hat{\beta}_2(1-\alpha) > 0,$$

$i=1,\dots,n$. Let $a_i = 0$ or 1 according as i satisfies (3.5) or not and denote A_n the diagonal matrix with diagonal $(a_1,\dots,a_n)$. Then

(3.6) $$L_n(\alpha) = (C_n' A_n C_n)^- (C_n' A_n X_n)$$

where

(3.7) $$X_n = (X_{n1},\dots,X_{nn})', \qquad C_n' = \begin{pmatrix} 1 & \dots & 1 \\ c_{n1} & \dots & c_{nn} \end{pmatrix}$$

and $(C_n' A_n C_n)^-$ is a generalized inverse of $(C_n' A_n C_n)$.

We shall study the asymptotic relations of M_n and $L_n(\alpha)$ under the following conditions :

(A) $$\lim_{n\to\infty} (n^{-1/2} \max_{1\leq i\leq n} |c_{ni}|) = 0 \ ; \quad \sum_{i=1}^n c_{ni} = 0 \ ;$$

$$\lim_{n\to\infty} n^{-1} \sum_{i=1}^n c_{ni}^2 = b^2 > 0, \quad b^2 < \infty;$$

$$\lim_{n\to\infty} n^{-1} \sum_{i=1}^n |c_{ni}|^3 = b'^2 > 0, \quad b'^2 < \infty;$$

$$n^{-1} \sum_{i=1}^n c_{ni}^4 = O(1), \quad \text{as} \quad n \to \infty .$$

Denote

(3.8) $$\underset{\sim}{Q} = \begin{pmatrix} 1 & 0 \\ 0 & b^2 \end{pmatrix} .$$

(B) The d.f. F has an absolutely continuous symmetric density f having a finite Fisher information $\mathcal{J}(f) = \int (f'/f)^2 dF < \infty$, and satisfying

$f(x) > a > 0$ for all x satisfying

$\alpha - \varepsilon \leq F(x) \leq 1-\alpha+\varepsilon$, $0 < \alpha \leq \frac{1}{2}$, $\varepsilon > 0$;

assume that f' exists in the interval $(F^{-1}(\alpha-\varepsilon),\ F^{-1}(1-\alpha+\varepsilon))$.

The asymptotic behavior of $\underset{\sim}{L}_n(\alpha)$ was studied by Ruppert and Caroll (1978), (1980), who proved that $\underset{\sim}{L}_n(\alpha)$ is asymptotically normally distributed,

(3.9) $$n^{1/2}(\underset{\sim}{L}_n(\alpha) - \underset{\sim}{\beta}) \xrightarrow{\mathcal{D}} N_2(\underset{\sim}{0},\ \sigma^2(\alpha,F)\cdot \underset{\sim}{Q}^{-1}), \quad \text{as} \quad n \to \infty ,$$

with $\sigma^2(\alpha,F)$ being the asymptotic variance of the α-trimmed mean. The same authors also derived the following asymptotic representation of $\underset{\sim}{L}_n(\alpha)$:

(3.10)
$$n^{1/2}(L_{n1}(\alpha) - \beta_1) = n^{-1/2}(1-2\alpha)^{-1} \sum_{i=1}^{n} \psi(e_i) + o_p(1)$$
$$n^{1/2}(L_{n2}(\alpha) - \beta_2) = n^{-1/2}(1-2\alpha)^{-1} \sum_{i=1}^{n} c_i\, \psi(e_i) + o_p(1)$$

where

(3.11) $$e_i = e_{ni} = X_{ni} - \beta_1 - c_{ni}\ \beta_2$$

is the i-th residual, $i=1,\ldots,n$, and ψ is the Huber ψ-function given in (2.4) with $c = F^{-1}(1-\alpha)$. Combining (3.10) with Theorem 4.1 of Jurečková (1977) implies that

(3.12) $$n^{1/2} \| \underset{\sim}{L}_n(\alpha) - \underset{\sim}{M}_n \| \xrightarrow{P} 0 , \quad \text{as} \quad n \to \infty$$

holds for the M-estimator $\underset{\sim}{M}_n$ generated by ψ of (2.4) with $c = F^{-1}(1-\alpha)$. Thus, there is a close relation of $\underset{\sim}{L}_n(\alpha)$ and $\underset{\sim}{M}_n$, and we are interested in the order of convergence in (3.12). Depending on the (generally unknown) d.f. F, this relation does not enable to calculate the value of $\underset{\sim}{M}_n$ once we have calculated the value of the corresponding $\underset{\sim}{L}_n(\alpha)$ and vice versa; the relation rather indicates the closeness of two special classes of estimators and shows that $\underset{\sim}{L}_n(\alpha)$ is a worthwhile alternative to Huber's M-estimator.

We shall show that the rate of convergence in (3.12) is $O_p(n^{-1/4})$, so that $\|\underset{\sim}{L}_n(\alpha) - \underset{\sim}{M}_n\| = O_p(n^{-3/4})$. This order differs from that proved in the location case and the proof does not indicate that it is the best possible; but the author surmises that it is the case due to the fact that $\underset{\sim}{L}_n(\alpha)$ is, unlike in the location case, a two-step estimator.

THEOREM 3.1. Let $X_{n1},\dots,X_{nn}$, $n=1,2,\dots$ be the triangular array of independent observations with X_{ni} distributed according to the d.f. $F(x-\beta_1-c_{ni}\beta_2)$, $i=1,\dots,n$. Then, under the conditions (A) and (B),

$$\|\underset{\sim}{L}_n(\alpha) - \underset{\sim}{M}_n\| = O_p(n^{-3/4}), \quad \text{as} \quad n\to\infty, \tag{3.13}$$

where $\underset{\sim}{L}_n(\alpha)$ is the α-trimmed least-squares estimator and $\underset{\sim}{M}_n$ is Huber's estimator generated by ψ of (2.4) with $c = F^{-1}(1-\alpha)$; $0 < \alpha \leq \frac{1}{2}$.

P r o o f . The proof is based on a refinement of that of Theorem 3 of Ruppert and Caroll (1978). It follows from Jurečková and Sen (1981b), Corollary 3.2, that

$$n^{1/2}(M_{nj}-\beta_j) = n^{-1/2}(1-2\alpha)^{-1}\sum_{i=1}^{n} d_{ij}\psi(e_i)+O_p(n^{-1/4}), \tag{3.14}$$

as $n\to\infty$, with e_i defined in (3.11) and $d_{i1}=1$, $d_{i2}=c_i/b^2$, $i=1,\dots,n$. Moreover, it follows from the proof of Theorem 3 in Ruppert and Caroll (1978) that

$$n^{-1}(\underset{\sim}{C}_n' \underset{\sim}{A}_n \underset{\sim}{C}_n) \xrightarrow{p} (1-2\alpha)\underset{\sim}{Q}, \quad \text{as} \quad n\to\infty \tag{3.15}$$

and it follows from Theorem 2 of the same paper that

$$\begin{aligned}&\hat{\beta}_1(\eta) - \beta_1 + c_i(\hat{\beta}_2(\eta) - \beta_2)= \\ &= F^{-1}(\eta)+\left[f(F^{-1}(\alpha))\right]^{-1}\left[\eta-F_n(F^{-1}(\eta)-)-(c_i/b^2)F_n^*(F^{-1}(\eta)-)\right] \\ &+ o_p(n^{-1/2})\end{aligned} \tag{3.16}$$

holds for $\eta=\alpha$, $1-\alpha$, where

$$F_n(x) = n^{-1}\sum_{i=1}^{n} I\left[e_i\leq x\right],\quad F_n^*(x)=n^{-1}\sum_{i=1}^{n} c_i\, I\left[e_i\leq x\right]; \tag{3.17}$$

notice that $F_n(x)-F(x) = O_p(n^{-1/2})$ and $F_n^*(x) = O_p(n^{-1/2})$ for every fixed $x\in R^1$:

It follows from (3.6), (3.14) and (3.15) that it is sufficient to prove

$$n^{-1/2}\sum_{i=1}^{n} d_{ij}\left[a_ie_i-\psi(e_i)\right] = O_p(n^{-1/4}), \qquad j=1,2. \tag{3.18}$$

Let us consider the quantity

$$\begin{aligned} & T^{(j)}(\tilde{\Delta},\tilde{\Delta}') \\ (3.19) \quad & = n^{-1/2}\sum_{i=1}^{n} d_{ij}\, e_i\, I\left[F^{-1}(\alpha)+n^{-1/2}(\Delta_1+c_i\Delta_2)\le e_i \le \right. \\ & \left.\le F^{-1}(1-\alpha)+n^{-1/2}(\Delta_1'+c_i\Delta_2')\right] ;\ |\Delta_j|,|\Delta_j'|\le M ;\quad j=1,2. \end{aligned}$$

Then, for $n\ge n_0$,

$$\begin{aligned} (3.20) \quad & \operatorname{Var} n^{1/4}\left[T^{(2)}(\tilde{\Delta},\tilde{\Delta}')-T^{(2)}(\tilde{0},\tilde{0})\right] \\ & = b^{-2}n^{-1}\sum_{i=1}^{n} c_i^2(F^{-1}(\alpha))^2 f(F^{-1}(\alpha))\left[|\Delta_1|+|\Delta_1'|+|c_i|(|\Delta_2|+|\Delta_2'|)\right] \\ & + O(n^{-1/2}) = M.O(1) \end{aligned}$$

and an analogous relation holds for $\operatorname{Var} n^{1/4}\left[T^{(1)}(\tilde{\Delta},\tilde{\Delta}')-T^{(1)}(\tilde{0},\tilde{0})\right]$. Thus, for every fixed $\tilde{\Delta}$, $\tilde{\Delta}'$,

$$(3.21) \quad n^{1/4}|T^{(j)}(\tilde{\Delta},\tilde{\Delta}') - T^{(j)}(\tilde{0},\tilde{0})-E\left[T^{(j)}(\tilde{\Delta},\tilde{\Delta}')-T^{(j)}(\tilde{0},\tilde{0})\right]| = O_p(1)$$

as $n\to\infty$, $j=1,2$. For every positive integer k, put

$$(3.22) \quad \Delta^{(k)} = [k\Delta]/k.$$

Then $\Delta^{(k)}\le\Delta\le\Delta^{(k)}+\frac{1}{k}$ and, analogously as in (3.20), we get

$$\begin{aligned} & \sup_{\tilde{\Delta},\tilde{\Delta}'\in A}\left\{n^{1/4}|T^{(j)}(\tilde{\Delta},\tilde{\Delta}') - T^{(j)}(\tilde{\Delta}^{(k)},\Delta'^{(k)})\right. \\ (3.23) \quad & \left.-E\left[T^{(j)}(\tilde{\Delta},\tilde{\Delta}')-T^{(j)}(\Delta^{(k)},\Delta'^{(k)})\right]\right\} = k^{-1}\, O_p(1),\quad j=1,2 \end{aligned}$$

where $A=\left\{\tilde{t}=(t_1,t_2)' : |t_1|\le M, |t_2|\le M\right\}$. Combining (3.21) and (3.23), we get

$$\begin{aligned} (3.24) \quad & \sup_{\tilde{\Delta},\tilde{\Delta}'\in A}|T^{(j)}(\tilde{\Delta},\tilde{\Delta}')-T^{(j)}(\tilde{0},\tilde{0})-E\left[T^{(j)}(\tilde{\Delta},\tilde{\Delta}')-T^{(j)}(\tilde{0},\tilde{0})\right]| \\ & = O_p(n^{-1/4}),\quad j=1,2. \end{aligned}$$

Finally, we shall prove that

$$\begin{aligned} & n^{1/4}|E\left[T^{(j)}(\tilde{\Delta},\tilde{\Delta}')-T^{(j)}(\tilde{0},\tilde{0})-F^{-1}(1-\alpha)f(F^{-1}(\alpha))(\Delta_j+\Delta_j')\right]| \\ (3.25) \quad & \to 0,\ \text{ as } n\to\infty,\ \text{uniformly for } \tilde{\Delta},\tilde{\Delta}'\in A. \end{aligned}$$

The left-hand side of (3.25) takes on the form

$$\begin{aligned} & n^{-1/4}\left|\sum_{i=1}^{n} d_{ij}\left[\int_{a'}^{a'+\delta'}(xf(x)-F^{-1}(1-\alpha)f(F^{-1}(1-\alpha)))\,dx\right.\right. \\ (3.26) \quad & \left.\left. -\int_{a}^{a+\delta}(xf(x)-F^{-1}(\alpha)f(F^{-1}(\alpha)))\,dx\right]\right| \end{aligned}$$

with $a=F^{-1}(\alpha)$, $\delta=n^{-1/2}(\Delta_1+c_i\Delta_2)$, $a'=F^{-1}(1-\alpha)$, $\delta'=n^{-1/2}(\Delta_1'+c_i\Delta_2')$.

(3.26) is bounded from above by

$$C.n^{-5/4}\sum_{i=1}^{n}|d_{ij}|(\delta^2+\delta'^2) = M^2\, O(n^{-1/4})$$

and this gives (3.25). (3.24) and (3.25) give

$$(3.27)\quad \sup_{\Delta,\Delta'\in A}\left\{|T^{(j)}(\Delta,\Delta')-T^{(j)}(0,0)-F^{-1}(1-\alpha)f(F^{-1}(\alpha))(\Delta_j+\Delta'_j)|\right\} = O_p(n^{-1/4})$$

and this together with (3.16) yields

$$(3.28)\quad n^{-1/2}\sum_{i=1}^{n}d_{ij}\,a_ie_i = n^{-1/2}\sum_{i=1}^{n}d_{ij}\left\{e_i\, I\left[F^{-1}(\alpha)\le e_i\le F^{-1}(1-\alpha)\right] + F^{-1}(\alpha)I\left[e_i<F^{-1}(\alpha)\right]+F^{-1}(1-\alpha)I\left[e_i>F^{-1}(1-\alpha)\right]\right\}+O_p(n^{-1/4}),$$

$j=1,2$, and this completes the proof of the theorem. Q.E.D.

4. ASYMPTOTIC RELATION OF M-ESTIMATOR AND OF ITS ONE-STEP VERSION

Bickel (1975) introduced the one-step version M_n^* of the M-estimator M_n in the linear model, which is asymptotically equivalent to M_n in probability as $n\to\infty$. We shall study the order of the approximation of M_n by M_n^* in one-parameter regression model. The order depends on the smoothness of the ψ-function generating M_n and generally it is lower if ψ has jump-discontinuities than in the case that ψ is smooth.

I. Let us first consider the M-estimator generated by a smooth ψ-function. We assume that the following conditions are satisfied:

(A) $X_1,X_2,\ldots$ is a sequence of independent observations, X_i distributed according to the d.f. $F(x-\theta c_i)$, $i=1,2,\ldots$, such that F is symmetric and admits an absolutely continuous density f with a finite Fisher information, $\mathcal{I}(f)=\int(f'/f)^2\,dF<\infty$.

(B) $\psi: R^1\to R^1$ is nonconstant, nondecreasing, absolutely continuous on any bounded interval in R^1 and satisfies

$$(4.1)\qquad \psi(-x)=-\psi(x),\qquad x\in R^1 \text{ and}$$

$$(4.2)\qquad 0<\int\psi^2(x)\,dF(x)<\infty .$$

Let ψ' be the derivative of ψ; we assume that

$$(4.3)\qquad \gamma=\gamma(\psi,F)=\int\psi'(x)\,dF(x)\ne 0,$$

$$(4.4)\qquad \int(\psi'(x))^2\,dF(x)<\infty \quad\text{and}$$

$$(4.5)\qquad \lim_{t\to 0}\int\left[\psi'(x+t)-\psi'(x)\right]^2 dF(x)=0 .$$

Moreover, we assume either that ψ is constant outside of an interval $[-k,k]$, $0<k<\infty$ and has two bounded derivatives inside, or that ψ' is absolutely continuous on any bounded interval in R^1 and its derivative ψ'' satisfies

(4.6) $$\lim_{t\to 0}\int[\psi''(x+t)-\psi''(x)]\,dF(x) = 0.$$

(C) $\{c_n\}_{n=1}$ is a sequence of known constants; define $C_n^2=\sum_{i=1}^n c_i^2$ and assume that

(4.7) $$\lim_{n\to\infty}\left[\max_{1\le i\le n} c_i^4 \cdot(\sum_{j=1}^n c_j^4)^{-1}\right] = 0.$$

The M-estimator M_n of θ is defined as

(4.8) $$M_n = \tfrac{1}{2}(M_n^+ + M_n^-)$$
$$M_n^+ = \inf\left\{t: \sum_{i=1}^n c_i\psi(X_i-c_i\, t) < 0\right\}$$
$$M_n^- = \sup\left\{t: \sum_{i=1}^n c_i\psi(X_i-c_i\, t) > 0\right\}.$$

Let $\hat{\theta}_n$ be an initial estimator of θ satisfying

(4.9) $$C_n(\hat{\theta}_n-\theta) = O_p(1), \quad \text{as} \quad n\to\infty,$$

(4.10) $$\hat{\theta}_n(X_1+c_1t,\dots,X_n+c_nt) = \hat{\theta}_n(X_1,\dots,X_n)+t.$$

Define the one-step version of M_n as

(4.11) $$M_n^* = \hat{\theta}_n + C_n^{-2}\hat{\gamma}_n^{-1}\sum_{i=1}^n c_i\psi(X_i-c_i\,\hat{\theta}_n)$$

provided $\hat{\gamma}_n \neq 0$ and $M_n^* = \hat{\theta}_n$ otherwise, where

(4.12) $$\hat{\gamma}_n = C_n^{-1}(t_2-t_1)^{-1}\sum_{i=1}^n c_i[\psi(X_i+C_n^{-1}c_it_2)-\psi(X_i+C_n^{-1}c_it_1)]$$

with $t_1,t_2 (t_1<t_2)$ being fixed real numbers. The following theorem gives the order of approximation of M_n by M_n^*.

THEOREM 4.1. Under the assumptions (A),(B),(C), with M_n and M_n^* defined in (4.8)-(4.12),

(4.13) $$C_n(M_n-M_n^*) = O_p\left[C_n^{-2}(\sum_{i=1}^n c_i^4)^{1/2}\right], \quad \text{as } n\to\infty.$$

COROLLARY 4.1. Suppose that in addition to the conditions of Theorem 4.1,

(4.14) $$\lim_{n\to\infty}(C_n^2/n) = b_1^2>0, \qquad b_1^2<\infty,$$

(4.15) $$\lim_{n\to\infty} n^{-1}\sum_{i=1}^n c_i^4 = b_2^2>0, \qquad b_2^2<\infty.$$

Then

(4.16) $$n^{1/2}(M_n - M_n^*) = O_p(n^{-1/2}).$$

P r o o f . We may put $\theta=0$, without loss of generality. It follows from Jurečková and Sen (1981a), Theorem 2.2 (cf. also Jurečková (1980), Theorems 3.1 and 3.2) that

(4.17) $$c_n M_n = c_n^{-1} \gamma^{-1} \sum_{i=1}^n c_i \psi(X_i) + O_p(a_n),$$

with $a_n = c_n^{-2}(\sum_{i=1}^n c_i^4)^{1/2}$. It follows from (4.11) and (4.17) that, provided $\hat{\gamma}_n \neq 0$,

(4.18) $$\begin{aligned} c_n(M_n - M_n^*) &= -c_n\hat{\theta}_n + \gamma^{-1} c_n^{-1} \sum_{i=1}^n c_i \psi(X_i) \\ &- \hat{\gamma}_n^{-1} c_n^{-1} \sum_{i=1}^n c_i \psi(X_i - c_i\hat{\theta}_n) + O_p(a_n) \\ &= [(\gamma/\hat{\gamma}_n)-1][c_n\hat{\theta}_n - \gamma^{-1} c_n^{-1} \sum_{i=1}^n c_i \psi(X_i)] + (\gamma/\hat{\gamma}_n) O_p(a_n). \end{aligned}$$

It follows from Jurečková and Sen (1981a), Theorem 2.2, that $\gamma/\hat{\gamma}_n = 1+O_p(a_n)$, so that (4.13) follows from (4.18).

Q.E.D.

II. Let us consider the case that ψ consists of two components, $\psi = \psi_1 + \psi_2$, such that both ψ_1 and ψ_2 are nondecreasing and skew-symmetric, ψ_1 is absolutely continuous and satisfies the assumption (B) while ψ_2 is a step-function,

(4.19) $$\psi_2(x) = \beta_j \quad \ldots \quad x \in E_j, \qquad j=0,\ldots,p$$

where $E_j = (a_j, a_{j+1})$, $j=0,\ldots,p$; $a_0 = -\infty < a_1 < \ldots < a_p < a_{p+1} = \infty$ and $\psi_2(a_j) = \frac{1}{2}(\beta_{j-1} + \beta_j)$, $j=1,\ldots,p$.

Let $X_1, X_2, \ldots$ be a sequence of independent observations, X_i distributed according to $F(x - c_i\theta)$, $i=1,2,\ldots$ where F satisfies the assumption (A). Moreover, assume that f is bounded in neighborhood of $a_1,\ldots,a_p$ and

(4.20) $$0 < \sum_{j=1}^p (\beta_j - \beta_{j-1})^r f(a_j) < \infty \quad \text{for} \quad r=1,2 .$$

Let $\{c_n\}$ be a sequence of real numbers satisfying (C). Define M_n, M_n^* in the same way as in (4.8), (4.9)-(4.12). The following theorem shows that the approximation of M_n by M_n^* is generally of lower order than in the previous case.

THEOREM 4.2. Under the above assumptions,

(4.21) $$c_n(M_n - M_n^*) = O_p\left[c_n^{-3/2}(\sum_{i=1}^n |c_i|^3)^{1/2}\right],$$

as $n \to \infty$. If, in addition, $\{c_n\}$ satisfies (4.14) and

(4.22) $$\lim_{n\to\infty} n^{-1} \sum_{i=1}^n |c_i|^3 = b_3^2 > 0, \qquad b_3^2 < \infty,$$

then

(4.23) $$n^{1/2}(M_n - M_n^*) = O_p(n^{-1/4}).$$

P r o o f . The proof follows from Jurečková and Sen(1981b), Corollary 3.2, along the same lines as the proof of Theorem 4.1.

REFERENCES

BAHADUR R.R.(1966): A note on quantiles in large samples. Ann.Math. Statist. 37, 577-580.

BICKEL P.J.(1965): On some robust estimates of location.Ann.Math. Statist. 36, 847-858.

BICKEL P.J.(1973): On some analogues to linear combinations of order statistics in linear model. Ann.Statist.1, 597-616.

BICKEL P.J.(1975): One-step Huber estimates in the linear model. Journ.Amer.Statist.Assoc. 70, 428-433.

BICKEL P.J. and LEHMANN E.L.(1975): Descriptive statistics for non-parametric model. II.location. Ann.Statist.3, 1045-1069.

HODGES J.L. and LEHMANN E.L.(1963): Estimates of location based on rank tests: Ann.Math.Statist. 34, 598-611.

HUBER P.J.(1964): Robust estimation of a location parameter. Ann. Statist. 35, 73-101.

HUŠKOVÁ M.andJUREČKOVÁ J.(1981): Second order asymptotic relations of M-estimators and R-estimators in two-sample location model. Journ.Statist.Planning Infer. 5, 309-328.

JAECKEL L.A.(1971): Robust estimates of location: Symmetry and asym-metric contamination. Ann.Math.Statist.42, 1020-1034.

JUREČKOVÁ J.(1977): Asymptotic relations of M-estimates and R-estima-tes in linear regression model.Ann.Statist.5, 464-472.

JUREČKOVÁ J.(1980): Asymptotic representation of M-estimators of lo-cation. Math.Operationsforsch.Statist.,Ser.Statistics 11,61-73.

JUREČKOVÁ J.(1982): Robust estimators of location and their second order asymptotic relations. Submitted to ISI Centennial Vo-lume.

JUREČKOVÁ J.and SEN P.K.(1981a): Invariance principles for some sto-chastic processes relating to M-estimators and their role in in sequential statistical inference. Sankhya A 43, 190-210.

JUREČKOVÁ J.and SEN P.K.(1981b):Sequential procedures based on M-es-timators with discontinuous score functions.Journ.Statist. Planning Infer.5, 253-266.

KOENKER R. and BASSETT G.(1978): Regression quantiles. Econometrica 46, 33-50.

RUPPERT D. and CARROLL R.J. (1978): Robust regression by trimmed least squares estimation. Inst.of Statist.Mimeo Series No 1186, Univ.of North Carolina at Chapel Hill.

RUPPERT D., CARROLL R.J.(1980) : Trimmed least squares estimation in the linear model. Journ.Amer.Statist.Assoc.75, 828-838.

Charles University
Department of Probability and Statistics
Sokolovská 83
186 00 Praha 8
Czechoslovakia

GENERATORS OF SOME CLASSES OF PROBABILITY MEASURES ON BANACH SPACES

Zbigniew J. JUREK
Wrocław

Key words: Banach space, infinitely divisible measures, Lévy measure, self-decomposable measure, s-self-decomposable measure.

ABSTRACT

For the classes of s-self-decomposable and self-decomposable measures we find measures whose finite convolutions form dense subsets. These classes are also characterized as subsets of the class of all infinitely divisible measures. Complete results a given for measures on Hilbert spaces.

1. PRELIMINERIES AND NOTATIONS

Let E be a real separable Banach space with a norm $||\cdot||$. By $P(E)$ we shall denote the semi-group of all Borel probability measures on E with the convolution $*$. For $\mu \in P(E)$ and Borel mapping f from E into E, $f\mu$ is a measure given by the formula $(f\mu)(A) = \mu(f^{-1}(A))$ for all Borel subsets A. Let ID(E) denotes the class of all infinitely divisible measures, cf. Araujo, Gine (1981) p.136. It is easy to verify that ID(E) is closed sub-semi-group of $P(E)$ (in weak convergence topology). Since $\mu \in ID(E)$ are uniquely determined by a triplet $x \in E$, a Gaussian covariance operator R and Lévy measure M we will write that $\mu = [x,R,M]$. Let S be the unit sphere in E. For a Borel subset A of S and an interval I of positive reals we define $\langle A,I\rangle := \{x \in E: x/||x|| \in E, \ ||x|| \in I\}$. For $[x,R,M] \in ID(E)$ its Lévy spectral function L is defined as follows

$$L(A,r) := -M(\langle A,(r,\infty)\rangle) \tag{1.1}$$

for Borel subsets A of S and positive reals r.

2. SELF-DECOMPOSABLE PROBABILITY MEASURES

Let $T_a: E \to E$ is defined by means of formula $T_a x = ax$. We say $\mu \in P(E)$ is self-decomposable if

(2.1) $$\forall\ (0 < c < 1)\ \exists\ (\mu_c \in P(E)) \qquad \mu = T_c\mu * \mu_c \ .$$

By $L(E)$ we denote the class of all self-decomposable measures on E. It is well-known that $L(E)$ is closed sub-semi-group p of ID(E), cf. Kumar, Schreiber (1975). In particular from (2.1) we have $[x,R,M] \in L(E)$ iff for every $0 < c < 1$

(2.2) $$M \geq T_c M \ .$$

<u>PROPOSITION</u> 2.1. $\mu = [x,R,M] \in L(E)$ iff for each Borel subset B of S its Lévy spectral function L(B,r) has right and left derivatives with respect to r, for $0 < r < \infty$, and the function $r dL(B,r)/dr$ is non-increasing on $(0,\infty)$ for each B. [Here $dL(B,r)/dr$ denotes either the right or left derivatives, possibly different ones at different points].

<u>Proof</u>. If $\mu \in L(E)$ then (2.2) implies that for each B the function $f(t) := L(B,e^t)$ is continuous concave on the real line. Consequently f has the right and left derivatives and $f'(t) = e^t dL(B,e^t)/dt$ is non-increasing. Conversely, if for fixed B $L(B,\cdot)$ has the above mentioned properties then it is continuous and concave. Further, since $r dL(B,r)/dr$ is non-increasing on $(0,\infty)$ we have

$$dL(B,r)/dr \geq c^{-1} dL(B,r/c)/dr \ .$$

Hence, similarly as in Jurek (1978), we obtain

$$M(\langle B,(u/c,\ v/c]\rangle) \leq M(\langle B,(u,\ v]\rangle)$$

for $0 < u < v$ and $0 < c < 1$. So, M satisfies (2.2), i.e., $[x,R,M] \in L(E)$.

For a positive real number α and a finite Borel measure m on S we put

(2.3) $$M_{\alpha,m}(A) := \int_S \int_0^\alpha 1_A(tx)t^{-1}\, dt\, m(dx) \ , \qquad A \subseteq E\backslash\{0\} \ ,$$

and 1_A denotes the indicator of a set A. Since

$$\int_{\|x\| \leq 1} \|x\| M_{\alpha,m}(dx) = m(S)(\alpha \wedge 1) < \infty$$

then Theorem III.6.3 in Araujo, Gine (1981) implies that M is a Lévy measure. Let us denote by G a class of all measures of the form $[x,R,M_{\alpha,m}]$. Note that by (2.2) we obtain $G \subseteq L(E)$.

<u>PROPOSITION</u> 2.2. For $\mu = [x,0,M] \in L(E)$ there exist a subsequence

$\{k_n\}$ of positive integers, positive real numbers α_{nj} and positive finite Borel measures m_{nj} on S ($j=1,2,\ldots,k_n$; $n=1,2,\ldots$) such that

(a) $$N_n := \sum_{j=1}^{k_n} M_{\alpha_{nj},m_{nj}} \Longrightarrow M$$

outside every neighbourhood of zero in E. Moreover, if $\int(1\wedge\|x\|)^p M(dx)<\infty$ for some $p>0$ then

(b) $$\lim_{\varepsilon\to 0}\limsup_{n\to\infty} \int_{\|x\|\le\varepsilon} \|x\|^p N_u(dx) = 0$$

Proof. Let L be Lévy spectral function of μ. Put $b_{nk}(A) = L(A,k/2^n)$ and

(2.4) $$a_{nk}(A) = (b_{nk}(A) - b_{nk-1}(A))/\log k/(k-1)\ , \quad \text{for } k=2,3,\ldots,n2^n,$$

and $a_{n1}(A) = a_{n2}(A)$, $a_{n,n2^n+1}(A) = b_{n,n2^n+1}(A) \equiv 0$ for all A. Taking into account Proposition 1.1 and Lemma 2 in Kubik (1962) we infer that $a_{nk+1} \le a_{nk}$ for $k=1,2,\ldots,n2^n-1$. Hence

(2.5) $$m_{nj}(A) := a_{nj}(A) - a_{n,j+1}(A) \quad \text{for } j=1,2,\ldots,n2^n,$$

are finite non-negative measures. Further, let

(2.6) $$\alpha_{nj} = j2^{-n} \quad \text{for } j=1,\ldots,n2^n \quad \text{and } k_n = n2^n\ .$$

It is enough to prove (a) for sets of the form $\langle A,(t,\infty)\rangle$, $t>0$. Put $L_n(A,t) := -N_n(\langle A,(t,\infty)\rangle)$ for $t>0$. From (2.4) - (2.6) we infer that if $(i-1)2^{-n} \le t \le i2^{-n}$ for some $i \in \{2,3,\ldots,k_n\}$ then

$$L_n(A,t) = \sum_{k=i}^{k_n} m_{nk}(A)\log(k^{-1}2^n t) =$$

$$= a_{ni}(A)\log(2^n ti^{-1}) - \sum_{k=i}^{k_n} [a_{nk}(A) - a_{nk+1}(A)]\log k =$$

$$= a_{ni}(A)\log(2^n ti^{-1}) - \sum_{k=i+1}^{k_n} a_{nk}(A)\log(k/k-1) =$$

$$= a_{ni}(A)\log(2^n ti^{-1}) + b_{ni}(A) - b_{n,k_n}(A)\ .$$

Assuming for fixed $i\ge 2$ and $(i-1)2^{-n} \le t \le i2^{-n}$

$$L_{ni}(A,t) := a_{ni} \log(2^n ti^{-1}) + b_{ni}(A)$$

we shall get

(2.7) $$|L_{ni}(A,t) - L(A,t)| \le L(A,i2^{-n}) - L(A,(i-1)2^{-n})$$

because $L_{ni}(A,i2^{-n}) = L(A,i2^{-n})$, $L_{ni}(A,(i-1)2^{-n}) = L(A,(i-1)2^{-n})$ and the above functions are non-decreasing. Further, for $\varepsilon > 0$ there exists $n_o \in N$ such that for $n \ge n_o$ and $k=2,3,\dots,k_n$

$$L(A,k2^{-n}) - L(A,(k-1)2^{-n}) \le \varepsilon \quad , \quad -L(A,n) \le \varepsilon \ .$$

Therefore for all $t > 0$ and all Borel subset A of S, $\lim_{n\to\infty} L_n(A,t) = L(A,t)$, which proves the condition (a). Assume that $\int (1 \wedge \|x\|^p) M(dx) < \infty$ for some $p > 0$. From (2.4) - (2.6) we have

(2.8) $$\int_{\|x\| \le \varepsilon} \|x\|^p N_n(dx) = p^{-1} 2^{-p(n-1)} a_{n2}(S) +$$

$$+ p^{-1} a_{n,[\varepsilon 2^n]+1}(S)(\varepsilon^p - ([\varepsilon 2^{-n}]2^{-n})^p) + p^{-1} \sum_{k=3}^{[2^n \varepsilon]} a_{nk}(S)((k2^{-n})^p - ((k-1)2^{-n})^p) \ ,$$

where $[r]$ denotes the integer part of r. By (2.4) for $k \ge 2$ and $p > 0$ we have

$$p^{-1}((k2^{-n})^p - ((k-1)2^{-n})^p) \le k \int_{(k-1)/2^n}^{k/2^n} t^{p-1} dt \int_{(k-1)/2^n}^{k/2^n} dL(S,t) \le$$

$$\le (k/k-1)^{p \vee 1} \int_{(k-1)/2^n}^{k/2^n} t^p dL(S,t) \le 2^{p \vee 1} \int_{(k-1)/2^n \le \|x\| \le k/2^n} \|x\|^p M(dx) \ ,$$

and $a \vee b$ denotes the maximum of reals a and b. Hence the third component in (2.8) converges to zero. Similar arguments imply that the first and the second component tends to zero too, which completes the proof.

For a Hilbert space H from Proposition 2.2, with p=2, we obtain

<u>THEOREM</u> 2.1. A class $L(H)$ is equal to the class of finite convolutions of measures from G and of their limits.

Let $L_o(E) := \{\mu = [x,R,M] \in L(E) : \int (1 \wedge \|x\|) M(dx) < \infty\}$. Of course, $G \subseteq L_o(E)$ but stable measures with an exponent $0 < p < 1$ do not belong to $L_o(E)$, cf. Araujo, Giné (1981), Theorem III.6.15.

<u>THEOREM</u> 2.2. Each measure from $L_o(E)$ is a weak limit of finite convolutions of measures from the class G.

<u>Proof</u>. If $\mu = [x,0,M] \in L_o(E)$ and M is symmetric then our theorem is consequence of Proposition 2.2 and Theorem V.10.5 in Woyczyński (1978). If $M^-(A) := M(-A)$ and $[0,0,M] \in L_o(E)$ then $[0,0,M+M^-] \in L_o(E)$ and consequently

$$[0,0,N_n] * [0,0,N_n^-] \Longrightarrow [0,0,M+M^-] \ .$$

This together with Corollary III.4.6 in [1] implies that the sequence $\{[0,0,N_n]\}$ is conditionally compact. But from Proposition 2.2 and the exercise 2, page 133 in Araujo-Gine (1981) we infer that $[0,0,N_n] \Rightarrow$ $\Rightarrow [0,0,M]$.

3. s-SELF-DECOMPOSABLE PROBABILITY MEASURES

For $r \geq 0$ we define shrinking operation U_r from E into E as follows: $U_r x = 0$ if $||x|| \leq r$, and $U_r x = (1 - r/||x||)x$ if $||x|| > r$. Further, let

$$\mathcal{U}(E) := \{[x,R,M] \in ID(E): M \geq U_t M \quad \text{for all } t > 0\} .$$

In case of Hilbert space H the class $\mathcal{U}(H)$ coincides with a class of limit distribution of some sequences, cf. Jurek (1981). For $\mathcal{U}(E)$ we have analogous discriptions to that ones for $\mathcal{L}(E)$. Moreover, the proofs are similar and we will omit them. Cf. also Jurek (1978).

PROPOSITION 3.1. $\mu = [x,R,M] \in \mathcal{U}(E)$ iff for each Borel subset B of S its Lévy spectral function $L(B,r)$ has right and left derivatives with respect to r, for $0 < r < \infty$, and the function $dL(B,r)/dr$ is non-increasing on $(0,\infty)$ for each B.

The above fact and Proposition 2.1 give

COROLLARY. $\mathcal{L}(E) \subseteq \mathcal{U}(E)$.

Let $\mathcal{K}$ denotes the class of all measures of the form $[x,R,N_{\alpha,m}]$, where $\alpha > 0$, m is finite Borel measure on S and

$$N_{\alpha,m}(A) := \int_S \int_0^{\alpha} 1_A(tx)dt\, m(dx) .$$

Note that $N_{\alpha,m}$ are finite and $\mathcal{K} \subseteq \mathcal{U}(E)$. Further we also have

PROPOSITION 3.2. For $[x,R,M] \in \mathcal{U}(E)$ there exist a sequence $\{k_n\}$ of positive integers, positive real numbers α_{nj} and positive finite Borel measures m_{nj} on S ($j=1,\dots,k_n$; $n=1,2,\dots$) such that

$$\text{(a)} \qquad M_n := \sum_{j=1}^{k_n} N_{\alpha_{nj},m_{nj}} \Longrightarrow M$$

outside every neighbourhood of zero in E. Moreover, if $\int(1 \wedge ||x||^p)M(dx) < \infty$ for some $p > 0$ then

$$\text{(b)} \qquad \lim_{\varepsilon\to 0} \limsup_{n\to\infty} \int_{||x||\leq\varepsilon} ||x||^p M_n(dx) = 0 .$$

Hence for Hilbert space H we have the following

THEOREM 3.1. $\mathcal{U}(H)$ is the smallest closed sub-semi-group if $ID(H)$ containing the class $\mathcal{K}$.

If $\ddot{U}_o(E) := \{[x,R,M] \in U(E): \int(1\wedge\|x\|)M(dx) < \infty\}$ then we get

THEOREM 3.2. Each measure from $U_o(E)$ is a weak limit of finite convolutions of measures from the class K.

It seems to be true that the integrability conditions in Theorems 2.2 and 3.2 may be omitted, i.e., G and K generates the whole class $L(E)$ and $U(E)$ respectively, for arbitrary Banach space.

REFERENCES

Araujo A., Giné E. (1981): The central limit theorem for real and Banach valued random variables. John Wiley & Sons, New York.

Jurek Z.J. (1978): Some characterizations of the class of s-selfdecomposable distributions. Bull.Acad.Pol.Sci. 26, No.8, 719-725.

(1981): Limit distributions for sums of shrunken random variables. Dissertationes Math. (Rozprawy Matematyczne) 185, PWN Warszawa.

Kubik L. (1962): A characterization of the class L of probability measures, Studia Math. 21, 245-252.

Kumar A., Schreiber B.M. (1975): Self-decomposable probability measures on Banach spaces. Studia Math. 53, 55--71.

Woyczyński W.A. (1978): Geometry and martingales in Banach spaces, Part II: Independent increments. Advances in Probab. 4, J.Kuelbs editor, M.Dekker, New York.

University of Wrocław
Institute of Mathematics
Pl.Grunwaldzki 2/4, 50-384 Wrocław
Poland.

MONTE-CARLO METHODS FROM THE POINT OF VIEW OF ALGORITHMIC COMPLEXITY

Ivan Kramosil
Prague

Key words: Monte-Carlo methods, pseudo-random numbers, computational complexity, algorithmic complexity of binary strings, Turing machine.

ABSTRACT

The unknown measure of a measurable subset of the unit real interval is estimated using an appropriate Monte-Carlo method. The random sample is simulated by a binary sequence of high algorithmic complexity and the tested set is supposed to be effectively decidable by a Turing machine. Under such conditions, if the algorithmic complexity of the pseudo-random input in question exceeds a threshold value, then the described Monte-Carlo method yields the correct value of the estimated measure. Moreover, if the computational complexity of the indicator (characteristic function) of the tested set is uniformly bounded, then the mentioned threshold value can be effectively computed.

1. INTRODUCTION

Monte-Carlo methods consist in an application of laws of large numbers with the aim to replace the unknown expected value of a random variable by the average value obtained by a finite random sample. If $X_1, X_2, \ldots$ are equally distributed and statistically independent random variables, then under condition that $EX < \infty$, $EX_i < \infty$ exist $n^{-1} \sum_{i=1}^{n} X_i$ tends with the probability one to the expected value EX. We shall concentrate our attention to a special case when random variables X_i are of the form $In_E(Y_i)$, where In_E is the indicator (characteristic function) of a measurable set $E \subset I = \langle 0,1 \rangle$, and Y_i are statistically independent, equally distributed real-valued random variables defined on a probability space $\langle \Omega, \mathcal{S}, P \rangle$, such that for all semi-open intervals $\langle a,b) \subset I$ and for all $i=1,2,\ldots$,

$$(1) \qquad P(\{\omega:\omega\in\Omega,\ Y_i(\omega)\in\langle a,b)\}) = b - a = \mu(\langle a,b)),$$

where μ denotes the Lebesque measure on the real line. The strong law of large numbers then claims that

$$(2) \qquad P(\{\omega:\omega\in\Omega,\ \lim_{n\to\infty} n^{-1} \sum_{i=1}^{n} In_E(Y_i(\omega)) = \mu(E)\}) = 1,$$

hence, the average number of at random sampled points from I which belong to E tends to $\mu(E)$ with the probability one.

Some other well-known assertions of axiomatic probability theory enable to describe and to compute or estimate quantitatively the risk connected with the decision to replace $\mu(E)$ by $n^{-1} \sum_{i=1}^{n} In_E(Y_i(\omega))$; the Tchebyshev inequality may serve as an example. But it is not the aim of this paper to penetrate more deeply into the probabilistic theoretical foundations of Monte-Carlo methods and we refer the reader to numerous textbooks or monographies. Here we shall study the problem from two other points of view, namely:

(a) its computational effectiveness, i. e., from the point of view of the computability of the function $n^{-1} \sum_{i=1}^{n} In_E(Y_i(\omega))$,

(b) the possibility to replace the random sample $Y_1(\omega), Y_2(\omega), \ldots, Y_n(\omega)$, or to approximate it, by an appropriate pseudo-random input.

The aim of this paper is to show and formalize, in an appropriate way, some mutual connections between the two approaches.

2. EFFECTIVE SET INDICATORS

In this chapter we shall study the effectivity connected with the indicator function In_E. The theory of Monte-Carlo methods, and the mathematical statistics in general, use a simple phrase "sample at random $Y_i(\omega)\in\langle 0,1\rangle$ and decide, whether it belongs to E or not", not going into details. This approach is influenced by the platonistic viewpoint of classical mathematics which considers real numbers and functions as finite and completely accessible objects. As a matter of fact, discrete descriptions of real numbers (say, their binary or decadic expansions) are, in general, infinite, but only finite objects may enter an effective procedure which computes In_E. Hence, the idea must be abandoned, that the tested real number itself enters this procedure, and other solution must be found. The most general one can be described as follows.

Let L be a formalized language, say, the first-order predicate calculus with appropriate functional, predicate and individual constants and with individual indeterminates ranging over the space of reals. Hence, unary predicates of the form $A(x)$ correspond to various properties of reals which are expressible in L. An effective indicator is a triple $\langle V, Ax, Cn^* \rangle$, where V is a unary predicate from L, $Ax \subset L$ is a recursive set of axioms (including, e. g., logical and arithmetical axioms), and Cn^* is a recursive restriction of the consequence relation defined by the usual deduction rules. Hence, $Cn^*(Ax)$ is a recursive subset of all formulas derivable from axioms. The indicator accepts inputs of the form $A(x_o) \varepsilon L$ and sets $In_E(x_o) = 1$ (i.e., proclaims that x_o belongs to E), iff $\ulcorner A(x_o) \rightarrow V(x_o)\urcorner \varepsilon Cn^*(Ax)$, i. e., iff it is able to prove that x_o satisfies the characteristic formula V, using Cn^* and $A(x_o)$.

The indicators accepting real numbers in the form of an initial segment of their digital expansion can be seen as special cases of indicators described above, namely, for $x_o = x_1x_2x_3\ldots$, $A(x_o)$ is of the form $\wedge_{j=1}^{n(x)} (A_j(x_o) = x_j)$, no matter which the digital base may be. Such indicators will be called digital, and only digital indicators will be investigated below, it is why the adjective will be often omitted.

A digital indicator works as follows: it reads the first digit of the tested real x and either (1) decides that x belongs to E, i.e., $In_E(x) = 1$, or (2) decides that x does not belong to E, i. e., $In_E(x) = 0$, or (3) reads the following digit in the expansion and tries again to decide about x. The indicator is effective, if it finishes its work over x (by saying whether $x \varepsilon E$ or not) after a finite number of steps, i. e. using only a finite number of digits. So we may identify an effective digital indicator with a Turing machine working over a tape which is infinite in both directions and which contains an infinite sequence of digits. The machine begins by reading the first digit, the boxes left from this first digit are supposed to be blanks. For each real x the machine finishes eventually its work, writing 1 or 0 on the output tape. Because of the equivalence between Turing machines and recursive functions we arrive at the following definition.

Definition 1. Let A be a finite set of symbols (digits). A partially recursive function In, defined in the set A^* of all finite sequences (strings) of digits and taking its values in the two-element set $\{0,1\}$ is called an effective digital indicator w.r.t. A, if there is, for each infinite sequence $x = x_1x_2\ldots$ of digits, just one $n = n(x)$ such that In is defined for $x_1x_2\ldots x_n \varepsilon A^*$. Let r be a mapping ascribing to each $x \varepsilon A^\infty$ a real number $r(x)$, let E be the set of reals defined by $E = \{r(x): x\varepsilon A^\infty, \mathrm{In}(x)=1\}$. Then In is called an <u>indicator</u> for E, or E is called <u>to be indicated</u> by In, or In_E, to make the dependence clear. ¤

The condition means that In may be unambiguously extended to a function taking the space A^∞ of infinite digital sequences into $\{0,1\}$ setting, for each $x = x_1x_2\ldots \varepsilon A^\infty$, $\mathrm{In}(x) = \mathrm{In}(x_1x_2\ldots x_{n(x)})$ and we have taken profit of this possibility in the second part of the definition. As a rule, we shall take $\{0,1\}$, or $\{0,1,\ldots,9\}$, for A. It is possible that there are $x,y\varepsilon A^\infty$, $x \neq y$, such that $r(x) = r(y)$, e. g., $r(0111\ldots) = r(1000\ldots)$, if $A = \{0,1\}$. In such a case our definition assures that if $\mathrm{In}_E(x) \neq \mathrm{In}_E(y)$, then $r(x)\varepsilon E$. Not wanting to admit this possibility, we have to ensure the one-to-one character of the mapping r, but this cannot be achieved inside the formalism of effective digital indicators.

Let $N = \{0,1,2,\ldots\}$ denote the set of natural numbers, let $x[n]$ denote the initial segment of the length n for an $x\varepsilon A^\infty$, or $x\varepsilon A^*$ with $\ell(x)\geq n$, let $\mathrm{Dom(In)} \subset A^*$ denote the set of strings for which In is defined.

<u>Theorem 1.</u> Let In be an effective digital indicator over a finite alphabet A. Then

(3) $(\exists n_o \varepsilon N)\ (\forall x\varepsilon A^\infty)\ (\exists n(x)\varepsilon N)\ ((n(x)\leq n_o)\ \&\ (x[n]\varepsilon \mathrm{Dom(In)}))$. ¤

Hence, the lengths of initial segments, which uses In in its decision making, are uniformly bounded.

<u>Proof.</u> In can be expressed by a (card A)-ary tree with finite branches, each branch being labelled by a finite digital string from Dom(In). If there existed, in In, an infinite sequence of branches with unboundedly increasing lengths, then there would exist, in In, at least one infinite branch (according to the well-known König's lemma, cf. Ore (1962)). However, this contradicts the effectiveness of In. ¤

<u>Theorem 2.</u> Let A be a finite set of symbols, numbered by naturals 1,2,..., card A, let $i(a)$ be the index of a symbol $a\varepsilon A$. Let $x\varepsilon A^\infty$ be identified with the real number the card A-digital expansion of which is $i(x_1)i(x_2)i(x_3)\ldots$. Then a set $E\subset I$ can be indicated by an effective digital indicator w.r.t. A iff there exists a natural n_o such that

(4) $E = \bigcup_{j=1}^{K} \langle (k(j)-1)\ (\text{card } A)^{-n_o},\ k(j)\ (\text{card } A)^{-n_o})$,

where K, $k(j) \leq (\text{card } A)^{n_o}$ are natural numbers, and the intervals are disjoint. ¤

Proof. Let us prove the assertion for the case A = {0,1}, the generalization to the case when card A>2 is straighforward. Let In_E be an effective digital indicator w.r.t. A = {0,1} and for EcI. According to Theorem 1 there exists n_o N such that for all $x \in A^\infty$ there exists $n(x) \leq n_o$ for which $In_E(x[n(x)])$ is defined. If $In_E(x[n(x)])=1$, then for all $y \in A^\infty$ also $In_E(x[n(x)]*y) = 1$, where * denotes the operation of concatenation. Hence, $E_x = \{x[n(x)]*y: y \in A^\infty\} \subset E$, but E_x is nothing else than the set of reals the binary expansion of which begins with $x[n(x)]$. So $E_x = \langle x[n(x)]2^{-n(x)}, (x[n(x)]+1)2^{-n(x)})$, if we take $x[n(x)]$ as integer in binary code, hence E_x can be written as a union of intervals of the form $\langle k.2^{-n_o}, (k+1)2^{-n_o})$, as $n(x) \leq n_o$. The total number of such intervals is 2^{n_o}, hence, the assertion of Theorem 2 holds with $K \leq 2^{n_o}$. ¤

3. COMPUTATIONAL COMPLEXITY OF EFFECTIVE INDICATORS

We have finished the last chapter with two results which will be, combined with some results concerning the algorithmic complexity, of principial importance in the next chapters. However, we have to emphasize the fact that both these results are of principially non-constructive and existential character, namely, the uniform upper bound for the lengths of decisive initial segments is not an effectively computable function. It is caused by the existential character of König's lemma which is based on the well-known non-constructive topological assertion that it is possible to choose a finite covering from each open covering of a compact set. In this chapter we shall try to arrive at a constructive result enlarging the set of conditions. Namely, we shall assume that the computational complexity of the indicator In_E of the tested set EcI is uniformly bounded. Such a condition seems to be quite natural, as the decision problem whether a real x belongs to E or not is, in Monte-Carlo methods, supposed to be effectively and <u>easily</u> solvable (otherwise it would be beyond a sense to replace the original decision problem on $\mu(E)$ by a great number of such decision problems).

We shall suppose that to each effective digital indicator In_E a unique Turing machine $TM(In_E)$ is ascribed. $TM(In_E)$ computes In_E

and its existence follows immediately from Definition 1. Hence, the notions of time and space computational complexities, defined for Turing machines, may be used also for effective indicators. The <u>space complexity</u> sc(TM(x)) of a Turing machine TM w.r.t. input value x is defined by the number of cells on the operational tape on which TM operates given x (we may consider the simple case investigated by Davis (1958) when input, operational and output tapes are identical with the only tape of TM). The time complexity of TM w.r.t. x is defined by the number of actions taken by TM when operating on x; this complexity is denoted by tc(TM(x)). If TM(x) is not defined, we set $tc(TM(x))=\infty$, it is possible (not necessary), that in this case also $sc(TM(x)) = \infty$. Let us set, moreover, $sc(In_E(x)) = sc(TM(In_E)(x))$, $tc(In_E(x)) = tc(TM(In_E)(x))$, and

$$C(In_E) = \sup\{\min\{sc(In_E(x)), tc(In_E(x))\}: x \in Dom(In_E)\}. \tag{5}$$

Hence, for each real number in <0,1>, either the space complexity or the time complexity, connected with the decision whether this real belongs to $E \subset \langle 0,1 \rangle$ or not, is majorized by $C(In_E)$. If $C(In_E) < \infty$, we may obtain constructive variants of Theorems 1 and 2.

<u>Theorem 3.</u> Let In_E be an effective digital indicator of a set $E \subset I$ w.r.t. a finite alphabet A, let $C(In_E) < \infty$. Then the assertions of Theorems 1 and 2 hold with n_o replaced by $C(In_E)$. ¤

<u>Proof.</u> Again, the result will be proved for A = {0,1}; the simple generalization to the case when card A>2 is left to the reader. Take an $x \in \{0,1\}^*$ for which $In_E(x)$ is defined. All the digits in x must be read, i. e. all boxes in which x is written must be operated, so $sc(x) \geq \ell(x)$, $tc(x) \geq \ell(x)$, as it takes a unit of time complexity (one step or action) to move the head of the $TM(In_E)$ one box aside. If not all digits in x were used when $In_E(x)$ computed, then In_E would be defined for an $x[k]$, $k < \ell(x)$, and this contradicts the condition that there is, for each $y \in \{0,1\}^\infty$, just one j such that $y[j] \in Dom(In_E)$. Hence, $\min\{sc(x), tc(x)\} \geq \ell(x)$ and $C(In_E) \geq \ell(x)$ as well, so if $In_E(x)=1$, then $In_E(y)=1$ for all $y \in I$ such that $y[\ell(x)]=x$. These reals form an interval of the form $\langle k.2^{-\ell(x)}, (k+1)2^{-\ell(x)})$. As $\ell(x) \leq C(In_E)$, such an interval can be written as a finite union of intervals of the form $\langle k_1 2^{-C(In_E)}, (k_1+1)2^{-C(In_E)})$, the total number of such intervals is limited by $2^{C(In_E)}$. ¤

4. ALGORITHMIC COMPLEXITY OF PSEUDO-RANDOM BINARY STRINGS

The axiomatic probability theory attributes the adjective "random" just to generators or devices which produce some results according to given probabilistic laws, but this adjective cannot be used for the obtained results. Hence, "random (finite) sequence" is a finite sequence of random variables defined on a probability space and governed by a finite-dimensional probabilistic distribution. On the other hand, when $X_1, X_2, \ldots, X_n$ is such a random sequence of, say, binary-valued random variables, this probability theory is not able to say whether a particular result $\langle X_1(\omega), X_2(\omega), \ldots, X_n(\omega)\rangle \in \{0,1\}^n$ is random or not.

However, such a classification criterion is necessary when we try to replace a "true-random" sample by a pseudo-random, i. e., in fact, deterministic one. Such a replacement may be useful because of the fact that pseudo-random inputs may be obtained more easily and quickly than the true-random ones, and this fact is of great importance just when Monte-Carlo methods are to be applied.

An interesting and useful idea (presented for the first time by Kolmogorov (1965) and then developed by Fine (1973), Martin-Löf (1966), Schnorr (1971), Chaitin (1974) and others) is to identify the random sequences or strings with those which cannot be generated by short programs, namely, with those which cannot be generated by programs "substantially shorter" than the generated strings themselves. In order to make this idea more precise, consider a universal Turing machine U. If p,S,u are finite binary strings, we write $U(p,S) = u$ to describe the situation when U, given p and S in this order on the input tape, finishes eventually its work over p and S yielding u on the output tape.

Definition 2. Let U be a universal Turing machine (UTM), let x,S be finite binary strings. The conditional algorithmic complexity of x w.r.t. U and under the condition S is denoted by $K_U(x/S)$ and defined by

$$K_U(x/S) = \min\{\ell : \ell \in N,\ \ell = \ell(p),\ U(p,S) = x\}. \tag{6}$$

For $S = \Lambda$ (the empty string) we shall write $K_U(x)$ and omit the adjective "conditional". ☐

The function K_U may be also used to characterize the algorithmic complexity of subsets $E \subset I$. For $j,n \in N$, $0 < j \le n$, denote by $I(j,n)$ the set $\langle (j-1)n^{-1}, jn^{-1})$, so $I = \langle 0,1\rangle = I(1,1) \cup \{1\}$. A binary string

$e(n,E) \in \{0,1\}^n$ is uniquely defined by the condition that its j-th component $e(n,E)(j) = 1$ iff $I(j,n) \cap E \neq \emptyset$, $j \leq n$ (i. e. $e(n,E)(j) = 0$ iff $I(j,n) \subset E-I$). $K_U(E/n)$ will be defined by $K_U(e(n,E)/n)$.

In what follows we shall formulate and prove our results for the simple case of binary alphabet $A = \{0,1\}$, keeping in mind the fact that their generalization to a case when card $A>2$ is a matter of purely technical routine.

Theorem 4. Let In_E be an effective digital indicator of a set $E \subset I$ w.r.t. $A = \{0,1\}$, let U be a UTM. Then there exists a natural number $c_1(E,U)$, independent of n, such that $K_U(E/n) \leq c_1(E,U)$ for all $n \in N$. Moreover, if $C(In_E) < \infty$, then there exists a natural number $c(U)$, dependent only on U, such that $K_U(E/n) \leq 2^{C(In_E)} + c(U)$. ∎

Proof. According to Theorem 2 there exists, for each $E \subset I$ indicable by an effective digital indicator, a natural number $n_o = n_o(E)$ such that E can be written as a union of at most 2^{n_o} disjoint intervals of the form $I(j,2^{n_o})$. Hence, E is uniquely determined by the sequence $e(2^{n_o},E)$ which ascribes units just to the intervals belonging to E. So there exists a fixed program which computes $e(n,E)$ for each $n \in N$, given this n and the sequence $e(2^{n_o},E)$. Rewriting this fact in the terms of algorithmic complexities we can see that there exists $c^*(U) \in N$, independent of E and such that

$$K_U(E/n) = K_U(e(n,E)/n) \leq K_U(e(2^{n_o},E)/n) + c^*(U). \tag{7}$$

On the other hand, there exists $c_2(U) \in N$ such that, for all $x,S \in \{0,1\}^*$, $K_U(x/S) \leq \ell(x) + c_2(U)$. Clearly, it suffices to write a program consisting of x itself and of an instruction saying that x should be rewritten on the output tape (the length of this instruction does not depend on x or S). So $K_U(E/n) \leq 2^{n_o(E)} + c^*(U) + c_2(U)$ and the first assertion is proved by denoting the right-hand side of this inequality as $c_1(E,U)$. If $C(In_E) < \infty$, we may replace n_o by $C(In_E)$ and set $c(U) = c^*(U) + c_2(U)$ and the other assertion is proved as well. ∎

In the rest of this paper, our aim is to estimate the Borel measure $\mu(E)$ of measurable and effectively indicable subsets of $\langle 0,1 \rangle$ by pseudo-random Monte-Carlo method which uses, instead of true-random samples, binary strings of high algorithmic complexity.

Let us denote by $B(n,m)$, $n,m \in N$, the set of all binary strings $x \in \{0,1\}^n$ such that $\sum_{j=1}^{n} x(j) = w(x) \leq m$, set $c(n,m) = \text{card}(B(n,m))$. So $c(n,m) = \sum_{j=0}^{m} \binom{n}{j}$, and $B(n,m) = \{0,1\}^n$, $c(n,m) = 2^n$ for $n \leq m$.

We define $c(u,v)$ also for non-integer u,v, setting $c(u,v) = c(\mathrm{Int}(u),\mathrm{Int}(v))$, $\mathrm{Int}(u) = \max\{n: n\in N,\ n\le u\}$. A sequence $x\in B(n,m)$ is called T-random w.r.t. $B(n,m)$ for a $T\in N$, $T>0$, if $K_U(x/n,m) \ge \log_2 c(n,m) - T$.

<u>Theorem 5.</u> For each positive integers n,m,T there is at least one $x\in B(n,m)$ which is T-random w.r.t. $B(n,m)$. ⌑

<u>Proof.</u> Programs are binary strings, so there are at most 2^i programs of the length i and at most $2^{\log_2 c(n,m)-T+1} - 1 = c(n,m).2^{-T+1} - 1$ programs which are shorter than $\log_2 c(n,m) - T$. $T>0$ implies that $c(n,m)2^{-T+1} - 1 < c(n,m)$, so there is at least one $x\in B(n,m)$ which cannot be generated by a program shorter than $\log_2 c(n,m) - T$. Hence, x is T-random w.r.t. $B(n,m)$. ⌑

Denote by $\mathcal{G}$ the minimal algebra of subsets in $I = \langle 0,1\rangle$ generated by semi-open intervals. A sequence $x\in\{0,1\}^n$ is called <u>adequate</u> w.r.t. E, if $\sum_{j=1}^{n}(e(n,E)(j)).(x(j)) > 0$, i. e., if there is at least one $j\le n$ such that $x(j) = e(n,E)(j) = 1$.

<u>Theorem 6.</u> There exists $c\in N$ such that for all positive $n,m,T\in N$, all $E\in\mathcal{G}$ and all $x\in B(n,m)$ this implication holds: if

$$K_U(e(n,E)/n) < \log_2[c(n,m)(c(n-n\mu(E),\ m))^{-1}] - T - c, \tag{8}$$

and if x is T-random w.r.t. $B(n,m)$, then x is adequate w.r.t. E. ⌑

<u>Proof.</u> Let $x\in B(n,m)$ be non-adequate w.r.t. E, so $e(n,E)(j)=1$ implies $x(j) = 0$, $j\le n$. Hence, $x(j) = 1$ may occur only if $e(n,E)(j)=0$ and we may define x, given $e(n,E)$, by saying to which zeros in $e(n,E)$ correspond units in x. Among the sets E with the same $\mu(E)$ the number of zeros in $e(n,E)$ is maximal, if E is a union of intervals $I(j,n)$; this number of zeros is $n(1-\mu(E))$. Hence, the positions of units in x w.r.t. zeros in $e(n,E)$ can be defined by a string with $n(1-\mu(E))$ symbols and at most m units, i. e. by a string from $B(n(1-\mu(E)),m)$. In order to encode such a string we need a binary string of the length $\mathrm{Int}(\log_2(c(n(1-\mu(E)),m))+1)$.

Having $e(n,E)$ and a natural number not exceeding $c(n(1-\mu(E)),m)$, a fixed program of the length c_1 yields a non-adequate x in such a way that it ascribes zeros to all units in $e(n,E)$ and to all indices not belonging to the string of the length $n(1-\mu(E))$ encoded by the used integer. For $c' = c_1+1$ we obtain

$$K_U(x/n,m) \le K_U(e(n,E)/n) + \log_2 c(n(1-\mu(E)),m) + c', \tag{9}$$

so (8) implies, for all non-adequate $x\in B(n,m)$,

(10) $K_U(x/n,m) \leq \left(\log_2 c(n,m)(c(n(1-\mu(E)),m))^{-1}\right) - T - c +$
$+ \log_2 c(n(1-\mu(E)),m) + c' < \log_2 c(n,m) - T,$

if $c>c'$, so x is not T-random w.r.t. B(n,m). Hence, T-randomness of x w.r.t. B(n,m) assures that x is adequate w.r.t. E. ⌑

Definition 3. Let x,y be two binary strings. The j-th component $y(j)$, $j \leq \ell(y)$, of y is called to be sampled by x, if there is $i \leq \ell(x)$ such that $x(i)=1$ and $I(j,\ell(y)) \cap I(i,\ell(x)) \neq \emptyset$. A y(j) is called successful w.r.t. x, if it is sampled by x and if $y(j) = 1$. The (relative) weight $w(y/x)$ of y w.r.t. x is defined by $w(y/x) = \beta(x,y)(\alpha(x,y))^{-1}$, where

$$\alpha(x,y) = \text{card}\{j : j \leq \ell(y),\ y(j) \text{ is sampled by } x\},$$
(11)
$$\beta(x,y) = \text{card}\{j : j \leq \ell(y),\ y(j) \text{ is successful w.r.t. } x\}. \text{ ⌑}$$

Intuitively said, units in x "sample" components of y and $w(y/x)$ defines the relative frequency of units among the sampled components. If the "sample" made by x is "representative" enough, then $w(y/x)$ may serve as an estimate for the average weight $(\ell(y))^{-1}.w(y) = (\ell(y))^{-1} \sum_{j=1}^{\ell(y)} y(j)$ of the sequence y. Moreover, if $E \subset I$, $E \in G$, then $\lim_{n\to\infty} n^{-1} w(e(n,E)) = \mu(E)$, hence, $w(e(n,E)/x)$ may serve as an estimate of $\mu(E)$. The following theorem shows, that if E is indicable by an effective digital indicator and if the algorithmic complexity of x is high enough, then the estimation is perfect.

5. RESULTS CONCERNING ESTIMATIONS

Theorem 7. Let In_E be an effective digital indicator of a set $\emptyset \neq E \subset I$, $E \in G$, w.r.t. $A = \{0,1\}$, let U be a UTM. Then there exist $K_1 = K_1(E) \in N$, $K_2 = K_2(E) \in N$ such that, for all $x \in \{0,1\}^*$, if $\ell(x) > K_1.K_2$, $K_U(x/\ell(x)) \geq \ell(x) - T$, i. e., x is T-random w.r.t. $B(\ell(x),\ell(x))$ with the same $T \in N$ as in Theorem 6, then $w(e(K_2,E)/x) = \mu(E)$. ⌑

Proof. Let In_E, U satisfy the conditions, let $c \in N$ satisfy Theorem 6. Taking m=n, i. e. $c(m,n) = 2^n$, the condition (8) reads, for $E_1 \in G$:

(12) $K_U(e(n,E_1)/n) = K_U(E_1/n) < \log_2[c(n,n)(c(n(1-\mu(E_1)),n))^{-1}] - T - c =$
$= \log_2[c(n,n)(c(n(1-\mu(E_1)),n(1-\mu(E_1))))^{-1}] - T - c \leq$
$\leq \log_2[2^n(2^{n(1-\mu(E_1))})^{-1}] - T - c =$
$= \log_2(2^{n\mu(E_1)}) - T - c = n\mu(E_1) - T - c\ .$

Take the tested set $E \in G$, according to Theorem 2 there exists $n_o \in N$ such that E can be written as a finite union of intervals of the type $I(j,2^{n_o})$. Take $E_1 = I(j,2^{n_o})$ for a $j \le 2^{n_o}$, so $\mu(E_1) = 2^{-n_o}$. All information concerning E is contained in the string $e(2^{n_o},E)$. E_1 is completely defined by j and 2^{n_o} in the sense that, given n, a fixed program of the length, say $c_3 = c_3(U)$ computes, given $n \in N$, the string $e(n,E_1)$ from j, 2^{n_o} and n. Hence, $K_U(e(n,E_1)/n) \le 2n_o + c_3$, and the condition (12) will hold, if $2n_o + c_3 < (2^{-n_o}.n) - T - c$. It is satisfied for $n > (2n_o + c_3 + T + c).2^{n_o} = K_1.K_2$, setting $K_1 = 2n_o + c_3 + T + c$, $K_2 = 2^{n_o}$, so we may state that if x is a finite binary string which is T-random w.r.t. $B(\ell(x),\ell(x))$ and such that $\ell(x) > K_1.K_2$, then x is adequate w.r.t. $I(j,2^{n_o(E)})$.

The adequateness of x w.r.t. a fixed $I(j,2^{n_o})$ does not eliminate, but rather implies the adequateness of x w.r.t. <u>all</u> intervals $I(j,2^{n_o})$ simultaneously as the relations $K_U(e(n,E_1)/n) \le 2n_o + c_3$ and $\mu(E_1) = 2^{-n_o}$, which were the only used, hold for all such intervals. Hence, for each $j \le 2^{n_o}$ there is an $i_j \le \ell(x)$ such that $e(\ell(x), I(j,2^{n_o}))(i_j) = x(i_j) = 1$. But this means nothing else than that each $I(j,2^{n_o})$ is sampled by x, in symbols, $\alpha(x,e(2^{n_o},E)) = 2^{n_o}$. It follows immediately that each $e(2^{n_o},E)(j)$ which equals to one is classified as successful, so $\beta(x,e(2^{n_o},E)) = w(e(2^{n_o},E))$. Hence,

$$(13) \quad (\alpha(x,e(2^{n_o},E)))^{-1}\beta(x,e(2^{n_o},E)) = w(e(2^{n_o},E))(2^{-n_o}) = \mu(E),$$

as E can be written as a finite union of intervals of the length 2^{-n_o} and there is one-to-one correspondence between such intervals and the units in $e(2^{n_o},E)$. ⌑

<u>Theorem 8.</u> (A constructive version of Theorem 7.) Let the notations and conditions of Theorem 7 hold, let T,c satisfy Theorem 6, let $c_3 \in N$ be the same as in the proof of Theorem 7. Let x be a finite binary sequence which is T-random w.r.t. $B(\ell(x),\ell(x))$ and such that $\ell(x) > (2C(In_E)+c_3+T+c).2^{C(In_E)}$, supposing that $C(In_E)<\infty$. Then $w(e(2^{C(In_E)},E)/x) = \mu(E)$. ⌑

<u>Proof.</u> As shown in the proof of Theorem 3, we may take $n_o = C(In_E)$; repeating the proof of Theorem 7 under this condition, the assertion of Theorem 8 is immediately obtained. ⌑

As can be easily seen, the limitation to strings which are T-random w.r.t. the set of all strings of the same length, i. e. to such $x \in \{0,1\}^*$ for which $K_U(x/\ell(x)) \ge \ell(x) - T$, is not too restrictive. Or, we may enumerate all strings from $B(n,m)$ by binary strings corresponding to natural numbers not exceeding $c(n,m)$. Hence, using sequences which are T-random w.r.t. $B(n,m)$ we may replace them by sequences which are T'-random w.r.t. $B(\log_2 c(n,m)+1, \log_2 c(n,m)+1)$ and apply the results proved above with the threshold value possibly enlarged by a constant corresponding to $|T'-T|$ and to the length of the coding program. Let us close this section by mentioning briefly a more general result dealing with the case when the tested set is not effectively indicable in the sense defined above. Because of the reasons mentioned here we shall limit ourselves to the case when pseudo-random inputs are represented by strings which are T-random w.r.t. the set of all strings of the same length. More generally and in more details this case is investigated by Kramosil (1982) where also some informal comments can be found.

Theorem 9. Let $E \subset I = \langle 0,1 \rangle$, $E \in \mathcal{G}$, $E \neq \emptyset$, let U be a UTM, let $\mathcal{S} = \{S_1, S_2, \ldots\}$ be an infinite sequence of finite binary strings such that, for all $n \in N$, $\ell(S_n)=n$, $K_U(S_n/n) \geq n-T$ for the same T as in Theorem 6. Then there exists a total function $k: N \rightarrow N$ such that

$$\text{(14)} \qquad \lim_{n\to\infty} w(e(n,E)/S_{k(n)}) = \mu(E). \ \square$$

Proof. The proof of Theorem 7 yields that there exists, for each $n \in N$, $K(n) \in N$ such that if $x \in \{0,1\}^*$ and $K_U(x/\ell(x)) \geq K(n)$, then x is adequate w.r.t. all $I(j,n)$, $j \leq n$. Hence, we define $k(n)$ in such a way that $k(n) = \min\{n_1 : K_U(S_{n_1}/n_1) \geq K(n)\}$. Referring, once more, to the proof of Theorem 7 we may state that the adequateness of $S_{k(n)}$ w.r.t. all $I(j,n)$, $j \leq n$, implies that $w(e(n,E)/S_{k(n)}) = n^{-1} w(e(n,E))$. However, $\lim_{n\to\infty} w(e(n,E)) = \mu(E)$ for $E \in \mathcal{G}$, as such E can be written as a finite union of intervals and for all $E_1 = \langle a_1, b_1) \subset I$ the value $n^{-1} w(e(n,E_1))$ tends to $(b_1 - a_1) = \mu(E_1)$. Hence, $\lim_{n\to\infty} w(e(n,E)/S_{k(n)}) = \mu(E)$. $\square$

Theorems 7 and 9 show that we may use pseudo-random side inputs in Monte-Carlo methods supposing that these inputs are of high algorithmic complexity. The assertions are stronger than in the true-random case, but the conditions are also stronger and non-effective. No effective generator is able to produce only strings with algorithmic complexity, exceeding an a priori given threshold value, no effective general procedure exists to decide, for each $x \in \{0,1\}^*$, $k \in N$, whether $K_U(x) \geq k$ or not. From this point of view we may consider random number generators as effective devices providing us, with probability high enough, by strings of high algorithmic complexity. This reliability cannot be absolute, each effective device produces, from time to time, a string of low complexity; used as pseudo-random inputs such strings may cause an error the probability of which is studied by the means of classical probability theory. So we should take the approach investigated here rather as a tool which enables to penetrate more deeply into the nature of complexity, randomness and their interconnections than as a tool which could immediately replace the classical methods of probability theory and mathematical statistics.

The author thanks to A. Perez for valuable remarks and comments.

REFERENCES

Chaitin G. J. (1974): Information-Theoretic Limitations of Formal Systems. Journal of the Association for Computing Machinery 21, No. 3, 403-424.

Davis M. (1958): Computability and Unsolvability. McGraw-Hill Book Comp., New York - London.

Fine T. L. (1973): Theories of Probability. An Examination of Foundations. Academic Press, New York.

Kolmogorov A. N. (1965): Three Approaches to the Quantitative Definition of Information. Problemy peredači informacii 1, 4-7.

Kramosil I. (1982): Pseudo-Random Monte-Carlo Methods. To appear in Fundamenta Informaticae, PWN Warsaw.

Martin-Löf P. (1966): The Definition of Random Sequences. Information and Control 9, 602-619.

Ore O. (1962): Theory of Graphs. AMS Publications, Providence, Rhode Island. Russian translation: Nauka, Moscow, 1980.

Schnorr C. P. (1971): Zufälligkeit und Wahrscheinlichkeit. Lecture Notes in Mathematics 218, Springer-Verlag, Berlin-Heidelberg.

Czechoslovak Academy of Sciences
Institute of Information Theory
and Automation
Pod vodárenskou věží 4
182 08 Praha 8 - Libeň
Czechoslovakia

DISCRIMINATION RATE LEAST FAVORABLE PAIRS OF DISTRIBUTIONS FOR ε-CONTAMINATED STATISTICAL HYPOTHESES OR WITH f-DIVERGENCE LIKE NEIGHBORHOODS

Albert Perez

Prague

Key words: Composite statistical hypotheses, discrimination rate, least favorable pairs of distributions, f-divergence.

ABSTRACT

Based on our paper [1], concerning the discrimination rate of simple statistical hypotheses by unfitted decision procedures, we introduce in the case of composite statistical hypotheses considered here the concept of Discrimination Rate Least Favorable Pair (DR-LFP) of distributions for a given decision procedure and the concept of Maximal DR-LFP which is the analog of the well-known LFP concept. Their investigation is illustrated in two cases: (I) the ε-contamination Huber's model and (II) our f-divergence like neighborhood model of composite statistical hypotheses.

INTRODUCTION

Let $H_1 = \{Q_1\}$ and $H_2 = \{Q_2\}$ be two disjoint sets of probability measures on a measurable space $(X,\mathfrak{X})$. In the sequel, H_1 and H_2 will represent the two composite statistical hypotheses to be discriminated on the base of a growing number of i.i.d. observations on the sample space $(X,\mathfrak{X})$.

As examples we consider either the <u>ε-contamination Huber's model</u>:

(I) $$H_i = \{Q_i : Q_i = (1-\varepsilon_i)P_i + \varepsilon_i S_i\}, \quad i=1,2$$

(with $0 \le \varepsilon_1, \varepsilon_2 \le 1$; $P_1 \ne P_2$ given and S_1 and S_2 arbitrary probability measures on $(X,\mathfrak{X})$) or an <u>f-divergence like neighborhood model</u>:

(II) $$H_i = \{Q_i : \mathrm{H}_f(P_i, Q_i) := \int f\left(\frac{dP_i}{dQ_i}\right) dQ_i \le \varepsilon_i\}, \quad i=1,2$$

(with $\varepsilon_i \geq 0$; $P_1 \neq P_2$ given probability measures on $(X,\mathfrak{X})$; $f(u)$ continuous and convex function of $u \in (0,\infty)$ with $f(1) = 0$ and, by definition, $f(0) = \lim_{u \searrow 0} f(u)$ and $0 \circ f(0) = 0$; here u coincides with the Radon-Nikodym derivative $\frac{dP_i}{dQ_i}$ we assume existing; $H_f(P,Q)$ is the Csiszár's "f-divergence" in the case here. We call it in our papers also "generalized f-entropy" by extension of the generalized or relative Shannon's entropy $H(P,Q)$, obtained for $f(u) = u \log u$, we studied in a 1957 paper). One may use $H_f(Q_i,P_i)$ instead in (II).

In the present paper we largely use the extremal generalized f-entropy method first introduced in our paper [2] for minimizing f-divergences (f general) $H_f(P,Q)$ with respect to P (Q) given Q (P) under different constraints.

In order to introduce the concept of Discrimination Rate Least Favorable Pair (DR-LFP) of distributions for composite statistical hypotheses let us consider first the DR of a pair (Q_1,Q_2) of <u>simple</u> statistical hypotheses for a maximum-likelihood (ML) decision procedure adapted to some other pair $(\overline{Q}_1,\overline{Q}_2)$. This problem was studied in our paper [1]. From Corollary 1 of [1] we obtain in particular that the corresponding error probabilities for a growing number, n, of i.i.d. observations converge both exponentially to zero provided that the following conditions are fulfilled:

(i) $$H_{Q_1}(\overline{Q}_1,\overline{Q}_2) := \int \log\frac{d\overline{Q}_1}{d\overline{Q}_2}\, dQ_1 > 0; \quad H_{Q_2}(\overline{Q}_2,\overline{Q}_1) > 0$$

(ii) $$L_{Q_1}(\overline{Q}_2,\overline{Q}_1) := \int \frac{d\overline{Q}_2}{d\overline{Q}_1} \log \frac{d\overline{Q}_2}{d\overline{Q}_1}\, dQ_1 > 0; \quad L_{Q_2}(\overline{Q}_1,\overline{Q}_2) > 0,$$

implicitly assuming that $\overline{Q}_1 \equiv \overline{Q}_2$ $[Q_1]$ and $[Q_2]$.

Namely, it holds:

<u>THEOREM 1.</u>

Under conditions (i) and (ii), the following limits exist and are negative

(iii) $$\begin{cases} \lim\limits_{n\to\infty} \frac{1}{n} \log Q_1^n(d\overline{Q}_1^n < d\overline{Q}_2^n) = \min\limits_{\alpha} \log \int \left(\frac{d\overline{Q}_2}{d\overline{Q}_1}\right)^{\alpha} dQ_1 < 0 \\ \lim\limits_{n\to\infty} \frac{1}{n} \log Q_2^n(d\overline{Q}_2^n \leq d\overline{Q}_1^n) = \min\limits_{\alpha} \log \int \left(\frac{d\overline{Q}_1}{d\overline{Q}_2}\right)^{\alpha} dQ_2 < 0 \end{cases}$$

For the corresponding exponential rates it holds:

$$
\text{(iv)} \quad \left\{
\begin{array}{l}
-\min\limits_{\alpha} \log \int \left(\frac{d\overline{Q}_2}{d\overline{Q}_1}\right)^{\alpha} dQ_1 = \min\limits_{R} \mathrm{H}(R,Q_1) =: \mathrm{H}(R^{Q_1}_{\overline{Q}_1\overline{Q}_2},Q_1) \\
\qquad\qquad\qquad\qquad\qquad\qquad \text{under constraint } \mathrm{H}(R,\overline{Q}_1)=\mathrm{H}(R,\overline{Q}_2) \\
-\min\limits_{\alpha} \log \int \left(\frac{d\overline{Q}_1}{d\overline{Q}_2}\right)^{\alpha} dQ_2 = \min\limits_{R} \mathrm{H}(R,Q_2) =: \mathrm{H}(R^{Q_2}_{\overline{Q}_1\overline{Q}_2},Q_2) \\
\qquad\qquad\qquad\qquad\qquad\qquad \text{under constraint } \mathrm{H}(R,\overline{Q}_1)=\mathrm{H}(R,\overline{Q}_2)
\end{array}
\right.
$$

where R is an arbitrary probability measure on $(X,\mathfrak{X})$ and $\mathrm{H}(R,Q)$ is, as above, the relative Shannon's entropy of R with respect to Q (if $R \not\ll Q$, $\mathrm{H}(R,Q):=\infty$).

<u>Proof.</u> (iii) is proved in [1], Corollary 1. As to (iv), (first equality), we have for the minimizing α

$$\frac{d}{d\alpha} \int \left(\frac{d\overline{Q}_2}{d\overline{Q}_1}\right)^{\alpha} dQ_1 = \int \log\left(\frac{d\overline{Q}_2}{d\overline{Q}_1}\right) \left(\frac{d\overline{Q}_2}{d\overline{Q}_1}\right)^{\alpha} dQ_1 = 0$$

Let us search for the probability measure R minimizing $\mathrm{H}(R,Q_1)$ under constraint $\mathrm{H}(R,\overline{Q}_1) = \mathrm{H}(R,\overline{Q}_2)$ by applying the Lagrange multipliers method. Parting from the expression

$$\int \log\left(\frac{dR}{dQ_1}\right) \left(\frac{dR}{dQ_1}\right) dQ_1 - \lambda_1 \int \log\left(\frac{d\overline{Q}_2}{d\overline{Q}_1}\right) \left(\frac{dR}{dQ_1}\right) dQ_1 - \mu_1 \int \frac{dR}{dQ_1} dQ_1$$

we obtain the equation

$$\log\left(\frac{dR}{dQ_1}\right) + 1 - \lambda_1 \log\left(\frac{d\overline{Q}_2}{d\overline{Q}_1}\right) - \mu_1 = 0$$

The solution is $R = R^{Q_1}_{\overline{Q}_1\overline{Q}_2}$ with

$$\text{(1)} \qquad \frac{dR^{Q_1}_{\overline{Q}_1\overline{Q}_2}}{dQ_1} = \left(\frac{d\overline{Q}_2}{d\overline{Q}_1}\right)^{\lambda_1} \Big/ \int \left(\frac{d\overline{Q}_2}{d\overline{Q}_1}\right)^{\lambda_1} dQ_1$$

where λ_1 is given by the same condition as that for the minimizing α. Similarly, we obtain

$$\text{(2)} \qquad \frac{dR^{Q_2}_{\overline{Q}_1\overline{Q}_2}}{dQ_2} = \left(\frac{d\overline{Q}_1}{d\overline{Q}_2}\right)^{\lambda_2} \Big/ \int \left(\frac{d\overline{Q}_1}{d\overline{Q}_2}\right)^{\lambda_2} dQ_2$$

where λ_2 coincides with the α minimizing $\int \left(\frac{d\overline{Q}_1}{d\overline{Q}_2}\right)^{\alpha} dQ_2$. From (1) and (2) we obtain easily (iv). Q.E.D.

From the above it follows that the discrimination rate of Q_1 and Q_2 by a ML decision procedure adapted to $(\overline{Q}_1,\overline{Q}_2)$ is given by $\min_i H(R^{Q_i}_{\overline{Q}_1\overline{Q}_2}, Q_i)$.

DISCRIMINATION RATES FOR COMPOSITE STATISTICAL HYPOTHESES

Passing now to the case of composite statistical hypotheses H_1 and H_2, we shall assume in the sequel that for any two pairs (Q_1,Q_2) and $(\overline{Q}_1,\overline{Q}_2)$ of $H_1 \times H_2$ the discrimination rate of Q_1 and Q_2 by a ML decision procedure adapted to $(\overline{Q}_1,\overline{Q}_2)$ is given, as above, by $\min_i H(R^{Q_i}_{\overline{Q}_1\overline{Q}_2}, Q_i)$ where $R^{Q_i}_{\overline{Q}_1\overline{Q}_2}$ $(i=1,2)$ are such that

$$H(R^{Q_i}_{\overline{Q}_1\overline{Q}_2}, Q_i) = \min_R H(R, Q_i) \tag{3}$$

under constraint $H(R,\overline{Q}_1)=H(R,\overline{Q}_2)$

In the case that, for some $i=1,2$, $Q_i=\overline{Q}_i$ we shall write $R_{\overline{Q}_1\overline{Q}_2}$ instead of $R^{Q_i}_{\overline{Q}_1\overline{Q}_2}$.

Let $Q_i^* = Q_i^*(\overline{Q}_1,\overline{Q}_2)$ $(i=1,2)$ be defined (if they exist) by

$$H(R^{Q_i^*}_{\overline{Q}_1\overline{Q}_2}, Q_i^*) := \inf_{Q_i \in H_i} H(R^{Q_i}_{\overline{Q}_1\overline{Q}_2}, Q_i) \tag{4}$$

The pair (Q_1^*,Q_2^*) is said to be a Discrimination Rate Least Favorable Pair (DR-LFP) of distributions for discriminating H_1 and H_2 by a ML decision procedure adapted to $(\overline{Q}_1,\overline{Q}_2)$.

Correspondingly, the quantity

$$DR^* = DR^*(\overline{Q}_1,\overline{Q}_2) := \min_i \inf_{Q_i \in H_i} H(R^{Q_i}_{\overline{Q}_1\overline{Q}_2}, Q_i) \tag{5}$$

is called Discrimination Rate of H_1 and H_2 by a ML decision procedure adapted to $(\overline{Q}_1,\overline{Q}_2)$. It coincides with $\min_i H(R^{Q_i^*}_{\overline{Q}_1\overline{Q}_2}, Q_i^*)$ if the pair (Q_1^*,Q_2^*) exists.

Let $\hat{Q}_i$ $(i=1,2)$ be defined (if they exist) by

$$H(R_{\hat{Q}_1\hat{Q}_2}, \hat{Q}_1) = H(R_{\hat{Q}_1\hat{Q}_2}, \hat{Q}_2) := \sup_{\overline{Q}_1\in H_1, \overline{Q}_2 \in H_2} \min_i \inf_{Q_i \in H_i} H(R^{Q_i}_{\overline{Q}_1\overline{Q}_2}, Q_i) \tag{6}$$

$$= \sup_{\overline{Q}_1\in H_1, \overline{Q}_2 \in H_2} \min_i H(R^{Q_i^*}_{\overline{Q}_1\overline{Q}_2}, Q_i^*)$$

Remark that if the latter supremum is obtained for some pair $(\hat{Q}_1,\hat{Q}_2)$ then one can take $Q_i^*(\hat{Q}_1,\hat{Q}_2) = \hat{Q}_i$ since it holds in general

$$\min_i H(R^{Q_i}_{\overline{Q}_1\overline{Q}_2}, Q_i) \leq H(R_{Q_1Q_2}, Q_1) = H(R_{Q_1Q_2}, Q_2) \tag{7}$$

due to the fact that the discrimination rate of Q_1 and Q_2 obtains its maximum if we apply a ML decision procedure adapted to (Q_1,Q_2).

The pair $(\hat{Q}_1,\hat{Q}_2)$ is said to be Maximum Discrimination Rate Least Favorable Pair (MDR-LFP) of distributions for discriminating the composite statistical hypotheses H_1 and H_2.

Correspondingly, the quantity

$$\hat{\mathrm{DR}} := \sup_{\overline{Q}_1\in H_1,\overline{Q}_2\in H_2} \min_i \inf_{Q_i\in H_i} H(R^{Q_i}_{\overline{Q}_1\overline{Q}_2}, Q_i) \tag{8}$$

is called Maximum Discrimination Rate of H_1 and H_2. It coincides with $H(R_{\hat{Q}_1\hat{Q}_2},\hat{Q}_1)$ if the pair $(\hat{Q}_1,\hat{Q}_2)$ exists.

Provided that the *min* figuring below exist, the discrimination rate $\mathrm{DR}^*(\overline{Q}_1,\overline{Q}_2)$ resp. the corresponding DR-LFP (Q_1^*,Q_2^*) may be approximated arbitrarily from above by means of:

$$\begin{aligned} H(R_{\overline{Q}_1\overline{Q}_2},\overline{Q}_1) &\geq H(R_{\overline{Q}_1\overline{Q}_2}, Q_i^o) := \min_{Q_i\in H_i} H(R_{\overline{Q}_1\overline{Q}_2}, Q_i) \geq \\ &\geq H(R^{Q_i^o}_{\overline{Q}_1\overline{Q}_2}, Q_i^o) \geq H(R^{Q_i^o}_{\overline{Q}_1\overline{Q}_2}, Q_i^1) := \min_{Q_i\in H_i} H(R^{Q_i^o}_{\overline{Q}_1\overline{Q}_2}, Q_i) \geq \\ &\geq \ldots \geq H(R^{Q_i^{k-1}}_{\overline{Q}_1\overline{Q}_2}, Q_i^k) \geq H(R^{Q_i^k}_{\overline{Q}_1\overline{Q}_2}, Q_i^k) \xrightarrow[k\to\infty]{} H(R^{Q_i^*}_{\overline{Q}_1\overline{Q}_2}, Q_i^*) \end{aligned} \tag{9}$$

As it concerns $\hat{\mathrm{DR}}$ and $(\hat{Q}_1,\hat{Q}_2)$, the following results hold.

LEMMA 1. For any pair $(Q_1,Q_2)\in H_1 \times H_2$,

$$H(R_{Q_1Q_2}, Q_1) \geq \hat{\mathrm{DR}} \geq \min_i H(R^{Q_i^*}_{Q_1Q_2}, Q_i^*), \tag{10}$$

where $Q_i^* = Q_i^*(Q_1,Q_2)$.

Proof. The second inequality (10) follows immediately from the definition (8) of $\hat{\mathrm{DR}}$. As to the first inequality (10) it results from the fact that for any $\varepsilon > 0$ there exists a pair $(Q_1^\varepsilon,Q_2^\varepsilon) \in H_1 \times H_2$ such that (cf. (8) and (7))

$$\hat{DR} \le \min_i H(R^{Q_i^*(Q_1^\varepsilon,Q_2^\varepsilon)}_{Q_1^\varepsilon Q_2^\varepsilon}, Q_i^*(Q_1^\varepsilon,Q_2^\varepsilon)) + \varepsilon \le$$

$$\le \min_i H(R^{Q_i}_{Q_1^\varepsilon Q_2^\varepsilon}, Q_i) + \varepsilon \le H(R_{Q_1Q_2}, Q_1) + \varepsilon$$

Q.E.D.

LEMMA 2. Let $\{Q_{1j}, Q_{2j}\}_{j\ge 1}$ be a sequence of pairs $(Q_{1j}, Q_{2j}) \in H_1 \times H_2$ such that

$$\lim_{j\to\infty}\left[H(R_{Q_{1j}Q_{2j}}, Q_{1j}) - \min_i H(R^{Q_i^*(Q_{1j},Q_{2j})}_{Q_{1j}Q_{2j}}, Q_i^*(Q_{1j},Q_{2j}))\right] = 0 \tag{11}$$

Then

$$\hat{DR} = \lim_{j\to\infty} H(R_{Q_{1j}Q_{2j}}, Q_{1j})$$

and conversely.

Proof. It results immediately from the assumption and from Lemma 1.

THEOREM 2. A necessary and sufficient condition for a pair $(Q_1,Q_2) \in H_1 \times H_2$ to be MDR-LFP is

$$H(R_{Q_1Q_2}, Q_1) = \min_{Q_i' \in H_i} H(R_{Q_1Q_2}, Q_i') \quad (i=1,2) \tag{12}$$

Proof. *Necessity*: According to the remark following (6), it holds

$$Q_i^*(\hat{Q}_1, \hat{Q}_2) = \hat{Q}_i \quad (i=1,2)$$

so that

$$H(R_{\hat{Q}_1\hat{Q}_2}, \hat{Q}_1) = H(R^{Q_i^*(\hat{Q}_1,\hat{Q}_2)}_{\hat{Q}_1\hat{Q}_2}, Q_i^*(\hat{Q}_1,\hat{Q}_2)) \quad (i=1,2)$$

But (9) for $(\overline{Q}_1,\overline{Q}_2) = (\hat{Q}_1,\hat{Q}_2)$ then implies (12) for $(Q_1,Q_2) = (\hat{Q}_1,\hat{Q}_2)$.

Sufficiency: If $Q_i^o(Q_1,Q_2)$ is the minimizing Q_i' in (12), one can take

$$Q_i^o(Q_1,Q_2) = Q_i \quad (i=1,2)$$

The sequence (9) of inequalities by taking $(\overline{Q}_1,\overline{Q}_2) = (Q_1,Q_2)$ is then reduced to a sequence of equalities since

$$R_{Q_1Q_2}^{Q_i^o} = R_{Q_1Q_2}^{Q_i} = R_{Q_1Q_2} \quad \text{and, thus}$$

$$Q_i^1(Q_1^o, Q_2^o) = Q_i^1(Q_1, Q_2) = Q_i^o(Q_1, Q_2) = Q_i \quad (i=1,2)$$

and so on. As a consequence

$$H(R_{Q_1Q_2}, Q_1) = H(R_{Q_1Q_2}^{Q_i^*(Q_1,Q_2)}, Q_i^*(Q_1,Q_2))$$

By Lemma 1 then

$$H(R_{Q_1Q_2}, Q_1) = \hat{DR} = H(R_{\hat{Q}_1\hat{Q}_2}, \hat{Q}_1)$$

Thus, one may take $\hat{Q}_1 = Q_1$ and $\hat{Q}_2 = Q_2$.

Q.E.D.

THEOREM 3. A necessary and sufficient condition for a pair $(Q_1, Q_2) \in H_1 \times H_2$ to be a DR-LFP corresponding to a ML decision procedure adapted to $(\bar{Q}_1, \bar{Q}_2)$ is

$$H(R_{\bar{Q}_1\bar{Q}_2}^{Q_i}, Q_i) = \min_{Q_i' \in H_i} H(R_{\bar{Q}_1\bar{Q}_2}, Q_i') \tag{13}$$

Proof. It is based on (9) and it is an obvious simplified version of the Proof of Theorem 2.

Q.E.D.

Based on (9), Lemma 2 and Theorem 2, a *conjecture* for approximating arbitrarily the maximum discrimination rate $\hat{DR}$ of H_1 and H_2 resp. the corresponding MDR-LFP $(\hat{Q}_1, \hat{Q}_2)$ is the following.

Provided that the *min* figuring below exist, we construct the sequence $\{Q_{1j}, Q_{2j}\}_{j\geq 1}$ of Lemma 2 as follows beginning from some suitable pair (P_1, P_2) (for instance, in the case of models (I) or (II) one may take the corresponding P_i's as starting point):

$$(14) \quad \begin{cases} Q_{i1} : H(R_{P_1P_2}, Q_{i1}) := \min\limits_{Q_i \in H_i} H(R_{P_1P_2}, Q_i) & (i=1,2) \\ Q_{i2} : H(R_{Q_{11}Q_{21}}, Q_{i2}) := \min\limits_{Q_i \in H_i} H(R_{Q_{11}Q_{21}}, Q_i) & (i=1,2) \\ \cdots\cdots & \\ Q_{ij} : H(R_{Q_{1j-1}Q_{2j-1}}, Q_{ij}) := \min\limits_{Q_i \in H_i} H(R_{Q_{1j-1}Q_{2j-1}}, Q_i) & (i=1,2) \end{cases}$$

As expected, the following Theorem holds.

THEOREM 4. If the composite statistical hypotheses H_1 and H_2 are replaced by a reduced ones H_1' and H_2' with $H_i' \subset H_i$ $(i=1,2)$, then

(15) $\hat{DR}(H_1,H_2) \equiv H(R_{\hat{Q}_1\hat{Q}_2},\hat{Q}_1) \leq \hat{DR}(H_1',H_2') \equiv H(R_{\hat{Q}_1'\hat{Q}_2'},\hat{Q}_1')$,

where $(\hat{Q}_1,\hat{Q}_2)$ is a MDR-LFP for (H_1,H_2) and $(\hat{Q}_1',\hat{Q}_2')$ for (H_1',H_2').

<u>Proof.</u> For any $(\overline{Q}_1,\overline{Q}_2)$ it holds

$$H(R^{Q_i^*(\overline{Q}_1,\overline{Q}_2;H_i)}_{\overline{Q}_1\overline{Q}_2},Q_i^*(\overline{Q}_1,\overline{Q}_2;H_i)) \leq \text{(since } H_i'\subseteq H_i\text{)}$$

$$\leq H(R^{Q_i^*(\overline{Q}_1,\overline{Q}_2;H_i')}_{\overline{Q}_1\overline{Q}_2},Q_i^*(\overline{Q}_1,\overline{Q}_2;H_i')) \leq H(R^{Q_i'}_{\overline{Q}_1\overline{Q}_2},Q_i')$$

for any $Q_i'\in H_i'$ (by definition of Q_i^*)

Take, now, $\begin{cases} \overline{Q}_1 = \hat{Q}_1, & \overline{Q}_2 = \hat{Q}_2 \\ Q_1' = \hat{Q}_1', & Q_2' = \hat{Q}_2' \end{cases}$

$\Longrightarrow$ (a) $H(R_{\hat{Q}_1\hat{Q}_2},\hat{Q}_1) \leq H(R^{\hat{Q}_i'}_{\hat{Q}_1\hat{Q}_2},\hat{Q}_i')$ $(i=1,2)$

But, in general, it holds

$$\min_i H(R^{Q_i}_{\overline{Q}_1\overline{Q}_2},Q_i) \leq \min_i H(R^{Q_i}_{Q_1Q_2},Q_i) = H(R_{Q_1Q_2},Q_1)$$

Take here: $\overline{Q}_1 = \hat{Q}_1$; $\overline{Q}_2 = \hat{Q}_2$; $Q_i = \hat{Q}_i'$ $(i=1,2)$

$\Longrightarrow$ (b) $\min_i H(R^{\hat{Q}_i'}_{\hat{Q}_1\hat{Q}_2},\hat{Q}_i') \leq H(R_{\hat{Q}_1'\hat{Q}_2'},\hat{Q}_1')$

Combining (a) with (b) we obtain (15).

Q.E.D.

In this context, let us remark that the H_i of model (I), H_i^{I}, may be "embeded" in a suitable H_i of model (II), H_i^{II} $(i=1,2)$:

(16) $$H_i^{I} \subset H_i^{II} := \{Q_i : H_f(P_i,Q_i) \leq \delta_i\}$$

with

(17) $$\delta_i := \max_{Q_i\in H_i^{I}} H_f(P_i,Q_i) = f(\frac{1}{1-\varepsilon_i})(1-\varepsilon_i)+\varepsilon_i f(0),$$

this maximum being here obtained for any $S_i \perp P_i$ provided that, for $\alpha>0$, $\frac{d}{du}[uf(\frac{\alpha}{u})]<0$. For instance, this holds for $f(u) = u\log u$;

$f(u) = u^{\beta}-1$ with $\beta>1$, etc.

Thus, estimates of DR in the frame of one model may be sometimes applied for estimating DR in the other model. For instance, according to Theorem 4, $\widehat{\mathrm{DR}}^{\mathrm{II}} \leq \widehat{\mathrm{DR}}^{\mathrm{I}}$ since $H_i^{\mathrm{I}} \subseteq H_i^{\mathrm{II}}$.

SOME SPECIAL RESULTS CONCERNING MODELS (I) AND (II)

That the composite statistical hypotheses of the ε-contaminated model (I) are *convex* sets it is immediate. As to those of the f-divergence like neighborhood model (II), their convexity may be proved as follows:

Let $Q_{i1}, Q_{i2}, \ldots, Q_{in} \in H_i := \{Q_i : H_f(P_i, Q_i) \leq \varepsilon_i\}$ and let $Q := \sum_{k=1}^{n} \alpha_k\, Q_{ik}$ for some arbitrary probability vector $\{\alpha_k\}_{k=1}^{n}$. Then

$$H_f(P_i, Q) \leq \sum_{k=1}^{n} \alpha_k\, H_f(P_i, Q_{ik}) \tag{18}$$

since the function $g(u) := uf(\frac{1}{u})$ is convex for $u \in (0,\infty)$: Assuming that f has a second derivative we, namely, obtain

$$g'(u) = f(\tfrac{1}{u}) - f'(\tfrac{1}{u})\tfrac{1}{u}$$

$$g''(u) = -f'(\tfrac{1}{u})\tfrac{1}{u^2}+f'(\tfrac{1}{u})\tfrac{1}{u^2}+\tfrac{1}{u}f''(\tfrac{1}{u})\tfrac{1}{u^2} = \tfrac{1}{u^3}f''(\tfrac{1}{u}) \geq 0$$

since $f(v)$ is, by assumption, convex.

Q.E.D.

<u>Model (I)</u>. Equations for DR-LFP (Q_1^*, Q_2^*) and for MDR-LFP $(\hat{Q}_1, \hat{Q}_2)$

For establishing these equations we are based on Theorems 3 and 2, respectively. The minima there figuring may be obtained by an extremal approach based on Lagrange multipliers method. Thus, one finds that the solution for some probability measure R on $(X, \mathfrak{X})$ of

$$Q_i^R : \mathrm{H}(R, Q_i^R) := \min_{Q_i \in H_i} \mathrm{H}(R, Q_i),$$

given that $H_i = \{Q_i : Q_i = (1-\varepsilon_i)P_i + \varepsilon_i S_i\}$, is:

$$\left\{ \begin{array}{lll} dQ_i^R = \dfrac{\varepsilon_i + (1-\varepsilon_i)P_i(A_i^R)}{R(A_i^R)}\, dR & \text{on} & A_i^R \\ dQ_i^R = (1-\varepsilon_i)\, dP_i & \text{on} & A_i^{Rc} \end{array} \right. \qquad (i=1,2) \tag{19}$$

with

$$A_i^R = \{x \in X : \frac{dR}{dP_i} > \frac{(1-\varepsilon_i)R(A_i^R)}{\varepsilon_i+(1-\varepsilon_i)P_i(A_i^R)}\}$$

For $R = R_{\frac{Q_i^*}{Q_1 Q_2}}$ in (19) we obtain the equations for (Q_1^*, Q_2^*) by replacing Q_i^R by Q_i^*.

Similarly, for $R = R_{\hat{Q}_1\hat{Q}_2}$ in (19) we obtain the equations for $(\hat{Q}_1, \hat{Q}_2)$ by replacing Q_i^R by $\hat{Q}_i$.

Namely, the equations for $\hat{Q}_i$ $(i=1,2)$ in model (I) are

$$(20)\quad \begin{cases} d\hat{Q}_i = \dfrac{\varepsilon_i+(1-\varepsilon_i)P_i(\hat{A}_i)}{R_{\hat{Q}_1\hat{Q}_2}(\hat{A}_i)}\, dR_{\hat{Q}_1\hat{Q}_2} & \text{on } \hat{A}_i := A_i^{R_{\hat{Q}_1\hat{Q}_2}} \\ d\hat{Q}_i = (1-\varepsilon_i)dP_i & \text{on } \hat{A}_i^c \end{cases}$$

Since $dR_{\hat{Q}_1\hat{Q}_2} = \hat{c}\, d\hat{Q}_1^{\hat{\alpha}} d\hat{Q}_2^{1-\hat{\alpha}}$ with $\hat{c}^{-1} = \int d\hat{Q}_1^{\hat{\alpha}} d\hat{Q}_2^{1-\hat{\alpha}} = \min_{\alpha} \int d\hat{Q}_1^{\alpha} d\hat{Q}_2^{1-\alpha}$, one obtains on $\hat{A}_1$:

$$\frac{d\hat{Q}_1}{d\hat{Q}_2} = \frac{\varepsilon_1+(1-\varepsilon_1)P_1(\hat{A}_1)}{R_{\hat{Q}_1\hat{Q}_2}(\hat{A}_1)}\, \hat{c}\left(\frac{d\hat{Q}_1}{d\hat{Q}_2}\right)^{\hat{\alpha}} \Longrightarrow \frac{d\hat{Q}_1}{d\hat{Q}_2} = c_1$$

Similarly, on $\hat{A}_2$: $\frac{d\hat{Q}_1}{d\hat{Q}_2} = c_2$, where in general the constants c_1 and c_2 are different, i. e. $\hat{A}_1 \cap \hat{A}_2 = \emptyset$.

Finally, on $\hat{A}_1^c \cap \hat{A}_2^c$: $\frac{d\hat{Q}_1}{d\hat{Q}_2} = \frac{(1-\varepsilon_1)}{1-\varepsilon_2}\frac{dP_1}{dP_2}$ (cf. with Theorem 5.2. of [3]).

Instead of (20) one may write now

$$(21)\quad \begin{cases} d\hat{Q}_1 = c_1(1-\varepsilon_2)dP_2 & \text{on } \hat{A}_1 \\ \quad\;\; = (1-\varepsilon_1)dP_1 & \text{on } \hat{A}_1^c \\ d\hat{Q}_2 = c_2^{-1}(1-\varepsilon_1)dP_1 & \text{on } \hat{A}_2 \\ \quad\;\; = (1-\varepsilon_2)dP_2 & \text{on } \hat{A}_2^c \end{cases}$$

Thus, $dR_{\hat{Q}_1\hat{Q}_2} = \hat{c}\, d\hat{Q}_1^{\hat{\alpha}} d\hat{Q}_2^{1-\hat{\alpha}} = \hat{c}c_1(1-\varepsilon_2)dP_2$ on $\hat{A}_1$

$= \hat{c}c_2^{-1}(1-\varepsilon_1)dP_1$ on $\hat{A}_2$

As a consequence

$$\hat{A}_1 = \{x \in X : \hat{c}c_1(1-\varepsilon_2)\frac{dP_2}{dP_1} > \frac{(1-\varepsilon_1)\hat{c}c_1(1-\varepsilon_2)P_2(\hat{A}_1)}{\varepsilon_1+(1-\varepsilon_1)P_1(\hat{A}_1)}\} =$$

$$= \{x \in X : \frac{dP_2}{dP_1} > \frac{(1-\varepsilon_1)P_2(\hat{A}_1)}{\varepsilon_1+(1-\varepsilon_1)P_1(\hat{A}_1)}\} = A_1^{P_2}$$

Similarly, $\hat{A}_2 = A_2^{P_1}$. In other words, $(\hat{Q}_1, \hat{Q}_2)$ for model (I):

$$(22) \quad \begin{cases} \hat{Q}_1 : H(P_2, \hat{Q}_1) := \min_{Q_1 \in H_1} H(P_2, Q_1) \\ \hat{Q}_2 : H(P_1, \hat{Q}_2) := \min_{Q_2 \in H_2} H(P_1, Q_2) \end{cases}$$

Thus, we proved the following theorem.

THEOREM 5. A MDR-LFP $(\hat{Q}_1, \hat{Q}_2)$ for the ε-contamination Huber's model (I) is given by (19), where for $i=1$ we put $R = P_2$ and for $i=2$ we put $R = P_1$; in other words, (22) holds; H may be replaced by H_f.

This solution coincides with the solution (6.16) for a LFP obtained in [3].

Remark. In [3], however, as we know it is nothing said on the interesting property (22) of the solution obtained for the LFP.

Model II. Equations for DR-LFP (Q_1^*, Q_2^*) and for MDR-LFP $(\hat{Q}_1, \hat{Q}_2)$.

As in the case of model (I), for establishing these equations we are based on Theorems 3 and 2, respectively. As said in the introduction, for the constrained minimizations needed the extremal generalized f-entropy method, introduced in [2], is applied. Thus, instead of (19), one finds for

$$Q_i^R : H(R, Q_i^R) := \min_{Q_i \in H_i} H(R, Q_i)$$

given that

$$H_i = \{Q_i : H_f(P_i, Q_i) \le \varepsilon_i\}$$

with f differentiable, $R \ll P_i$ and $u_i := \frac{dP_i}{dQ_i^R}$ $(i=1,2)$

$$(23) \quad u_i \frac{dR}{dP_i} + \lambda_i \left[u_i f'(u_i) - f(u_i)\right] = \mu_i,$$

(provided that (23) has a solution $u_i \ge 0$), where the constants λ_i and μ_i are solutions obtained from the constraints:

$$H_f(P_i, Q_i^R) = \int f(u_i)\, dQ_i^R = \varepsilon_i \; ; \; \int dQ_i^R = 1 = \int u_i^{-1} dP_i .$$

If, for instance, $f(u) = u \log u$ one, thus, finds

$$(24) \quad dQ_i^R = \beta_i^R\, dR + (1 - \beta_i^R)\, dP_i$$

where (provided that $H(P_i, R) \ge \varepsilon_i$) $0 \le \beta_i^R \le 1$; β_i^R given by constraint $H(P_i, Q_i^R) = \varepsilon_i$.

If $f(u) = u^\beta - 1$, $(\beta>1)$, from (23) we obtain

$$u_i \frac{dR}{dP_i} + \lambda_i(\beta-1)u_i^\beta = \mu_i \tag{25}$$

with

$$\mu_i = 1 + \lambda_i(\beta-1)(\varepsilon_i+1)$$

For $R = R_{\overline{Q}_1\overline{Q}_2}^{Q_i^*}$ one obtains the equations for the DR-LFP (Q_1^*, Q_2^*) by replacing Q_i^R by Q_i^*.

Similarly, for $R = R_{\hat{Q}_1\hat{Q}_2}$ one obtains the equations for the MDR-LFP $(\hat{Q}_1, \hat{Q}_2)$ by replacing Q_i^R by $\hat{Q}_i$.

<u>THEOREM 6.</u> Let $P_1 \perp P_2$ in model (II) with $P_1(A) = P_2(A^c) = 1$. Then a MDR-LFP $(\hat{Q}_1, \hat{Q}_2)$ for the f-divergence like neighborhood model is given by

$$\left\{\begin{aligned} d\hat{Q}_1 &= \gamma_1 dP_2 + (1-\gamma_1)dP_1 = \begin{cases} (1-\gamma_1)dP_1 & \text{on } A \\ \gamma_1 dP_2 & \text{on } A^c \end{cases} \\ d\hat{Q}_2 &= \gamma_2 dP_1 + (1-\gamma_2)dP_2 = \begin{cases} \gamma_2 dP_1 & \text{on } A \\ (1-\gamma_2)dP_2 & \text{on } A^c \end{cases} \end{aligned}\right. \tag{26}$$

where $0<\gamma_i<1$ is given by constraint $H_f(P_i, \hat{Q}_i) = \varepsilon_i$ $(i=1,2)$, i.e. from

$$\gamma_i f(0) + (1-\gamma_i) f\left(\frac{1}{1-\gamma_i}\right) = \varepsilon_i \quad (i=1,2) \tag{27}$$

It holds (cf. Theorem 5)

$$\left\{\begin{aligned} H(P_2, \hat{Q}_1) &= \min_{Q_1 \in H_1} H(P_2, Q_1) \\ H(P_1, \hat{Q}_2) &= \min_{Q_2 \in H_2} H(P_1, Q_2) \end{aligned}\right. \tag{28}$$

whereas H in (28) may be replaced by any H_f.

For $f(u) = u\log u$ in (27), (26) coincides with (24) where we put $R = P_2$ and $R = P_1$, respectively.

The maximal discrimination rate $\hat{\mathrm{DR}}$ is given by

$$\hat{\mathrm{DR}} = -\log H_{\hat{\alpha}}\left(\begin{bmatrix} 1-\gamma_1 \\ \gamma_1 \end{bmatrix}, \begin{bmatrix} \gamma_2 \\ 1-\gamma_2 \end{bmatrix}\right) = -\min_{\alpha} \log\left[(1-\gamma_1)^\alpha \gamma_2^{1-\alpha} + \gamma_1^\alpha (1-\gamma_2)^{1-\alpha}\right] \tag{29}$$

where γ_1 and γ_2 solutions of (27).

<u>Proof.</u> For proving that $(\hat{Q}_1,\hat{Q}_2)$ as given by (26) and (27) is a MDR-LFP it is sufficient to prove that equations (23) for $R=R_{\hat{Q}_1\hat{Q}_2}$ and $u_i=\frac{dP_i}{dQ_i}$ are satisfied by $\hat{Q}_1$ and $\hat{Q}_2$ for suitable values of the constants λ_i and μ_i $(i=1,2)$. Indeed,

$$dR_{\hat{Q}_1\hat{Q}_2} = \hat{c}\,d\hat{Q}_1^{\hat{\alpha}}d\hat{Q}_2^{1-\hat{\alpha}} = \begin{cases} \hat{c}(1-\gamma_1)^{\hat{\alpha}}\gamma_2^{1-\hat{\alpha}}dP_1 & \text{on } A \\ \hat{c}\gamma_1^{\hat{\alpha}}(1-\gamma_2)^{1-\hat{\alpha}}dP_2 & \text{on } A^c \end{cases}$$

Thus,

$$\frac{dR_{\hat{Q}_1\hat{Q}_2}}{dQ_1} = \begin{cases} \hat{c}\left(\frac{\gamma_2}{1-\gamma_1}\right)^{1-\hat{\alpha}} & \text{on } A \\ \hat{c}\left(\frac{1-\gamma_2}{\gamma_1}\right)^{1-\hat{\alpha}} & \text{on } A^c \end{cases}$$

$$u_1 = \frac{dP_1}{dQ_1} = \begin{cases} \frac{1}{1-\gamma_1} & \text{on } A \\ 0 & \text{on } A^c \end{cases}$$

By substitution in (23) for $i=1$ we obtain the system

$$\begin{cases} \mu_1 = \hat{c}\left(\frac{\gamma_2}{1-\gamma_1}\right)^{1-\hat{\alpha}} + \lambda_1\left[\frac{1}{1-\gamma_1}f'\left(\frac{1}{1-\gamma_1}\right)-f\left(\frac{1}{1-\gamma_1}\right)\right] & \text{on } A \\ \mu_1 = \hat{c}\left(\frac{1-\gamma_2}{\gamma_1}\right)^{1-\hat{\alpha}} - \lambda_1 f(0) & \text{on } A^c \end{cases}$$

from where one derives the values of μ_1 and λ_1. Similarly for $i=2$.

For proving (28) it is sufficient to prove that equations (23) for $R = P_2$ ($R = P_1$) and $u_1 = \frac{dP_1}{dQ_1}$ $(u_2 = \frac{dP_2}{dQ_2})$ are satisfied by $\hat{Q}_1$ and $\hat{Q}_2$ for suitable values of the constants λ_1 (λ_2) and μ_1 (μ_2).

Indeed, (cf.(26))

$$\frac{dR}{dQ_1^R} = \frac{dP_2}{dQ_1} = \begin{cases} 0 & \text{on } A \\ \frac{1}{\gamma_1} & \text{on } A^c \end{cases}$$

By substitution in (23) for $i=1$ we obtain the system

$$\begin{cases} \mu_1 = \lambda_1\left[\frac{1}{1-\gamma_1}f'\left(\frac{1}{1-\gamma_1}\right)-f\left(\frac{1}{1-\gamma_1}\right)\right] & \text{on } A \\ \mu_1 = \frac{1}{\gamma_1}-\lambda_1 f(0) & \text{on } A^c \end{cases}$$

from where one derives the corresponding values of μ_1 and λ_1. Similarly for $i=2$. Thus, (28) is proved. The fact that in (28) H may be replaced by any H_f, say H_{f_1}, it results from the following equations, corresponding to (23),

$$(30)\qquad \mu_i = \frac{dR}{dQ_i^R} f_1'\left(\frac{dR}{dQ_i^R}\right) - f_1\left(\frac{dR}{dQ_i^R}\right) + \lambda_i\left[u_i f'(u_i) - f(u_i)\right]$$

Here, for $i=1$, $\frac{dR}{dQ_i^R} = \frac{dP_2}{dQ_1} = \begin{cases} 0 & \text{on } A \\ \frac{1}{\gamma_1} & \text{on } A^c \end{cases}$. Thus, μ_1 and λ_1 are here solution of the system

$$\begin{cases} \mu_1 = -f_1(0) + \lambda_1\left[\frac{1}{1-\gamma_1} f'\left(\frac{1}{1-\gamma_1}\right) - f\left(\frac{1}{1-\gamma_1}\right)\right] & \text{on } A \\ \mu_1 = \frac{1}{\gamma_1} f_1'\left(\frac{1}{\gamma_1}\right) - f_1\left(\frac{1}{\gamma_1}\right) - \lambda_1 f(0) & \text{on } A^c \end{cases}$$

Similarly for $i=2$. Thus, in (28) one may replace H by H_{f_1}.

As to (29), it is an immediate consequence of $\hat{D}R = H(R_{\hat{Q}_1\hat{Q}_2}, \hat{Q}_1) = -\log \min_\alpha \int d\hat{Q}_1^\alpha d\hat{Q}_2^{1-\alpha}$ (cf. Theorem 1).

Q.E.D.

REFERENCES

[1] Perez A. (1982): Discrimination Rate Loss of Simple Statistical Hypotheses by Unfitted Decision Procedures (To appear in Festschrift dedicated to professor Onicescu).

[2] Perez A. (1967): Information-Theoretic Risk Estimates in Statistical Decision. Kybernetika 3(1967), No.1, 1-21.

[3] Rieder H. (1977): Least Favorable Pairs for Special Capacities. The Annals of Statistics, 5(1977), No.5, 909-921.

Czechoslovak Academy of Sciences
Institute of Information Theory
and Automation
Pod vodárenskou věží 4
182 08 Praha 8 - Libeň
Czechoslovakia

ON THE LOCAL TIME OF BROWNIAN BRIDGE

P. Révész

Budapest

Key words: Local time, Brownian bridge, empirical process, strong invariance principle.

ABSTRACT

Let $\alpha_n(t)$ be the empirical process of a sequence of independent uniformly distributed r. v. 's. The main result says that the local time of α_n is close to the local time of a Brownian bridge B_n if n is big enough.

1. INTRODUCTION

Let $\{W(t), t \geq 0\}$ be a Wiener process and for any Borel set A let

$$H(A,t) = \lambda\{s: s \leq t, W(s) \in A\}$$

be the occupation time of W where λ is the Lebesgue measure. It is well-known that $H(A,t)$ (for any fixed t) is a random measure absolutely continuous with respect to λ with probability 1. The Radon-Nikodym derivative of H is called the local time of W and it will be denoted by η i.e. $\eta(x,t)$ is defined by

$$H(A,t) = \int_A \eta(x,t)dx \ . \tag{1.1}$$

The properties of the local time of a Wiener process are widely studied (see e.g. D. Geman - J. Horowitz, 1980). Here we only mention some of the continuity properties of η for later reference. The modulus of continuity in x is described by the following two results of H. P. McKean Jr. (1962) and D. B. Ray (1963).

$$(1.2)\quad \limsup_{h\to 0}\ \sup_{-\infty<x<\infty} \frac{|\eta(x+h,t)-\eta(x,t)|}{(h\log h^{-1})^{1/2}} = 2(\max_{-\infty<x<\infty} \eta(x,t))^{1/2} \quad \text{a.s.}$$

and

$$(1.3)\quad \limsup_{h\to 0} \frac{|\eta(h,t)-\eta(0,t)|}{(h\log h^{-1})^{1/2}} = 2(\eta(0,t))^{1/2} \quad \text{a.s.}$$

As far as the modulus of continuity in t is being concerned we have (Csáki, Csörgő, Földes, Révész, 1982)

$$(1.4)\quad \lim_{h\to 0} \frac{\sup_{0\leq t\leq 1-h} (\eta(0,t+h)-\eta(0,t))}{(h\log h^{-1})^{1/2}} = 1 \quad \text{a.s.}$$

and

$$(1.5)\quad \lim_{h\to 0} \frac{\sup_{0\leq t\leq 1-h}\ \sup_{-\infty<x<\infty} (\eta(x,t+h)-\eta(x,t))}{(2h\log h^{-1})^{1/2}} = 1 \quad \text{a.s.}$$

Comparing the modulus of continuity in x and that in t one can realize that the results giving the modulus in x are not so complete as the corresponding results in t. In fact one can ask whether the limsup of (1.2) can be replaced by lim. Also one can ask whether (1.2) and (1.3) hold uniformly in t. We formulate the following

CONJECTURE.

$$(1.2^*)\quad \lim_{h\to 0}\ \sup_{-\infty<x<\infty}\ \sup_{0\leq s\leq t} \frac{|\eta(x+h,s)-\eta(x,s)|}{(h\log h^{-1})^{1/2}} = 2(\max_{-\infty<x<\infty} \eta(x,t))^{1/2}$$

$$(1.3^*)\quad \limsup_{h\to 0}\ \sup_{0\leq s\leq t} \frac{|\eta(h,s)-\eta(0,s)|}{(h\log\log h^{-1})^{1/2}} = 2(\eta(0,t))^{1/2}$$

with probability one for any $t>0$.

Later on we return to this conjecture proving a much weaker result. In fact we will prove our

LEMMA 1. *We have*

$$(1.6)\quad \lim_{h\to 0}\ \sup_{-\infty<x<\infty}\ \sup_{0\leq s\leq t} \frac{|\eta(x+h,s)-\eta(x,s)|}{h^{1/2-\rho}} = 0 \quad \textit{a.s.}$$

for any fixed $t>0$ *and* $0<\rho<1/2$.

An analogue of the local time can be introduced for the partial sum sequence of integer valued i.i.d.r.v.'s. Let $X_1, X_2, \ldots$ be a sequence of integer valued i.i.d.r.v.'s. Put $S_0=0$, $S_n=X_1+X_2+\ldots+X_n$ $(n=1,2,\ldots)$,

$$\xi(x,n) = N\{k: k\leq n,\ S_k=x\} \qquad (x=0,\pm 1,\pm 2,\ldots;\ n=1,2,\ldots) \tag{1.7}$$

here and in what follows $N\{\ldots\}$ is the cardinality of the set in brackets. Several theorems state that the limit properties $(n\to\infty)$ of $\xi(x,n)$ (under some regularity conditions on the distribution of X_1) are the same as those of $\eta(x,n)$. Csáki and Révész (1982) proved a strong invariance principle saying that the sequence $\{X_i\}$ and the process W can be defined on a common probability space $\{\Omega,S,\mathbb{P}\}$ such that the corresponding local times ξ and η are close enough to each other. Such an invariance principle clearly implies the theorems stating that the limit properties of ξ and η are the same. Here we formulate only a special case of the general theorem.

THEOREM a. (Révész, 1981). *Let* $\{W(t),\ t\geq 0\}$ *be a Wiener process defined on a probability space* $\{\Omega,S,\mathbb{P}\}$. *Then on the same probability space* Ω *one can define a sequence* $X_1^{(o)}, X_2^{(o)},\ldots$ *of i.i.d.r.v.'s with*

$$\mathbb{P}(X_i^{(o)}=1) = \mathbb{P}(X_i^{(o)}=-1) = 1/2$$

such that

$$\lim_{n\to\infty} n^{-1/4-\rho} \sup_x |\xi(x,n)-\eta(x,n)| = 0 \quad a.s.$$

and

$$\lim_{n\to\infty} n^{-1/4-\rho} |S_n-W(n)| = 0 \quad a.s.$$

for any $\rho>0$ *where the* sup *is taken over all integers,* n *is running over the positive integers,* ξ *is defined by* (1.7) *(where now* $S_k = \sum_{i=1}^{k} X_i^{(o)}$ *) and* η *is the local time of* W .

In the case when $\{X_i\}$ is a sequence of i.i.d.r.v.'s with continuous distribution (say) then the natural concept of the local time is not so clear. This question was studied by Révész (1982). Here we formulate a special case of the general result.

Let $E_1, E_2, \ldots$ be a sequence of i.i.d.r.v.'s with $\mathbb{P}(E_1<x)=1-e^{-x}$ $(x\geq 0)$. Put $T_0=0$, $T_k=E_1+E_2+\ldots+E_k-k$ $(k=1,2,\ldots)$,

$$Z(A,n) = N\{k: k\leq n,\ T_k\in A\} \quad (n=1,2,\ldots)$$

for any Borel set A,

$$T(t)=T_k+(t-k)(E_{k+1}-1) \quad \text{if} \quad k\leq t\leq k+1 \quad (k=0,1,2,\ldots)$$

$$\zeta(x,t) = N\{s: s\leq t,\ T(s)=x\} .$$

Then we have

THEOREM B. *The sequence* $\{E_i\}$ *and the Wiener process* W *can be defined on a probability space* $\{\Omega, S, \mathbb{P}\}$ *such that for any* $\varepsilon>0$ *we have*

$$(1.8) \qquad \lim_{n\to\infty} n^{-1/3-\varepsilon}|\eta(0,\sigma^2 n)-\sigma^2 Z((-1/2,1/2),n)| = 0 \ \textit{a.s.}$$

$$(1.9) \qquad \lim_{n\to\infty} n^{-1/3-\varepsilon}|\eta(0,\sigma^2 n)-\mu_1^{-1}\sigma^2\zeta(0,n)| = 0 \ \textit{a.s.}$$

and

$$(1.10) \qquad \lim_{n\to\infty} n^{-1/4-\varepsilon}|T_n-W(n)| = 0 \ \textit{a.s.}$$

where $\mu_2=\mathbb{E}(|E_1-1|)=2e^{-1}$ *and* $\sigma^2=\mathbb{E}(E_1-1)^2=1$.

Clearly this theorem again says that the limit properties of Z and ζ are the same as those of η .

The properties of the local time of the Brownian bridge are not well-investigated. This lack is not only a theoretical one. It can be expected that the limit properties of the suitably defined local time of an empirical process and that of the quantile process are the same as the properties of the local time of a Brownian bridge. If it is the case then the study of the local time of the Brownian bridge can produce some results applicable in statistics. The present paper is devoted to this problem. Only the main ideas of the proofs will be sketched. The author intends to return to the details elsewhere in a survey paper of the question of local time and invariance principle.

The author is indebted to Prof. E. Csáki for his valuable remarks during the preparation of this paper.

2. THE EXISTENCE OF THE LOCAL TIME OF A BROWNIAN BRIDGE

Let $\{B(t),\ 0\leqq t\leqq 1\}$ be a Brownian bridge i.e. B is a continuous Gaussian process with covariance function

$$\mathbb{E}\,B(t_1)B(t_2) = \min(t_1\,,\,t_2)-t_1t_2\ .$$

Define the local time $\{\beta(x,t);\ -\infty<x<\infty\ ,\ 0\leqq t\leqq 1\}$ of B by

$$\beta(x,t) = \lim_{\varepsilon\to 0}\frac{1}{2\varepsilon}\,\lambda\{s: s\leqq t,\ |B(s)-x|\leqq\varepsilon\}\ .$$

In order to prove the existence of β (i.e. the fact that the above limit exists with probability 1) we present the following

LEMMA 2. *Let* $\{W(t),\ t\leqq 0\}$ *be a Wiener process and define the r.v.* τ *by*

$$\tau = \sup\{t:\ 0\leqq t\leqq 1\ ,\ W(t)=0\}\ .$$

Put

$$B(t) = \tau^{-1/2}W(t\tau)\qquad (0\leqq t\leqq 1)\ . \tag{2.1}$$

Then $\{B(t),\ 0\leqq t\leqq 1\}$ *is a Brownian bridge.*

The proof of this lemma can be obtained by a standard calculation. Hence it will be omitted.

This lemma clearly implies the existence of $\beta(x,t)$ for all $-\infty<x<+\infty\ ,\ 0\leqq t\leqq 1$ with probability 1. In fact we have

LEMMA 3. *Let the Brownian bridge* B *be defined by* (2.1). *Then*

$$\beta(x,t) = \tau^{-1/2}\eta(x\tau^{1/2},\ t\tau)\ .$$

PROOF. For any $0\leqq t\leqq 1$ we have

$$\lim_{\varepsilon\to 0}\frac{1}{2\varepsilon}\lambda\{s: |B(s)-x|\leqq\varepsilon\ ,\ 0\leqq s\leqq t\} =$$

$$= \lim_{\varepsilon\to 0}\frac{1}{2\varepsilon}\lambda\{s: |\tau^{-1/2}W(s\tau)-x|\leqq\varepsilon\ ,\ 0\leqq s\leqq t\} =$$

$$= \lim_{\varepsilon\to 0}\frac{1}{2\varepsilon}\lambda\{s: |W(s\tau)-x\tau^{1/2}|\leqq\varepsilon\tau^{1/2},\ 0\leqq s\leqq t\} =$$

$$= \lim_{\varepsilon\to 0}\frac{1}{2\varepsilon\tau}H((x\tau^{1/2}-\varepsilon\tau^{1/2}, x\tau^{1/2}+\varepsilon\tau^{1/2}), t\tau) = \tau^{-1/2}\eta(x\tau^{1/2}, t\tau).$$

3. ON THE CONTINUITY OF β

(1.5) and Lemma 3 together clearly imply

THEOREM 1.

$$\lim_{h\to 0} \frac{\sup_{0\leq t\leq 1-h} \sup_{-\infty<x<\infty} (\beta(x,t+h)-\beta(x,t))}{(2h \log h^{-1})^{1/2}} = 1 \quad a.s. \tag{3.1}$$

In the study of the continuity of β in x we have to apply our Lemma 1. Hence, at first we sketch its proof.

The random walk analogue of Lemma 1 can be obtained by a simple combinatorial argumentation. (The details will be omitted, they are close to the ideas used in the proof of Lemma 1.1 of Révész, 1981).

LEMMA 1*. *Let* $\xi(x,n)$ *be the local time of the simple random walk (cf. Theorem A). Then for any* $\rho>0$ *and* $\varepsilon>0$ *we have*

$$\mathbb{P}\{\sup_{-\infty<x<+\infty} \sup_{0\leq k\leq n} \frac{|\xi(x+1,k)-\xi(x,k)|}{n^{1/4+\rho}} \geq \varepsilon\} \leq Cn^{-4} \tag{3.2}$$

with some $C>0$.

Lemma 1* and Theorem A together imply

LEMMA 1**. (3.2) *remains true replacing* ξ *by* η *in it.*

Applying the well-known transformation

$$\{W(t),\ 0\leq t\leq 1\} \stackrel{\mathcal{D}}{=} \{T^{-1/2}W(tT),\ 0\leq t\leq 1\}$$

(for any $T>0$) Lemma 1** implies Lemma 1.

As a consequence of Lemmas 1 and 3 we obtain

THEOREM 2. *We have*

$$\lim_{h\to 0} \sup_{-\infty<x<\infty} \sup_{0\leq t\leq 1} \frac{|\beta(x+h,t)-\beta(x,t)|}{h^{1/2-\rho}} = 0 \quad a.\ s. \tag{3.3}$$

for any $0<\rho<1/2$.

This Theorem also implies

CONSEQUENCE 1.

$$\lim_{\varepsilon\to 0} \sup_{-\infty<x<\infty} \sup_{0\leq t\leq 1} \varepsilon^{-1/2+\rho}\left|\frac{1}{2\varepsilon}\lambda\{s: s\leq t,\ |B(s)-x|\leq\varepsilon\}-\beta(x,t)\right| = 0 \tag{3.4}$$

with probability 1 , *for any* $0<\rho<1/2$.

Let $B(t),\ 0\leq t\leq 1\}$ be a Brownian bridge and for any real y define the process

$$B_y(t) = B(t)+yt \quad (0\leq t\leq 1) .$$

It is easy to see that the local time of $B_y(t)$ (i.e. the process

$$\beta_y(x,t)=\lim_{\varepsilon\to 0}\frac{1}{2\varepsilon}\lambda\{s:s\leq t,\ |B_y(s)-x|\leq\varepsilon\}$$

$(-\infty<x<\infty\ ,\ 0\leq t\leq 1\ ,\ -\infty<y<\infty\ ))$ exists with probability 1 .

For later reference we present a Theorem stating the continuity of $\beta_y(x,t)$ in y . In fact we have

THEOREM 3.

$$\lim_{y\to 0} y^{-1/2+\rho}\sup_{-\infty<x<\infty}\sup_{0\leq t\leq 1}|\beta_y(x,t)-\beta(x,t)| = 0 \qquad (3.5)$$

with probability 1 , *for any* $0<\rho<1/2$.

This theorem can be easily obtained from Consequence 1.

4. TWO INVARIANCE PRINCIPLES

Let $\{Y_i^{(n)}\}_{i=1}^{2n}$ $(n=1,2,\ldots)$ be a double array with

$$\{Y_i^{(n)},\ i=1,2,\ldots,2n\}\overset{\mathcal{D}}{=}\{X_i^{(o)},\ i=1,2,\ldots,2n\Big|\sum_{i=1}^{2n}X_i^o=0\}. \qquad (4.1)$$

(The sequence $\{X_i^{(o)}\}$ was defined in Theorem A.) Put

$$U_k=U_k^{(n)}=Y_1^{(n)}+Y_2^{(n)}+\ldots+Y_k^{(n)} \quad (k=1,2,\ldots,2n\ ;\ n=1,2,\ldots),$$

$\gamma(x,k)=\gamma^{(n)}(x,k)=N\{j:j\leq k,\ U_j^{(n)}=x\}$.

Now, we can formulate our

THEOREM 4. *For each* n $(n=1,2,\ldots)$ *the sequence* $\{Y_i^{(n)}\}_{i=1}^{2n}$ *and a Brownian bridge* $\{B_n(t),\ 0\leq t\leq 1\}$ *can be defined on a probability space* $\{\Omega_n,S_n,\mathbb{P}_n\}$ *such that for any* $\varepsilon>0$ *we have*

$$\lim_{n\to\infty} n^{-1/4-\varepsilon}\sup_{k\leq n}\sup_{-\infty<x<\infty}|\gamma(x,k)-n^{1/2}\beta_n(xn^{-1/2},kn^{-1})| = 0 \ a.s. \qquad (4.2)$$

where β_n *is the local time of* B_n .

Before proving this theorem we give a few lemmas.

LEMMA 4. *Applying the notations of Theorem A we have*

$$\mathbb{P}\{\sup_{-\infty<x<\infty}\sup_{k\leq n}|\xi(x,k)-\eta(x,k)|\geq\varepsilon n^{1/4+\rho}\}\leq Cn^{-4}$$

and

$$\mathbb{P}\{\sup_{k\leq n}|S_k-W(k)|\geq\varepsilon n^{1/4+\rho}\}\leq Cn^{-4}$$

for any $\varepsilon>0$, $\rho>0$ *and for some* $C>0$.

This Lemma is slightly sharper than Theorem A. However the proof of Theorem A without any change is able to produce Lemma 4. The details are omitted.

LEMMA 5. *Applying the notations of Theorem A we have*

$$\mathbb{P}\{\sup_{-\infty<x<\infty}\sup_{k\leq 2n}|\xi(x,k)-\eta(x,k)|\geq\varepsilon n^{1/4+\rho}|S_{2n}=0\}\leq Cn^{-3}$$

and

$$\mathbb{P}\{\sup_{k\leq 2n}|S_k-W(k)|\geq\varepsilon n^{1/4+\rho}|S_{2n}=0\}\leq Cn^{-3}$$

for any $\varepsilon>0$, $\rho>0$ *and for some* $C>0$.

This lemma is a simple consequence of Lemma 4.

For a Wiener process W_1 define the sequence $0=\tau_o<\tau_1<\tau_2$ ... of stopping times by

$$\tau_1=\inf\{t:\ t>0,\ |W_1(t)|=1\}\ ,$$

$$\tau_2=\inf\{t:\ t>\tau_1\ ,\ |W_1(t)-W_1(\tau_1)|=1\}\ ,$$

$$.\ .\ .\ .\ .\ .\ .\ .\ .\ .\ .\ .\ .\ .$$

$$\tau_n=\inf\{t:\ t>\tau_{n-1}\ ,\ |W_1(t)-W_1(\tau_{n-1})|=1\}\ ,$$

$$.\ .\ .\ .\ .\ .\ .\ .\ .\ .\ .\ .\ .\ .$$

Put

$$A_n=\{\omega:\ \omega\in\Omega\ ,\ W_1(\tau_{2n})=0\}\ ,$$

$$W_1(\tau_k)=U_k\ ,\qquad Y_k=U_k-U_{k-1}\ .$$

Clearly the joint distribution of $U_1, U_2, \ldots, U_{2n}$ given A_n satisfies (4.1). Considering the local time of W_1 in the interval $(0,\tau_{2n})$ it will be as close to the local time of U_k as (4.2) says. (Apply Lemma 5.) The only problem is the fact that $W_1(t)$ in $(0,\tau_{2n})$ (more precisely the process $B^*(t)=\{\tau_{2n}^{-1/2}W_1(t\tau_{2n})\ ,\ 0\leq t\leq 1\}$) is *not* a Brownian bridge. The "only" reason of this negative statement is the fact that there is an interval $(1-\delta, 1)$ $(\delta>0)$ where $B^*(t)$ does not have roots. However, one can modify the process $B^*(t)$ a bit such that it becomes a Brownian bridge without loosing the closeness (4.2).

Let $\{W_o(t),\ t\geq 0\}$ be a Wiener process independent from W_1 . On the set A_n define:

$$\tau_1^{(o)} = \inf\{t:\ t>0,\ |W_o(t)|=1\}\ ,$$

$$\Delta_n = \sup\{t:\ t<\tau_{2n}\ ,\ |W_1(t)| = 1\}\ ,$$

$$W_2(t) = \begin{cases} W_1(t) & \text{if}\quad 0\leq t\leq \Delta_n\ , \\ W_o(\Delta_n+\tau_1^{(o)}-t) & \text{if}\quad \Delta_n\leq t\leq \Delta_n+\tau_1^{(o)}\ , \end{cases}$$

$$B_n(t) = (\Delta_n+\tau_1^{(o)})^{-1/2}\ W_2(t(\Delta_n+\tau_1^{(o)}))\qquad (0\leq t\leq 1)\ .$$

One can easily see that $\{B_n(t)\ ,\ 0\leq t\leq 1\}$ is a Brownian bridge for each $n=1,2,\ldots$. Applying Lemmas 4 and 5 and the method of proof of Theorem A one can see that the local time of U_n and that of B_n satisfies (4.2).

Let $V_1\ ,V_2\ ,\ldots$ be a sequence of uniformly distributed independent r.v.'s. Denote its empirical density function by

$$F_n(t) = n^{-1}N\{\,k:\ k\leq n\ ,\ V_k<t\,\}\ .$$

Further let

$$\alpha_n(x) = n^{1/2}(F_n(t)-t)$$

be the empirical process. The local time of α_n is denoted by

$$\delta_n(0,t) = N\{k:\ k\leq tn\ ,\ \alpha_n(V_{k-1})\alpha_n(V_k)<0\}\ .$$

Applying Theorem B instead of Theorem A one can prove

THEOREM 5. *For each* n $(n=1,2,\ldots)$ *one can define the sequence* $V_1\ ,V_2\ ,\ldots,V_n$ *of independent, uniformly distributed r.v.'s and a Brownian bridge* $\{B_n(x),\ 0\leq x\leq 1\}$ *on a probability space* $\{\Omega_n\ ,S_n\ ,\mathbf{P}_n\}$ *such that for any* $\varepsilon>0$ *we have*

$$\lim_{n\to\infty}\ \sup_{0\leq t\leq 1}\ n^{-1/4-\varepsilon}|\delta_n(0,t)-n^{1/2}\beta_n(0,t)|=0\quad \textit{a.s.}$$

5. AN APPLICATION

By a combinatorial argument one can see

$$\mathbf{P}\,(\gamma^{(n)}(0,2n)=\ell)=2^{\ell}\,\frac{\ell\binom{2n-\ell}{n}}{(2n-\ell)\binom{2n}{n}}\qquad (\ell=0,1,2,\ldots,\ n)\ . \tag{5.1}$$

By the Stirling formula (5.1) implies

$$\lim_{n\to\infty}\mathbf{P}\,((2n)^{-1/2}\,\gamma^{(n)}(0,2n)<y) = 1-e^{-y^2/2}\qquad (0\leq y<\infty)\ . \tag{5.2}$$

Applying Theorems 4 and 5 by (5.2) one can get that the exact distribution of the local time of a Brownian bridge and the limit distribution of the local time of an empirical process is $1-e^{-y^2/2}$ $(y\geq 0)$.

REFERENCES

Csáki, E. - Csörgő, M. - Földes, A. - Révész, P. (1982): How big are the increments of the local time of a Wiener process. Technical report, Carleton Math. Lecture Note. No. 37.

Csáki, E. - Révész, P. (1982): Strong invariance for local times. Technical report, Carleton Math. Lecture Note. No. 37.

Geman, D. - Horowitz, J. (1980): Occupation densities. *Ann. Probability* 8, 1-67.

McKean, H, P. Jr. (1962): A Hölder condition for Brownian local time. *J. Math. Kyoto Univ.* 1, 195-201.

Ray, D. B. (1963): Sojourn times of a diffusion process. *Illinois J. Math.* 7, 615-630.

Révész, P. (1981): Local time and invariance. *Analytical Methods in Probability Theory. Lecture Notes in Math.* 861, Springer-Verlag, Berlin.

Révész, P. (1982): A strong invariance principle of the local time of r.v.'s with continuous distribution. Technical report, Carleton Math. Lecture Note. No. 37.

Mathematical Institute of the
Hungarian Academy of Sciences
Budapest, Reáltanoda u. 13-15
H-1053
Hungary

ROBUST ESTIMATION OF ONE REAL PARAMETER WHEN NUISANCE PARAMETERS ARE PRESENT

Helmut Rieder
Bayreuth

Key words: influence curve, orthogonality condition, Hellinger balls, ε-contamination, total variation, asymptotic minimax bias, minimum asymptotic variance subject to bounded influence

---

ABSTRACT

One real parameter is to be estimated robustly in the presence of a finite-dimensional nuisance parameter. The parametrization of the underlying family of probability measures is assumed to be smooth. The laws of the independent observations may be subject to infinitesimal perturbations which are formulated in terms of Hellinger distance and L^1-neighborhoods. A class of regular estimators is considered whose influence curves must satisfy a certain orthogonality condition. While Hellinger balls lead to the classically optimal influence curve, the L^1-optimal influence curve is obtained by truncating the not quite orthogonalized first component of the logarithmic derivative. The construction of such estimators is addressed.

ESTIMATION FRAMEWORK

Let $(\Omega, \mathcal{B})$ be a measurable space, p a nonnegative integer, Θ a nonempty open subset of $\mathbb{R}^{1+p}$, and $\{P_\theta : \theta \in \Theta\}$ a family of probability measures (p.m.s) on $\mathcal{B}$.

The problem is to estimate the first component θ_1 of the parameter $\theta = (\theta_1, \theta_2) \in \Theta$, $\theta_1 \in \mathbb{R}^1$, $\theta_2 \in \mathbb{R}^p$, on the basis of n stochastically independent observations $x_1,\ldots,x_n \in \Omega$ whose laws equal P_θ, approximately. Our study will be asymptotic, as the sample size n tends to infinity.

For the purpose of convenient asymptotics we assume a smooth parametrization: $\{P_\theta : \theta \in \Theta\}$ is dominated by some σ-finite measure μ on $\mathcal{B}$, and for every $\theta \in \Theta$ there is a $\Lambda_\theta \in L^2(P_\theta)^{1+p}$ such that for one (hence any) choice of μ-densities $p_{\theta+h} \in dP_{\theta+h}/d\mu$

(1) $$\int (2(p_{\theta+h}^{1/2} - p_\theta^{1/2}) - \Lambda_\theta h^T p_\theta^{1/2})^2 d\mu = o(|h|^2)$$

as $|h|^2 = h\,h^T \longrightarrow 0$. Then, for every $\theta \in \Theta$, every convergent sequence $h_n \longrightarrow h$ in $\mathbb{R}^{1+p}$, and $\theta_n = \theta + n^{-1/2} h_n$, any version $L_n(\theta_n : \theta)$ of the loglikelihood $\log dP_{\theta_n}^n / dP_\theta^n$ of n-fold product measures can be expanded as follows

(2) $$L_n(\theta_n : \theta) = n^{-1/2} \Sigma \Lambda_\theta h^T - \frac{1}{2} h I_\theta h^T + o_{P_\theta^n}(1) \quad (n \to \infty)$$

where $\Sigma \Lambda_\theta = \sum_{i=1}^{n} \Lambda_\theta(x_i)$, and $I_\theta = \int \Lambda_\theta^T \Lambda_\theta \, dP_\theta$ denotes the Fisher information matrix. We assume I_θ has full rank.

For the estimation of θ_1, a class of regular estimators T, i.e. sequences of measurable mappings $T_n : \Omega^n \longrightarrow \mathbb{R}^1$, will be considered. For every $\theta \in \Theta$ let a set of functions Ψ_θ be defined by

(3) $$\Psi_\theta = \{\psi \in L^2(P_\theta) : \int \psi \, dP_\theta = 0, \int \psi \Lambda_{\theta,1} \, dP_\theta = 1, \int \psi \Lambda_{\theta,2} \, dP_\theta = 0\}$$

where, corresponding to the partition of $\theta = (\theta_1, \theta_2)$, $\Lambda_\theta = (\Lambda_{\theta,1}, \Lambda_{\theta,2})$, $\Lambda_{\theta,1} \in L^2(P_\theta)$, $\Lambda_{\theta,2} \in L^2(P_\theta)^p$. Then an estimator $T = (T_n)$ is called regular iff for every $\theta \in \Theta$ there exists a $\psi_\theta \in \Psi_\theta$ such that

(4) $$n^{1/2}(T_n - \theta_1) = n^{-1/2} \Sigma \psi_\theta + o_{P_\theta^n}(1) \quad (n \to \infty).$$

In a formal sense only, we call ψ_θ the influence curve of T at P_θ. When compared with the case $p = 0$, these influence curves must in general fulfill the additional orthogonality requirement $\int \psi_\theta \Lambda_{\theta,2} \, dP_\theta = 0$.

An estimator may be generated by some real-valued functional τ, which is defined on $\{P_\theta: \theta \in \Theta\}$ and the molecular p.m.s on $\mathcal{B}$, such that T_n is the value of τ at the empirical measure of sample size n. If τ is differentiable in the sense that for every $\theta \in \Theta$ there is a $\psi_\theta \in L^2(P_\theta)$, $\int \psi_\theta \, dP_\theta = 0$, such that

(5) $$\tau(P_{\theta+h}) = \tau(P_\theta) + \int \psi_\theta \Lambda_\theta \, dP_\theta \, h^T + o(|h|) \quad (h \to 0),$$

then the additional property $\int \psi_\theta \Lambda_\theta \, dP_\theta = (1,0)$ is enforced by the requirement of Fisher consistency, i.e.

(6) $$\tau(P_\theta) = \theta_1 \quad (\theta \in \Theta).$$

In general, given a $\psi_\theta \in L^2(P_\theta)$, $\int \psi_\theta \, dP_\theta = 0$, and the expansion (4), an application of LeCam's Third Lemma to (4) and (2) proves: The property $\int \psi_\theta \Lambda_\theta \, dP_\theta = (1,0)$ is equivalent to uniformity over h-compacts of the weak convergence of laws

(7) $$n^{1/2}(T_n - \theta_n(h)_1)(P_{\theta_n(h)}^n) \underset{w}{\longrightarrow} \text{normal}\,(0, \int \psi_\theta^2 \, dP_\theta)$$

where $\theta_n(h) = \theta + n^{-1/2} h$, $\theta_n(h)_1$ its first component; which is Hajek's (1970) notion of regularity.

From the idealistic viewpoint we assume that there is a true, unknown value of the parameter $\theta \in \Theta$. This value is now fixed and (mostly) dropped from notation. Then regular estimators shall be studied under infinitesimal perturbations of $P = P_\theta$.

Let $\delta, \varepsilon \in [0, \infty)$. Define neighborhoods (nbd.s) $\mathcal{P}_h = \mathcal{P}_h(P;\delta)$, $\mathcal{P}_{cv} = \mathcal{P}_{cv}(P;\varepsilon,\delta)$ by

(8) $$\mathcal{P}_h = \{Q \text{ p.m. on } \mathcal{B} : h(Q,P) \le \delta\}$$
$$\mathcal{P}_{cv} = \{Q \text{ p.m. on } \mathcal{B} : (1-\varepsilon)P - \delta \le Q \le (1-\varepsilon)P + \varepsilon + \delta \text{ on } \mathcal{B}\}.$$

The symbol h denotes Hellinger distance, $h^2(Q,P) = \frac{1}{2}\int (dQ^{1/2} - dP^{1/2})^2$, and cv stands for the superposition of ε-contamination and total variation nbd.s, which $\mathcal{P}_{cv}$ represents.

Take sequences $\delta_n, \varepsilon_n \in [0, \infty)$ such that

(9) $$n^{1/2}\delta_n \longrightarrow \delta, \quad n^{1/2}\varepsilon_n \longrightarrow \varepsilon \quad (n \to \infty),$$

and replace the radii δ, ε in (8) by δ_n, ε_n to get shrinking nbd.s $\mathcal{P}_{j,n}(j = h, cv)$

Given a loss function ℓ,

(10) $$\ell : [0, \infty) \longrightarrow [0, \infty), \text{ isotone}, \quad \ell(0+) < \ell(\infty-),$$

put $\ell_c = \ell \wedge c \, (0 < c < \infty)$. The full nbd. risks of a regular estimator T at θ are defined by

(11) $$\overline{r}_j(T,\theta) = \sup_c \ \limsup_n \ \sup \int \ell_c(|n^{1/2}(T_n - \theta_1)|)dw_n$$
$$\underline{r}_j(T,\theta) = \sup_c \ \liminf_n \ \sup \int \ell_c(|n^{1/2}(T_n - \theta_1)|)dw_n$$

where the last sup extends over all product measures $w_n = \prod_{i=1}^{n} Q_i$, $Q_1, \ldots, Q_n \in \mathcal{P}_{j,n}$ $(j = h, cv)$. The second version is relevant for lower bounds, the first for the attainment of such bounds.

A convenient asymptotic nbd. submodel is formulated using the following sets of functions $\mathcal{F}_j(j = h, cv)$

(12) $$\mathcal{F}_h = \{f \in L^\infty(P) : \int f \, dP = 0, \ \int f^2 \, dP \le 8 \, \delta^2\}$$
$$\mathcal{F}_{cv} = \{f_1 + f_2 : \ f_i \in L^\infty(P), \int f_i \, dP = 0 \ (i = 1,2), \ \inf_P f_1 \ge -\varepsilon, \int |f_2| dP \le 2\delta\}$$

where $\inf_P$ means P-essential infimum (P-essential supremum will be denoted by $\sup_P$). For $f \in \mathcal{F}_j$ let

(13) $$Q_n(f) = (1 + n^{-1/2} f) P .$$

Then, eventually, $Q_n(f) \in \mathcal{P}_h(P ; \delta_n')$ where $n^{1/2}\delta_n' \longrightarrow \delta$, respectively $Q_n(f) \in \mathcal{P}_{cv}(P; n^{-1/2}\varepsilon, n^{-1/2}\delta)$.

The corresponding submodel risk of a regular estimator T at θ is given by

(14) $$\rho_j(T, \theta) = \sup_c \ \sup_{f \in \mathcal{F}_j} \ \lim_n \int \ell_c(|n^{1/2}(T_n - \theta_1)|)dw_n(f),$$

where $w_n(f) = Q_n(f)^n$ $(f \in \mathcal{F}_j ; \ j = h, cv)$.

RISK EVALUATION

The last limit indeed exists, for, by familiar arguments (L^2-differentiability of $\{dQ_n(f)^{1/2}\}$, expansion (4), and LeCam's Third Lemma), $n^{1/2}(T_n-\theta_1)(w_n(f))$ is asymptotically normal $(b(\psi,f),\ \sigma^2(\psi))$ where $\psi = \psi_\theta$ and

(15) $$b(\psi,f) = \int \psi\, f\, dP\,, \quad \sigma^2(\psi) = \int \psi^2\, dP\,.$$

Since $\ell_c \circ |\ \ |$ is bounded, continuous Lebesgue a.e., and $\sigma^2(\psi) > 0$, the above limit equals $\int \ell_c(|b(\psi,f) + \sigma(\psi)\, y\,|)\, \nu\,(dy)$ $(\nu = \text{normal}(0,1))$. Let $\beta_j(\psi)$ be defined by

(16) $$\beta_j(\psi) = \sup\{\,|b(\psi,f)|:\ f \in F_j\} \quad (j = h, cv).$$

PROPOSITION 1. Let T be a regular estimator with influence curve $\psi = \psi_\theta$ at P_θ. Then

(17) $$\rho_j(T,\theta) = \int \ell(|\beta_j(\psi) + \sigma(\psi) y|)\, \nu\,(dy) \quad (j = h, cv)$$

(18) $$\beta_h(\psi) = 8^{1/2}\, \delta\, \sigma(\psi)$$
$$\beta_{cv}(\psi) = \varepsilon\, \sup_P|\psi| + \delta(\sup_P \psi - \inf_P \psi)\,.$$

NOTATION $\rho_j(\psi) = \rho_j(T\,,\,\theta)\ (j = h\,,\, cv)$.

PROOF Observe that, for $b \geq 0$, $t \geq 0$, $\nu\{y:\ |b+y| > t\}$ is differentiable with respect to b and has derivative $\dot\nu(b-t) - \dot\nu(-b-t) \geq 0$ ($\dot\nu$ the standard normal density). Therefore, $|b_2 + \sigma\, y\,|$ is stochastically larger than $|b_1 + \sigma\, y\,|$ under ν iff $|b_2| \geq |b_1|$ $(b_1, b_2 \in \mathbb{R}^1,\ \sigma \geq 0)$. As ℓ is isotone, (17) follows by dominated and monotone convergence.

To prove (18) put $b_j(\psi) = \sup\{\,b(\psi,f):\ f \in F_j\}$; then $\beta_j(\psi) = b_j(\psi) \vee b_j(-\psi)$. In case $j = h$ approximate $\psi \neq 0$ in $L^2(P)$ on one hand by bounded functions f, $\int f\, dP = 0$, then rescale f such that $\int f^2\, dP = 8\,\delta^2$. On the other hand, apply Cauchy-Schwarz inequality. Thus $b_h(\psi) = b_h(-\psi) = 8^{1/2}\, \delta\, \sigma(\psi)$.

In case $j = cv$, if $f_1 + f_2 \in F_{cv}$, then $b(\psi,f_1) = b(\psi,f_1+\varepsilon) \leq \sup_P \psi$ and $b(\psi,f_2) = b(\psi,f_2^+) - b(\psi,f_2^-) \leq \delta(\sup_P \psi - \inf_P \psi)$. To approximate the bound let $\inf_P \psi < c_1 \leq c_2 < \sup_P \psi$, put $C_1 = \{\psi < c_1\}$, $C_2 = \{\psi > c_2\}$, $\gamma_1 = P(C_1)$, $\gamma_2 = P(C_2)$, and define $f_1 = \varepsilon(\gamma_2^{-1} 1_{C_2} - 1)$, $f_2 = \delta(-\gamma_1^{-1} 1_{C_1} + \gamma_2^{-1} 1_{C_2})$. Then $f_1 + f_2 \in F_{cv}$ and $b(\psi, f_1 + f_2) \geq \varepsilon c_2 + \delta(c_2 - c_1)$. If c_1, c_2 tend to the essential extrema of ψ we get $b_{cv}(\psi) = \varepsilon \sup_P \psi + \delta(\sup_P \psi - \inf_P \psi)$. Maximizing with $b_{cv}(-\psi)$ yields the desired expression. □

HELLINGER OPTIMAL ψ_θ

Let $\begin{pmatrix} I_{11} & I_{12} \\ I_{21} & I_{22} \end{pmatrix} = I$ be the partitioned Fisher information matrix (at θ), where I_{11} is 1×1, I_{22} is $p \times p$, and define

$$\begin{aligned} \Delta &= \Lambda_1 - I_{12} I_{22}^{-1} \Lambda_2 \\ \hat{\sigma}^2 &= I_{11} - I_{12} I_{22}^{-1} I_{21} \\ \hat{\psi} &= \hat{\sigma}^{-2} \Delta ; \end{aligned} \tag{19}$$

note that $\hat{\sigma} > 0$ and $\hat{\psi} \in \Psi$.

THEOREM 2. $\rho_h(\psi)$ is minimized over Ψ iff $\psi = \hat{\psi}$ a.e. P; and $\rho_h(\hat{\psi}) = \int \ell(\hat{\sigma}^{-1} | 8^{1/2} \delta + y |) \, \nu(dy)$.

PROOF As ℓ is isotone, not constant a.e. ν, the minimization of $\rho_h(\psi) = \int \ell(\sigma(\psi) | 8^{1/2} \delta + y|) \, \nu(dy)$ over Ψ turns out to be equivalent to the classical task of minimizing $\sigma(\psi)$. Equivalently, maximize $\sigma^{-1}(\psi) \, b(\psi, \Lambda_1)$ over $L^2(P)$ subject to $\int \psi \, dP = 0$, $\int \psi \Lambda_2 \, dP = 0$. For every such ψ, e.g. $\psi = \Delta$, Cauchy-Schwarz inequality yields $\sigma^{-1}(\psi) \, b(\psi, \Lambda_1) = \sigma^{-1}(\psi) \, b(\psi, \Delta) \leq \sigma(\Delta) = \sigma^{-1}(\Delta) \, b(\Delta, \Lambda_1)$, with equality holding iff ψ is proportional to Δ. □

Thus the classically (i.e. for $\delta = 0$) optimal estimator is still optimal in the Hellinger submodel. Its influence curve may be unbounded.

BOUNDED INFLUENCE CURVES

The second risk $\rho_{cv}(\psi)$ involves both $\sigma(\psi)$ and the extrema of ψ, which suggests a two-step minimization procedure: First, put bounds on ψ, secondly, minimize $\sigma(\psi)$ subject to these bounds.

For $c=(c_o,c_1)\in\mathbb{R}^2$, $c_o<0<c_1$, and $d=(d_o,d_1,d_2)\in\mathbb{R}^{2+p}$ with d_o, $d_1\in\mathbb{R}^1$, $d_2\in\mathbb{R}^p$, define

(20) $\Psi_c = \{\psi\in L^\infty(P):\ \inf_P\psi\ \ge c_o,\ \sup_P\psi\ \le c_1\}$

(21) $\Delta(d) = d_o + d_1\Lambda_1 + d_2\Lambda_2^T$

(22) $\psi_s(c,d) = c_o 1(\Delta(d)<0) + c_1 1(\Delta(d)>0)$

(23) $\psi_t(c,d) = c_o \vee \Delta(d) \wedge c_1$

(24) $B_d(\psi) = (\sup_P\psi)\int\Delta(d)^+dP - (\inf_P\psi)\int\Delta(d)^-dP \quad (\psi\in\Psi)$.

Influence curves of type ψ_s are optimal with respect to bias, in the following sense.

THEOREM 3. Assume that $\psi_s(c,d)\in\Psi$. Then

(i) $\psi_s(c,d)$ minimizes B_d over Ψ; and $B_d(\psi_s(c,d)) = d_1$.

(ii) If $\psi\in\Psi$ minimizes B_d over Ψ and $\int\psi 1(\Delta(d)=0)\,dP=0$, then $\psi=\psi_s(c,d)$ a.e.P.

PROOF If $\psi\in\Psi$ then $d_1=\int\psi\Delta(d)\,dP \le (\sup_P\psi)\int\Delta(d)^+dP - (\inf_P\psi)\int\Delta(d)^-dP$, with equality holding iff, a.e. P, $\Delta(d)>0$ implies $\psi=\sup_P\psi$, and $\Delta(d)<0$ implies $\psi=\inf_P\psi$. Moreover, since the determinant

$$\begin{vmatrix} P(\Delta(d)<0) & -\int\Delta(d)^-dP \\ P(\Delta(d)>0) & \int\Delta(d)^+dP \end{vmatrix}$$

is positive, uniqueness follows. □

REMARK In case $p=0$ the function $\psi_s=\alpha^{-1}(P(\Lambda<0)1(\Lambda>0) - P(\Lambda>0)1(\Lambda<0))$ with $\alpha=P(\Lambda\neq 0)\int\Lambda^+dP$ is in Ψ, and $B_d(\psi)=B(\psi)=(\sup_P\psi-\inf_P\psi)\int\Lambda^+dP\ (\psi\in\Psi)$. The corresponding (M)-estimator may be regarded as a generalized version of the median, which is now applied to transformed observations, with possibly different weights for positive and negative values. Similarly, in the nuisance parameter case. □

Influence curves of type ψ_t, and ψ_s as a limiting case, minimize the asymptotic variance subject to bounds.

THEOREM 4. Assume that $\psi_t(c,d) \in \Psi$. Then $\sigma(\psi)$ is minimized over $\Psi \cap \Psi_c$ iff $\psi = \psi_t(c,d)$ a.e. P .

PROOF If $\psi \in \Psi$ then $\int (\psi - \Delta(d))^2 dP = \int \psi^2 dP + \int \Delta(d)^2 dP - 2d_1$. Over Ψ_c, the first integral is uniquely minimized by $\psi_t(c,d)$. □

Besides including functions ψ_s, the following extension of Theorem 4 also ensures existence of $\psi_t, \psi_s \in \Psi$.

THEOREM 5. Let $c = (c_o, c_1) \in \mathbb{R}^2$, $c_o \leq c_1$. Then either

(i) $\Psi \cap \Psi_c = \emptyset$

or

(ii) there exists a $d \in \mathbb{R}^{2+p}$ such that $\psi_* \in \Psi \cap \Psi_c$ for $\psi_* = \psi_s(c,d)$ or $\psi_* = \psi_t(c,d)$.

In case (ii), $\sigma(\psi)$ is minimized over $\Psi \cap \Psi_c$ iff $\psi = \psi_*$ a.e. P .

COROLLARY 6. If $\psi_1, \psi_2 \in \psi$ each are of type ψ_s or ψ_t with the same c then $\psi_1 = \psi_2$ a.e. P .

PROOF The corollary follows from the uniqueness statement of the theorem. To prove the theorem, observe that both Ψ and Ψ_c are convex, closed subsets of the Hilbert space $L^2(P)$. Hence, if $\Psi \cap \Psi_c \neq \emptyset$, there exists a unique element ψ_* of smallest norm in $\Psi \cap \Psi_c$.

This ψ_* can be determined by means of Lagrange multipliers: In view of the convexity of Ψ_c and the convexity, respectively linearity, of the mappings $\psi \longrightarrow \int \psi^2 \, dP$, $\psi \longrightarrow \int \psi(1,\Lambda) dP$ on $L^2(P)$, Theorem 3 on page 82 of Neustadt (1976) ensures the existence of $\beta_o \geq 0$ and

$\alpha = (\alpha_o, \alpha_1, \alpha_2) \in \mathbb{R}^{2+p}$ such that ψ_* minimizes the value of the P-integral of $\beta_o \psi^2 + \Delta(\alpha)\psi$ with respect to $\psi \in \Psi_c$. Minimizing the integrand pointwise, subject to $c_o \le \psi \le c_1$, shows that in case $\beta_o > 0$ necessarily

$$\psi_* = \psi_t(c, -(2\beta_o)^{-1}\alpha) \quad \text{a.e. P}$$

and in case $\beta_o = 0$ necessarily

$$\psi_* = \psi_s(c, -\alpha) \quad \text{a.e. P}. \qquad \square$$

L^1 - OPTIMAL ψ_θ

After the preceding first-step optimality result, it is now shown in a second step that the risk ρ_{cv} is minimized over Ψ by influence curves of type ψ_s or ψ_t. In contrast to the Hellinger optimal $\hat{\psi}$, the L^1-solution in general depends on the loss function and the nbd. radii.

THEOREM 7. Let $\delta + \varepsilon > 0$. Assume that ℓ, in addition to (10), is convex. Assume $\inf \rho_{cv}(\Psi) < \ell(\infty -)$.

(i) Then there exist $c = (c_o, c_1) \in \mathbb{R}^2$, $c_o < 0 < c_1$, $d \in \mathbb{R}^{2+p}$ and ψ_* of type $\psi_s(c,d)$ or $\psi_t(c,d)$ in Ψ such that $\rho_{cv}(\psi_*) = \inf \rho_{cv}(\Psi)$.

(ii) If $\psi \in \Psi$, $\rho_{cv}(\psi) = \inf \rho_{cv}(\Psi)$, then $\psi = \psi_*$ a.e. P.

(iii) In case $\delta = 0$ one can choose $c_o = -c_1$.

PROOF First, as ℓ is isotone, $\int \ell(|\beta_2 + \sigma y|)\, \nu(dy) \ge \int \ell(|\beta_1 + \sigma y|)\, \nu(dy)$ $(\sigma \ge 0,\ \beta_2 \ge \beta_1 \ge 0)$, by a previous argument. Secondly, since also $\ell_1 = \ell \circ |\ \ |$ is convex, $\ell_1(\beta + \sigma_2 y) - \ell_1(\beta + \sigma_1 y) \ge (\sigma_2 - \sigma_1) y\, \dot{\ell}_1(\beta + \sigma_1 y)$ for $y, \beta \in \mathbb{R}^1$, $\sigma_2 \ge \sigma_1 \ge 0$, where $\dot{\ell}_1$ can be chosen isotone and finite, and then $\int y\, \dot{\ell}_1(\beta + \sigma_1 y)\, \nu(dy) = \int y\, (\dot{\ell}_1(\beta + \sigma_1 y) - \dot{\ell}_1(\beta))\, \nu(dy) \ge 0$; thus $\int \ell(|\beta + \sigma_2 y|)\, \nu(dy) \ge \int \ell(|\beta + \sigma_1 y|)\, \nu(dy)$.

Choose a sequence $\psi_n \in \Psi$ such that $\ell(\infty -) > \rho_{cv}(\psi_n) \longrightarrow \inf \rho_{cv}(\Psi)$ $(n \to \infty)$. Then $\{\psi_n\}$ must be bounded in $L^\infty(P)$. Otherwise, $\beta_{cv}(\psi_k) + \sigma(\psi_k) y \longrightarrow \infty$ for every $y \in \mathbb{R}^1$, along some subsequence, and as ℓ_1 is continuous a.e. ν and nonnegative, Fatou's Lemma would yield the contradiction: $\liminf_k \rho_{cv}(\psi_k) \geq \ell(\infty-)$.

Boundedness in $L^\infty(P)$ implies boundedness in $L^2(P)$, i.e. weak sequential compactness (Dunford-Schwartz, Vol. I, IV. 4.7). Thus there exists a weakly convergent subsequence $\{\psi_m\}$ with limit $\psi \in \Psi$ (weakly closed). The convergences $\int \psi\,\psi_m\,dP \longrightarrow \int \psi^2\,dP$ and $\int_B \psi_m\,dP \longrightarrow \int_B \psi\,dP$ $(B \in \mathcal{B})$ tell us that necessarily $\sigma^2(\psi) \leq \liminf_m \sigma^2(\psi_m)$, $\inf_P \psi \geq \limsup_m \inf_P \psi_m$, and $\sup_P \psi \leq \liminf_m \sup_P \psi_m$. Therefore, due to the initially observed two isotony properties,

$$\int \ell(|\beta_{cv}(\psi) - \eta + (\sigma(\psi) - \eta)\,y|)\,\nu(dy)$$

$$\leq \liminf_m \rho_{cv}(\psi_m) = \inf \rho_{cv}(\Psi)$$

for every $0 < \eta < \beta_{cv}(\psi) \wedge \sigma(\psi)$. Let $\eta \longrightarrow 0$. Then, as ℓ_1 is continuous a.e. ν and nonnegative, it follows again by Fatou that $\rho_{cv}(\psi) \leq \inf \rho_{cv}(\Psi)$.

Put $c_0 = \inf_P \psi$, $c_1 = \sup_P \psi$; $c_0 < 0 < c_1$ as $\psi \in \Psi$, and c_0, c_1 finite as $\beta_{cv}(\psi) < \infty$. By Theorem 5 there exist a $d \in \mathbb{R}^{2+p}$ and ψ_* of type $\psi_s(c,d)$ or $\psi_t(c,d)$ in Ψ such that $\beta_{cv}(\psi_*) \leq \beta_{cv}(\psi)$ and $\sigma(\psi_*) \leq \sigma(\psi)$; hence $\rho_{cv}(\psi_*) \leq \rho_{cv}(\psi)$. If $\delta = 0$, since then $\beta_{cv}(\psi) = \varepsilon \sup_P |\psi|$, we may choose $c_0 = -c_1 = -\sup_P |\psi|$ at this point.

Finally, as ℓ is not constant a.e. ν, $\sigma(\psi_*) < \sigma(\psi)$ cannot hold. Therefore, by the uniqueness statement of Theorem 5, $\psi = \psi_*$ a.e. P. □

CONSTRUCTION OF ESTIMATORS OVER FULL NBD.S

The construction of a regular estimator corresponding to a prescribed family of influence curves $\psi_\theta \in \Psi_\theta$ $(\theta \in \Theta)$ requires certain additional assumptions, but is a classical and solved problem. As for full nbd.s, since the submodel risk of a regular estimator has turned out not to depend on the particular sequences $\{\delta_n\}$, $\{\varepsilon_n\}$ as long as they satisfy (9), we already obtain the lower bound:

$\underline{r}_j(T,\theta) \geq \rho_j(T,\theta)$. If, moreover, we want to extend the submodel optimality result to full nbd.s, it suffices to construct regular estimators with the property: $\overline{r}_j(T,\theta) = \rho_j(T,\theta)$; which, in view of possible noncontiguity of $\{w_n\}$ and $\{P_\theta^n\}$, is a nontrivial task.

Under the additional assumption that the parametrization of $\{P_\theta\colon \theta \in \Theta\}$ be one-to-one there exist estimators $\zeta_n\colon \Omega^n \longrightarrow \mathbb{R}^{1+p}$ which are suitably discretized and such that, for each $\theta \in \Theta$ and every $w_n = \prod_{i=1}^n Q_i$, $Q_1,\ldots,Q_n \in \mathcal{P}_{h,n} \cup \mathcal{P}_{cv,n}$, the sequence of laws $n^{1/2}(\zeta_n-\theta)\,(w_n)$ is tight; cf. LeCam (1969). Consider $\{\psi_\theta\colon \theta \in \Theta\}$ such that, for every $\theta \in \Theta$, $\psi_\theta \in \Psi_\theta$, $\sup_{x\in\Omega} |\,\psi_\theta(x)\,| < \infty$, and, for some $k_\theta \in \mathbb{R}^1$,

$$\sup_{x\in\Omega} |\,\psi_{\theta+h}(x) - \psi_\theta(x)\,| \leq k_\theta |h| + o(|h|) \quad (h \to 0). \tag{25}$$

Define $T_n\colon \Omega^n \longrightarrow \mathbb{R}^1$ by

$$T_n = \zeta_{n,1} + n^{-1}\,\Sigma\,\psi_{\zeta_n}. \tag{26}$$

THEOREM 8. Under the preceding assumptions, for every $\theta \in \Theta$, $\{T_n\}$ satisfies

$$n^{1/2}(T_n-\theta_1) = n^{-1/2}\,\Sigma\,\psi_\theta + o_{w_n}(1) \qquad (n\to\infty) \tag{27}$$

for all sequences $w_n = \prod_{i=1}^n Q_i$, $Q_1,\ldots,Q_n \in \mathcal{P}_{h,n} \cup \mathcal{P}_{cv,n}$, $n \in \mathbb{N}$.

PROOF Write

$$n^{1/2}(T_n-\theta_1) - n^{-1/2}\,\Sigma\,\psi_\theta = n^{1/2}(\zeta_{n,1}-\theta_1) + n^{-1/2}\,\Sigma\,(\psi_{\zeta_n} - \psi_\theta).$$

Due to $n^{1/2}$-consistency and discretization, ζ_n may be replaced by a nonstochastic sequence $\theta + n^{-1/2} h_n$, $h_n \longrightarrow h \in \mathbb{R}^{1+p}$; then $n^{1/2}(\zeta_{n,1}-\theta_1) \longrightarrow h_1$. Moreover, by (25),

$$\begin{aligned} &w_n\{\,|n^{-1/2}\,\Sigma\,(\psi_{\zeta_n} - \psi_\theta) - \textstyle\int(\psi_{\zeta_n} - \psi_\theta)\,dQ_i\,| > \eta\,\} \\ &\leq\; n^{-1}\,\eta^{-2}\,\Sigma \textstyle\int(\psi_{\zeta_n} - \psi_\theta)^2\,dQ_i = O(n^{-1}) \end{aligned}$$

and also: $n^{-1/2}\,\Sigma\int(\psi_{\zeta_n}-\psi_\theta)\,d\,(Q_i-P_\theta) = O(n^{-1/2})$, while: $n^{1/2}\int(\psi_{\zeta_n}-\psi_\theta)\,dP_\theta = -n^{1/2}\int \psi_\theta\, d(P_{\zeta_n}-P_\theta) - n^{1/2}\int(\psi_{\zeta_n}-\psi_\theta)\,d\,(P_{\zeta_n}-P_\theta) \longrightarrow -\int \psi_\theta\,\Lambda_\theta\,h^T dP_\theta = -h_1$. □

Define, with inf and sup denoting pointwise extrema,

$$\bar{b}_{cv}(\psi_\theta) = (\varepsilon + \delta)\sup \psi_\theta - \delta \inf_{P_\theta} \psi_\theta \tag{28}$$

$$\bar{\beta}_{cv}(\psi_\theta) = \bar{b}_{cv}(\psi_\theta) \vee \bar{b}_{cv}(-\psi_\theta) \quad .$$

THEOREM 9. The uniform expansion (27) implies that

$$\bar{r}_{cv}(T,\theta) = \underline{r}_{cv}(T,\theta) = \int \ell(|\bar{\beta}_{cv}(\psi_\theta) + \sigma(\psi_\theta)y|)\, \nu(dy), \tag{29}$$

respectively,

$$\bar{r}_h(T,\theta) = \underline{r}_h(T,\theta) = \rho_h(T,\theta) . \tag{30}$$

PROOF (29) follows by the arguments of the proof to Proposition 1, inv.. ng Proposition 3.1, Lemma 3.2, Lemma 3.3, and Theorem 3.5 of Rieder (1978).

In the second case, without restriction, consider product measures $w_n = Q_n{}^n$. In view of the uniform asymptotic normality ensured by Proposition 3.1 of Rieder (1978), we only need to bound the centering sequence $n^{1/2}\int \psi_\theta \, dQ_n$ by $8^{1/2}\,\delta\,\sigma(\psi_\theta)$. Indeed, employing the appropriate Hilbert space H of all $\xi R^{1/2}$ (R a p.m. on $\mathcal{B}$, $\xi \in L^2(R)$), $\{n^{1/2}(Q_n{}^{1/2} - P_\theta{}^{1/2})\}$ is bounded in H, hence weakly sequentially compact: There exists an element $\xi R^{1/2}$ such that, along some subsequence, $m^{1/2}(Q_m{}^{1/2} - P_\theta{}^{1/2}) \longrightarrow \xi R^{1/2}$ weakly; necessarily, $\int \xi^2 \, dR \le 2\,\delta^2$. Since $m^{1/2}\int \psi_\theta \, dQ_m \longrightarrow 2\int \psi_\theta\, \xi \, dR^{1/2}\, dP_\theta{}^{1/2}$, the assertion follows. (As for unbounded ψ_θ, cf. Beran (1981).) □

SUMMARY

The paper deals with problems of the infinitesimal theory of robust estimation, as developed e.g. by Hampel (1968), Beran (1981), Bickel (1981), and Rieder (1980); the presentation closely follows Bickel (1981), pp. 15 - 27. Essentially, the equivalence of minimum asymptotic risk over shrinking nbd.s to minimum asymptotic variance subject to bounded influence, as well as the distinguished rôle of Hellinger nbd.s, are carried over to the nuisance parameter case.

The L^1-Problem leads to an interesting class of nonlinearly orthogonalized, truncated influence curves. In addition, an asymptotic minimax bias property of sign-type influence curves is proved which corresponds to Huber's (1981) optimality of the median in location models with constant nbd.s .

REFERENCES

Beran, R.J. (1981): Efficient robust estimation in parametric models. Z. Wahrscheinlichkeitstheorie u. Verw. Gebiete 55 , 91-108.

Bickel, P.J. (1981): Quelques Aspects de la Statistique Robuste. Lecture Notes in Mathematics, Vol. 876, pp. 1-72; Springer Verlag, Berlin - Heidelberg - New York.

Hajek, J. (1970): A characterization of limiting distributions of regular estimates. Z. Wahrscheinlichkeitstheorie u. Verw. Gebiete 14 , 323-330.

Hampel, F.R. (1968): Contributions to the theory of robust estimation. Ph. D. Thesis, University of California, Berkeley.

Huber, P.J. (1981): Robust Statistics. Wiley, New York.

LeCam, L. (1969): Theorie Asymptotique de la Décision Statistique. Les Presses de l'Université de Montréal.

Neustadt, L.W. (1976): Optimization. Princeton University Press.

Rieder, H. (1978): A robust asymptotic testing model. Ann. Statist. 6 , 1080-1094.

Rieder, H. (1980): Estimates derived from robust tests. Ann. Statist. 8 , 106-115.

MATHEMATISCHES INSTITUT
UNIVERSITÄT BAYREUTH
D-8580 BAYREUTH, WEST GERMANY

Work performed while the author was visiting the Department of Statistics, University of California, Berkeley, in spring 1982. The attendance at the conference as well as this visit were supported by travel grants of Deutsche Forschungsgemeinschaft, Bonn.

DISCRETE METHODS IN COOPERATIVE GAME THEORY

Joachim Rosenmüller
Bielefeld

ABSTRACT

Extreme point problems and the convergence of solution concepts (the core, the "competitive equilibrium") are topics of game theory that are frequently dealt with in a nonatomic framework (a measure space of players endowed with a nonatomic non-additive set function). This note as a survey presents two discrete analogues of "nonatomicity": Nondegeneracy and homogeneity of additive and nonadditive set functions. It turns out that these concepts lead to combinatorial and number theoretical problems (e.g. MINKOWSKI's second theorem) such that general principels of game theory (convergence problems etc.) maybe formulated in a discrete manner. Hence a statement like "the core and the competitive equilibrium coincide for large sets of players" is formulated rigorously by means of Geometric Number Theory instead of nonatomic measure theory.

§1 NONDEGENERATE AND HOMOGENEOUS MEASURES

Let $\Omega = \{1,\dots,n\}$ ("the set of players") and $\underline{\underline{B}} := \mathfrak{P}(\Omega)$ ("the coalitions"). An additive set function $m : \underline{\underline{B}} \to \mathbb{R}$ is tantamount to a vector $m = (m_1,\dots,m_n) \in \mathbb{R}^n$ via $m(S) = \sum_{i\in S} m_i$ $(S \in \underline{\underline{B}})$. $\mathbb{A} = \mathbb{A}_\Omega$ is the system of such functions and $\mathbb{A}^+, \mathbb{A}_1$ is used to indicate nonnegativity, normalization ($m(\Omega) = 1$), etc.

Call $m \in \mathbb{A}^+$ *homogeneous* w.r.t. $\alpha \in \mathbb{R}_+$ ("m hom α") if, for $T \in \underline{\underline{B}}$, $m(T) > \alpha$, there is $S \subseteq T$ such that $m(S) = \alpha$ (v.NEUMANN-MORGENSTERN [8]); this is a direct translation of nonatomicity from the continuous case (where e.g. $\Omega = [0,1]$, $\underline{\underline{B}}$ = {Borelian sets}, m σ-additive).

E.g. $m = (\frac{3}{7}, \frac{1}{7}, \frac{1}{7}, \frac{1}{7}, \frac{1}{7})$ is homogeneous w.r.t. $\alpha = \frac{4}{7}$ and uniform distribution ($m_i = \frac{1}{n}$, $i \in \Omega$) is homogeneous w.r.t. $\alpha = \frac{k}{n}$, $k = 0,1,\dots,n$. Counterexamples are straightforward.

Next, $m \in \mathbb{A}^+$ is said to be *nondegenerate* (cf. [15][12]) w.r.t. $\underline{\underline{Q}} \subseteq \underline{\underline{B}}$ ("m n.d. $\underline{\underline{Q}}$") if the linear system of equations in variables $y_1,\dots,y_n$, given by

$$(1) \qquad \sum_{i\in S} y_i = m(S) \quad (S \in \underline{\underline{Q}})$$

has the unique solution $m = (m_1,\dots,m_n)$. In particular, for $\alpha \in \mathbb{R}_+$, we may consider

$\underline{\underline{Q}} = \underline{\underline{Q}}_\alpha = \{S \in \underline{\underline{B}} \mid m(S) = \alpha\}$; in this case we say "m hom α" if m hom $\underline{\underline{Q}}_\alpha$. (1) translates to

(2) $$\sum_{i \in S} y_i = \alpha \quad (m(S) = \alpha)$$

and "n.d." means that m is uniquely defined by its values on its α-constancy sets. In the above mentioned examples, m is n.d. α as well. For $m = (\frac{1}{8}, \frac{1}{8}, \frac{1}{8}, \frac{2}{8}, \frac{3}{8})$ clearly m hom $\frac{7}{8}$ but m n.d. $\frac{7}{8}$ is wrong. The relations between "hom" and "n.d." are discussed in [13], CH.III. Note that a nonatomic probability is clearly n.d. α for $0 < \alpha < \frac{1}{2}$ (i.e. coincides with any σ-additive measure having identical values α on all α-constancy sets of m).

§2 GAMES AND SOLUTIONS

Generally, a *game* is a tripel $\Delta = (\Omega, \underline{\underline{B}}, v)$ where $v : \underline{\underline{B}} \to \mathbb{R}$, $v(\emptyset) = 0$, is a (non-additive) set function. Examples are

(3) the *weighted majority*; $v = 1_{[\alpha,1]} \circ m$ where $m \in \mathbb{A}_1^+$, $\alpha \in (0,1)$ (representing a voting committee);

(4) *production games* ("one factor"); $v = f \circ m$ where $m \in \mathbb{A}$ and $f : \mathbb{R} \to \mathbb{R}$, $f(0) = 0$;

(5) *L.P.-games* ("linear program"); defined by $v(S) = \max \{cx \mid x \in \mathbb{R}_+^l, Ax \leq b(S)\}$, $(S \in \underline{\underline{B}})$ where $c \in \mathbb{R}_+^l$, A an lxm-matrix and $b \in (\mathbb{A}^+)^m$ (all strictly positive);

(6) *market games*, given by $v(S) = \max \{ \sum_{i \in S} u^i(x^i) \mid (x^i)_{i \in \Omega} \in \mathbb{R}_+^{mxn}, \sum_{i \in S} x^i = \sum_{i \in S} a^i \}$ $(S \in \underline{\underline{B}})$, where $a^i \in \mathbb{R}_+^m$ ("initial allocation of player i∈Ω") and $u^i : \mathbb{R}_+^m \to \mathbb{R}$ (usually concave and continuous, "player i's utility function"); here players i∈S maximize their joint utility by "exchange of commodities". (cf. e.g. [6] [7])

Certain *classes of set functions* v (and of games Δ) are of particular interest. E.g. v is *superadditive* ("$v \in \mathcal{S}$") if

(7) $$v(S) + v(T) \leq v(S+T) \quad \text{for } S,T \in \underline{\underline{B}} ,$$

(e.g. example (3) for $\alpha > \frac{1}{2}$, S+T = S∪T iff S∩T = ∅);
v is *convex* ("$v \in \mathcal{C}$") if

(8) $$v(S) + v(T) \leq v(S \cap T) + v(S \cup T) \quad \text{for } S,T \in \underline{\underline{B}},$$

(e.g. example (4) for f convex). These functions are also called "alternating capacities of 2nd order" (CHOQUET [5]) or "supermodular functions" (EDMONDS-ROTA [11]) v is *balanced* ("$v \in \mathcal{B}$"), if, for every "partition of the unit"

$1_\Omega = \sum_{S \in \underline{\underline{S}}} c_S 1_S$ ($\underline{\underline{S}} \subseteq \underline{\underline{B}}$, $c_S \geq 0$), it follows that

(9) $$\sum_{S \in \underline{\underline{S}}} c_S\, v(S) \leq v(\Omega)\ .$$

Note that (5) and (6) both provide examples for balanced games and functions. The "positive parts" $\mathcal{S}^+$, $\mathcal{B}^+$, $\mathcal{C}^+$ are convex polyhedral cones (not trivial for $\mathcal{B}^+$, see SHAPLEY [18]). The extreme rays of $\mathcal{B}_+$ are unknown as yet. The extreme rays of $\mathcal{S}^+$ are partially known: $v \in \mathcal{S}^+$ is "canonically" representable as

(10) $$v = \max_{t=1,\dots,\tau} f^{\alpha^t}_{c^t} \circ m^t$$

where $m^t \in A^+$ and $f^{\alpha^t}_{c^t} = \sum_{\kappa=1}^{k_t} c^t_\kappa 1_{(\alpha^t_\kappa, \alpha^t_{\kappa+1}]}$ is a step function ($\mathbb{R}^+ \to \mathbb{R}^+$); essentially m^t is *nondegenerate* w.r.t. α^t_κ ($t=1\dots\tau$, $\kappa=1,\dots,k_t$). If f^{α^t} is extreme among the set of nonnegative *superadditive* functions $f : \mathbb{R}^+ \to \mathbb{R}^+$ with $f(0) = 0$ and if, in addition m^t is *homogeneous* w.r.t. all α^t_κ, then v is extreme in $\mathcal{S}^+$. (See [12] for the details.)

Within convex territory, the situation is similar: $v \in \mathcal{C}^+$ may be "canonically" represented by

(11) $$v = \max_{t=1,\dots,\tau} (m^t - \alpha_t)$$

where $m^t \in \mathbb{A}^+$ and $\alpha_t \in \mathbb{R}^+$. $(m^t - \alpha_t)$ may be regarded as a supporting "affine" set function of v. For $\tau = 2$, $\alpha_2 = 0$, $m^2 = 0$, consider

$$v = \max\,(m^1 - \alpha_1, 0) = (m^1 - \alpha_1)^+\ .$$

Then v is extreme in $\mathcal{C}^+$ if and only if m^1 is *nondegenerate* w.r.t. α_1 and this result generalizes to $\tau > 2$. (See [16] for the details.) Thus, non-degeneracy and homogeneity play an important rôle in the description of $\mathcal{S}^+$ and $\mathcal{C}^+$ by their extreme rays.

The "intermediate" balanced class (we have $\mathcal{C}^+ \subseteq \mathbf{B}^+ \subseteq \mathcal{S}^+$) can be characterized as follows. (SHAPLEY-SHUBIK [20]): v and all restrictions v_T on subsets $T \in \underline{\underline{B}}$ are balanced if and only if v is a market game in the sense of example (6).

Let us now turn to the concept of a solution. Generally speaking, a "solution concept" is a mapping defined on some subsets of functions v, say $\mathbb{V}^0$ (e.g. $= A, \mathcal{S}, \mathcal{C},\dots$) and taking values in $\mathfrak{P}(A)$.

Let us mention the *Core*, defined by

$$\mathcal{C}(v) = \{x \in \mathbb{A} \mid x \geq v,\ x(\Omega) = v(\Omega)\}$$

which is nonempty if and only if v is balanced (SHAPLEY [18], BONDAREVA [3]); thus $\mathfrak{C} : \mathbb{B} \to \mathfrak{P}(\mathbb{A})$. *Stable sets* or von Neumann-Morgenstern-solutions are somewhat more tedious to be defined: let us say that $x \in A$ *dominates* $y \in \mathbb{A}$ if there is $S \in \underline{\underline{B}}$ such that $x_i > y_i$ $(i \in S)$, $x(S) \leq v(S)$. Then a *von Neumann-Morgenstern-solution* (for some specified v) is a subset $\mathfrak{S} \subseteq \{x \in A \mid x_i \geq v(\{i\})\} =: \mathfrak{J}(v)$ such that

1. For $x, y \in \mathfrak{S}$, x does not dominate y,
2. For $y \notin \mathfrak{S}$, $y \in \mathfrak{J}(v)$, there is $x \in \mathfrak{S}$ such that x dominates y.

The class of functions v admitting at least one von Neumann-Morgenstern-solution is unknown, see e.g. v. NEUMANN-MORGENSTERN [8], LUCAS [6], SCHMITZ [17] for details.

A further solution concept, to be considered with respect to market games (example (6)) is the *competitive* ("Walrasian") *equilibrium*; an advanced treatment is offered by HILDENBRAND [6] [7]. For the purpose of this exposition, we shall only consider a rudimentary version, based on example (5), the *L.P.-game:* Let $v = v^{Abc}$ be defined by (5). For $S \in \underline{\underline{B}}$ the *dual* linear program is given by

$$v(S) = \min \{yb(S) \mid y \in \mathbb{R}^m_+, \; yA \geq c\} \tag{12}$$

and by the well known "main theorem" of linear programming the equation in (12) is established. If $\bar{y} \in \mathbb{R}^m_+$ is an "optimal Ω-solution for the dual program" - i.e. a minimizer in (12) for $S = \Omega$, then $\bar{m} := \bar{y}\, b(\cdot) \in A^+$ is the "competitive equilibrium" ("C.E."). (Recall that $b(\cdot)$ is a "vectorvalued measure").

Again by the "main theorem", $\bar{m} = \bar{y}b$ is an element of the core $\mathfrak{C}(v^{Abc})$, $\bar{y}\, b \in \mathfrak{C}(v)$ and thus the "equivalence theorem" of general equilibrium theory - the core shrinks to the C.E. as markets (games) increase - may be studied in a particularly detailed manner in this context. As it turns out, nondegeneracy is a tool which - at least in this context - may replace "nonatomicity" as a sufficient requirement. (See AUMANN [1], HILDENBRAND [6] for the nonatomic case.)

§3 EXISTENCE AND COINCIDENCE OF SOLUTIONS

Whether a certain solution concept is preferable compared to another one may very well depend on the class of games to be "solved". For voting games the core is frequently empty and it is felt that the v. Neumann-Morgenstern-solution is an appropriate concept.

v. NEUMANN-MORGENSTERN prove in [8] the following Theorem (on the "main simple solution"). Let $v = 1_{[\alpha,1]} \circ m$, assume that m hom α and v and m have equal carrier (in which case the representation is unique). If $\alpha > \frac{1}{2}$ and v is constant sum $(v(S) + v(S^c) = 1)$, then

(13) $\mathcal{S}^{\alpha} = \{\frac{m_T}{\alpha} \mid T \in \underline{\underline{Q}}_{\alpha}\}$

is a von Neumann-Morgenstern-solution. There are generalizations to this of the following type. Recall the step functions $f^{\alpha^t}_{c^t}$ (§2). For $c^t_{\kappa} = \frac{\kappa}{K}$ and $\alpha^t_{\kappa} = \alpha_{\kappa}$ (i.e. $\tau=1$ in the canonical representation) ($\kappa=1,\ldots,K$), consider

$\mathcal{S}^{\alpha_K} := \{\frac{m_T}{\alpha_K} \mid T \in \underline{\underline{Q}}_{\alpha_K}\}$

and

$\mathcal{S}_{\kappa} := \{\frac{m_U}{\alpha_K} + \frac{m_V}{\frac{K}{K-1}\alpha_K} \mid$

$U \in \underline{\underline{Q}}_{\alpha_K},\ V \in \underline{\underline{Q}}_{\frac{K}{K-1}(\alpha_K - \alpha_{\kappa})},\ U \cap V = \emptyset\}$

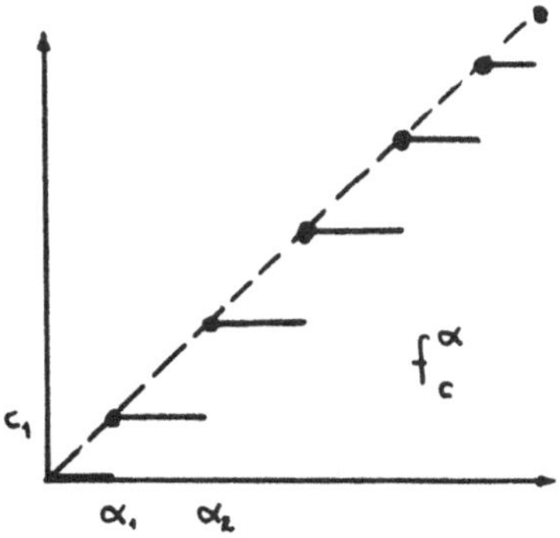

Define

(14) $\mathcal{S}^{\alpha} := \mathcal{S}^{\alpha_K} + \sum_{\kappa=1}^{K-1} \mathcal{S}_{\kappa}$.

Assume now further regularity properties of $\alpha = (\alpha_1,\ldots,\alpha_K)$, in particular assume that m hom α_K , m hom $K\alpha_{\kappa} - \kappa\alpha_K, \ldots$. Then $\mathcal{S}^{\alpha}$ is a v. Neumann-Morgenstern-solution.

There are various results on existence and nonexistence of v. Neumann-Morgenstern-solutions (see LUCAS [6], MUTO [9]). The above results are mentioned because they hinge mainly on homogeneity; the same property that renders $f^{\alpha}_{c} \circ m$ to be extreme in $\mathcal{S}^+$. Thus, it seems that via homogeneity "extreme games" and their solutions (which are extreme in a different sense) may be studied successfully.

A similar statement is true for the relation between extreme games in convex territory (i.e. $\mathcal{C}^+$) and the core (the appropriate concept for this class of games) if we replace homogeneity by nondegeneracy. As SHAPLEY observed, the extreme points of the core $\mathcal{C}(v)$ (which is also a convex compact polyhedron) may be obtained by a "bandwaggon process" (see [19]). If we have enough nondegeneracy of a specified function $v \in \mathcal{C}^+$ (w.r. to the "canonical" representation, see §2) the extreme points of the core correspond to certain "minimal winning coalitions" or "minimal profitable coalitions". But the same requirement ("enough nondegeneracy") renders the function v to be extreme in $\mathcal{C}^+$; thus again "extremality of v" and "extreme shape of $\mathcal{C}(v)$" is linked - this time via nondegeneracy (see [12] for the details).

Let us shortly touch a further application of the nondegeneracy concept: the "equivalence theorem" of the core and the C.E.; we shall treat this question within the framework of the L.P.-game.

Suppose $v = v^{Abc}$ is specified by example (5) and let $\bar{y} \in \mathbb{R}^m_+$ be an "optimal Ω-solution for the dual" (i.e. a minimizer of (12) for $S = \Omega$). Recall that $\bar{m} = \bar{y}b(\cdot) \in \mathbb{A}^+$ is the "C.E.". Now define

$$Q_0 := \{z \in \mathbb{R}^m_+ \mid \bar{y} \text{ is an optimal dual solution for the } (A,z,c)\text{-program}\} \tag{15}$$

$$= \{z \in \mathbb{R}^m_+ \mid \bar{y}z = \min \{yz \mid y \in \mathbb{R}^m_+, yA \geq c\}\},$$

and let

$$\underline{Q} = \{S \in \underline{B} \mid b(S) \in \underline{Q}_0, b(S^c) \in \underline{Q}_0\}. \tag{16}$$

Then it is not hard to see that $v = v^{Abc}$ is "additive on $\underline{Q}$", i.e.,

$$v(S) + v(S^c) = v(\Omega) \qquad (S \in \underline{Q}) . \tag{17}$$

Assume now that $\bar{m} = \bar{y}b(\cdot)$ is n.d. $\underline{Q}$. Then, if $\mu \in \mathcal{C}(v)$, it follows that for $S \in \underline{Q}$

$$\mu(\Omega) = \mu(S) + \mu(S^c) \geq v(S) + v(S^c) = v(\Omega) = \mu(\Omega) \tag{18}$$

i.e.

$$\mu(S) = v(S) = \bar{m}(S) \qquad (S \in \underline{Q}) \tag{19}$$

and $\mu = \bar{m}$ (by "n.d."), $\mathcal{C}(v) = \{\bar{m}\}$.

Hence, nondegeneracy implies the coincidence of core and C.E. (see OWEN [10] for the "replica" version and BILLERA-RAANAN [2] for the "nonatomic" version).

We are used to attach the "equivalence" theorem to a concept of "large economies" or markets. It is, therefore, desirable to translate the surrogates for nonatomicity in a way such that "large games" can be identified with "games enjoying n.d.-properties" (or "hom-properties", for that matter). This is the purpose of our last section.

§4 MANY PLAYERS IMPLY NONDEGENERACY

If m n.d. α ($m \in \mathbb{A}^+_1$, $\alpha \in (0,1)$), then it is not hard to see that m and α are rational [15]. By rescaling we may, therefore, assume that we are dealing with an integer valued "measure" $M = (M_1,\ldots,M_n)$ and a constant $\lambda \in \mathbb{N}$; we are looking for conditions sufficient to ensure that M n.d. λ. Next, let $g_1 < \ldots < g_r$ be integers denoting the different values M_i ($i \in \Omega$) can take; if we introduce

$$K_\rho := \{i \in \Omega \mid M_i = g_\rho\} \qquad (\rho=1,\ldots,r), \tag{20}$$

then

$$M(\cdot) = \sum_{\rho=1}^{r} \mid K_\rho \cap \cdot \mid g_\rho \quad . \tag{21}$$

Thus, K_ρ is the set of players of "type ρ" and g_ρ is the (common) "weight" of all players of type ρ. Now, nondegeneracy w.r.t. $\lambda \in \mathbb{N}$ is essentially a property of the pair $(g,k) \in \mathbb{N}^{2r}$, where

(23) $g = (g_1,\ldots,g_r)$; $k = (k_1,\ldots,k_r)$

with $k_\rho = | K_\rho |$ $(\rho=1,\ldots,r)$ (clearly, k_ρ is the size of the coalition of players of type ρ). Indeed, any equation

(24)
$$\sum_{i\in S} M_i = \lambda$$

is at once rewritten

(25)
$$\lambda = \sum_{\rho=1}^{r} \sum_{i\in K_\rho \cap S} M_i = \sum_{\rho=1}^{r} | K_\rho \cap S | g_\rho$$

i.e.

(26)
$$\lambda = \sum_{\rho=1}^{r} a_\rho^S g_\rho \qquad a_\rho^S = | S \cap K_\rho |$$

Now, nondegeneracy means that (24) enjoys a nonsingular coefficient matrix (for S varying in $\underline{\underline{Q}}_\lambda = \underline{\underline{Q}}_\lambda$ (M)) and, therefore, it is conceivable that in view of (26), "M n.d. λ" can be rephrased as a condition for (g,k) as follows.

(27)

M n.d. λ iff there are integers a_ρ^σ $(\rho,\sigma = 1,\ldots,r)$ such that

1. $0 \leq a_\rho^\sigma \leq k_\rho$ $\qquad (\sigma,\rho = 1,\ldots,r)$
2. $(a_\rho^\sigma)_{\rho,\sigma=1,\ldots,r}$ is nonsingular
3. $\sum_{\rho=1}^{r} a_\rho^\sigma g_\rho = \lambda$ $\qquad \sigma = 1,\ldots,r$

Parts of the problem specified by (27) were already delt with by SYLVESTER and FROBENIUS ($\sigma = 1$ and (27).2 omitted). Let us take a result from [15] : Given $g \in \mathbb{N}^r$, there are natural numbers R_r; L_ρ $(\rho = 1,\ldots,r)$ depending on $g_1,\ldots,g_r$ such that (27) can be solved (i.e., there are integers $(a_\rho^\sigma)\ldots$) provided $k_\rho \geq L_\rho$ $(\rho = 1,\ldots,r)$, $R_r \leq \lambda \leq M(\Omega) - R_r$ and $\lambda \equiv 0$ mod g.c.a. $(g_1,\ldots,g_r)$.

In other words: there is a nonempty "intervall" $[R_r, M(\Omega) - R_r]$ within the ideal spanned by g_ρ $(\rho = 1,\ldots,r)$ such that all λ's of this intervall allow for a solution of (27) provided *there are sufficiently many players of each type* ("$k_\rho \geq L_\rho$").

Via this result, extreme point problems, unique representations, existence theorems etc. may thus be solved in terms of "large sets of players". "If there are sufficiently many players of each type, then nondegeneracy takes place and thus a certain game is extreme or the core allows for extreme solutions" is the message conveyed.

A second example of the principle is based on the L.P.-game (example (5)).

According to section 3, the system $\underline{\underline{Q}}$ as defined by (15) and (16) is decisive for the coincidence of the "competitive equilibrium" $\bar{m} = \bar{y}b(\cdot)$ and the core of the L.P.-game. The introduction of "types" is done similarly to (20) and (21). We decompose

$$\Omega = \sum_{\rho=1}^{r} K_\rho$$

and assume the initial "distribution of resources" represented by the "vector valued measure" b satisfies

(28) $$b(\cdot) = \sum_{\rho=1}^{r} | K_\rho \cap \cdot | g_\rho$$

where $g_\rho \in \mathbb{R}_+^m$ denotes the common initial allocation of all players of type ρ. Again let $k_\rho := | K_\rho |$.

The next step is to switch from (28) to an analogue of (27) (less directly in this case). To this end, introduce a suitable matrix Λ such taht Q_0 (cf. (15)) is given by

$$Q_0 = \{z \in \mathbb{R}_+^m \mid \Lambda z \geq 0\}$$

In fact, Λ is the "tableau" associated by standard L.P. routine with the optimal solution $\bar{y}$ of the dual Ω-program (12). Then it turns out that

m n.d. $\underline{\underline{Q}}$ iff there are integers (a_ρ^σ) $(\rho, \sigma = 1,\ldots,r)$ such that

(29)

1. $0 \leq a_\rho^\sigma \leq k_\rho \qquad (\sigma, \rho = 1,\ldots,r)$
2. $(a_\rho^\sigma)_{\rho, \sigma = 1,\ldots,r}$ is nonsingular
3. $0 \leq \Lambda \sum_{\rho=1}^{r} a_\rho^\sigma g_\rho \leq \Lambda e$

"Lower bounds" for the numbers of players of each type may be (roughly) obtained as follows. By a transformation of bases assume that the g_ρ are unit vectors of $\mathbb{N}^r$. Then (29) reads:

Find r linearly independent $\mathbb{N}^r$ vectors a^σ $(\sigma=1,\ldots,r)$, bounded by $k \in \mathbb{N}^r$ and within a convex polyhedron determined by Λ.

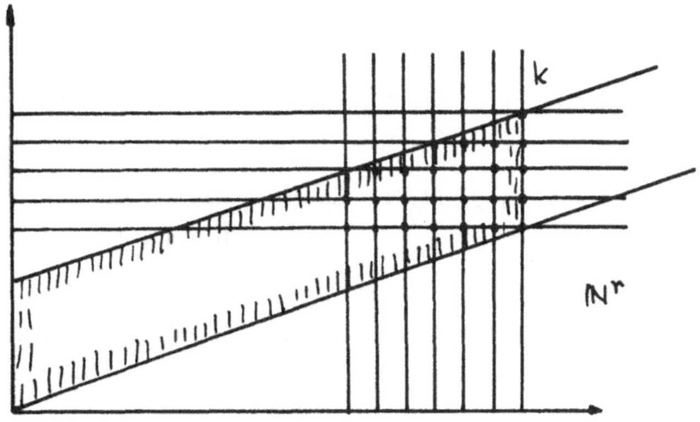

Given the data of the L.P.-game, (i.e., the tableau-matrix Λ) we may therefore specify regions in $\mathbb{N}^r$ such that if $k \in \mathbb{N}^r$ is an element of such a region then (30) can be solved, thus nondegeneracy takes place and the core and the C.E. coincide. The tools for this task are provided by Geometric Number Theory; in fact MINKOWSKI's 2nd Theorem (connecting lattice constants and the volume of convex polyhedra, see [4]) is essential for specifying regions of $k \in \mathbb{N}^r$ as mentioned above. As k resembles distributions of players over types, looking to our regions in $\mathbb{N}^r$ verifies again the basic result: "many" players induce the core and the C.E. to coincide - but how many can be said much more exactly compared to nonatomic theory if one is willing to apply Geometric Number Theory via nondegeneracy (see [14] for the details).

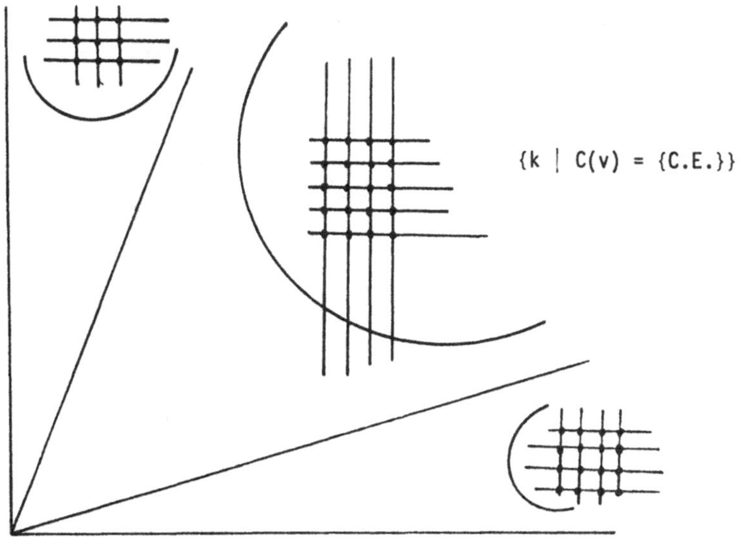

REFERENCES

[1] Aumann, J. and Shapley, L.S. (1974): Values of nonatomic games, Princeton University Press, Princeton, NJ..

[2] Billera, L.J. and Raanan, J. (1981): Cores of non-atomic linear production games, Math. of O.R., Vol. 6, 420-423.

[3] Bondareva, O.N. (1963): Some applications of linear programming methods to the theory of cooperative games, Problemy Kibernet. 10, 119-139 (in Russian).

[4] Cassels, J.W.S. (1959): An introduction to the geometry of numbers, Springer Verlag, Verlin, Göttingen, Heidelberg.

[5] Choquet, G. (1953-1954): Theory of capacities, Ann.Inst.Fourier 5, 131-295.

[6] Hildenbrand, W. (1974): Core and equilibria of a large economy, Princeton University Press, Princeton, NJ..

[7] Hildenbrand, W. (1982): Core of an economy, Handbook of Mathematical Economics, Vol.II, edited by K.J.Arrow and M.D.Intriligator, North-Holland Publishing Co..

[8] v.Neumann, J. and Morgenstern, O. (1944,1953): Theory of games and economic behavior, Princeton University Press, Princeton, NJ..

[9] Muto, S. (1982): Symmetric solutions for (n, n-2) Games with small values of v(n-2), International Journal of Game Theory, Vol. 11, 43-52.

[10] Owen, G. (1975): On the core of linear production games, Math. Programming, Vol. 9, 358-370.

[11] Rota, G.-C. and Edmonds, J. (1966): Submodular set functions (abstract, Waterloo Combinatorics Conference), University of Waterloo, Waterloo, Ontario.

[12] Rosenmüller, J. (1977): Extreme games and their solutions, Lecture Notes in Economics and Mathematical Systems 145, Springer, Heidelberg.

[13] Rosenmüller, J. (1981): The theory of games and markets, North-Holland Publishing Co..

[14] Rosenmüller, J. (about 1982): L.P.-games with sufficiently many players, to appear: International Journal of Game Theory.

[15] Rosenmüller, J. and Weidner, H.G.: (1973): A class of extreme convex set functions with finite carrier, Advances in Mathematics, Academic Press, Vol.10, No. 1, 1-38.

[16] Rosenmüller, J. and Weidner, H.G. (1974): Extreme convex set functions with finite carrier: general theory, Discrete Math.10, 343-382.

[17] Schmitz, N. (1976): Eine Klasse von Mehrpersonen-Spielen ohne von Neumann'sche Lösung, Internation Journal of Game Theory 5.

[18] Shapley, L.S. (1967): On balanced sets and cores, Naval Rest.Logist. Quart. 14, 453-460.

[19] Shapley, L.S. (1971): Cores of convex games, International Journal of Game Theory 1, 12-26.

[20] Shapley, L.S. and Shubik, M. (1966): Quasi cores in a monetary economy with non-convex preferences, Econometrica 34, 805-827.

Universität Bielefeld
Institute of
Mathematical Economics
Postfach 8640
4800 Bielefeld 1
W.Germany

A NEW GENERAL APPROACH TO MINIMUM DISTANCE ESTIMATION

Igor Vajda

Prague

Key words: f-divergence, minimum divergence estimators, consistency, equivariance, efficience, robustness.

ABSTRACT

Minimum distance methods based on various measures of "distance" of distributions have been considered in many papers and were shown to possess desirable properties. The author suggested a systemization of the various "distances" based on the general concept of f-divergence. Probabilistic models resulting from his f-estimators are required to minimize the divergence with a sample probability. Since the f-estimators decisions aim at meeting the interpretation of probability criterion itself, they can a priori be expected to be either "good" or at least interesting. On the other hand the f-estimators are regular enough to be tractable analytically. This opens possibilities to built up a general statistical theory covering most classical estimators as particular cases.

MINIMUM DIVERGENCE ESTIMATORS

Let $(\mathcal{X},\mathcal{B})$ be a measurable sample space and $\mathcal{P}$ the class of all probabilities on it. Each sample $\mathbf{x} = (x_1,\ldots,x_n)$ defines a sample probability (empirical probability) $P_n \in \mathcal{P}$ by

$$P_n(E) = \frac{1}{n} \sum_{i=1}^{n} I_E(x_i) \qquad \text{for } E \in \mathcal{B}.$$

The basic point of view of this paper is that the statistical estimation theory is interested in methods of selecting probabilistic models P from a theoretical family $\tilde{\mathcal{P}} \subset \mathcal{P}$ on the basis of an empirical evidence represented by the sample probability P_n.

A usual case is that $\tilde{\mathcal{P}} = \{P_\theta : \theta \in \Theta\}$ is parametrized and the "methods of selecting probabilistic models" can be reduced to point functions $T(P_n)$ taking on values from Θ so that $P_{T(P_n)} \in \{P_\theta : \theta \in \Theta\}$ represents the model selected.

Thus we define an estimator with projection family $\{P_\theta : \theta \in \Theta\}$ as a mapping $T: \mathcal{P} \to \Theta$. Since there is one-to-one correspondence between the contours in $\mathfrak{X}^n$ defined by the loss of order in the sample $\mathbf{x} = (x_1, \ldots, x_n)$ and the P_n, each sample order invariant estimator $T: \mathfrak{X}^n \to \Theta$ is in fact a point function $T(P_n)$ on the domain of all sample probabilities. This domain can at least formally be extended to the whole $\mathcal{P}$. For example, for any location projection family with $\mathfrak{X} = \Theta = R$, one can extend the sample mean $T(\mathbf{x}) = T(P_n) = \bar{\mathbf{x}}$ by

$$T(Q) = \begin{cases} \mathbf{E}_Q x & \text{if } \mathbf{E}_Q x \in R \\ 0 & \text{in the opposite case} \end{cases}$$

to all $Q \in \mathcal{P}$ (remark that $\mathbf{E}_\lambda \phi(x)$ denotes the Lebesque integral $\int \phi(x) d\lambda$ for any $\mathcal{B}$-measurable $\phi: \mathfrak{X} \to R^*$ and any σ-finite λ on $\mathcal{B}$). Analogically, for any scale projection family with $\mathfrak{X} = R$, $\Theta = (0,\infty)$, the sample deviation can be defined by the formula $T(Q) = \mathbf{E}_Q(x - \mathbf{E}_Q x)^2$ not only for sample probabilities $Q = P_n$ but for all $Q \in \mathcal{P}$ with $\mathbf{E}_Q x^2$ finite. This extension procedure fails with sample order depending estimators like the $T(\mathbf{x}) = x_1$ but the obvious sufficiency considerations imply that these counterexamples are not interesting at all.

An important idea has been brought into statistical estimation theory by Neyman (1949) and Wolfowitz (1957). They considered a concrete distance D on $\mathcal{P}$ and defined the so-called minimum distance estimator with projection family $\{P_\theta : \theta \in \Theta\}$ by the condition that $T(P_n)$ minimizes the function $D(P_\theta, P_n)$ on Θ. If $D(P_\theta, P_n)$ measures the accuracy with which the probabilities $P_n(E)$ approximate the $P_\theta(E)$ on a class of "all interesting events" $E \subset \mathfrak{X}$ then this idea is intuitively quite appealing as $P_n(E)$ is nothing but the empirical relative frequency of an event E playing the well-known role in the standard interpretation of a theoretical probability $P_\theta(E)$.

Before introducing a class of "information-type" distances satisfying the above stated approximation property w.r.t. the σ-algebra $\mathcal{B}$ of events E, let us remark that, even before the minimum distance estimators have been introduced, the perhaps greatest achievement of statistical estimation theory - the maximum likelihood estimator-has been shown to be efficient under the assumption that the sample generating family of probabilities coincides with the projection family $\{P_\theta : \theta \in \Theta\}$. As well known, the MLE with the given projection family is defined for every $Q \in \mathcal{P}$ by the condition

$$T(Q) \text{ maximizes } \mathbb{E}_Q \ln p_\theta \text{ on } \Theta$$

where, from now on, we suppose that the projection family is dominated by a σ-finite λ on $\mathcal{B}$ and denote $p_\theta = dP_\theta/d\lambda$. The minimum distance estimators have been revived in statistical estimation theory when Huber (1964) found that the MLE may loss its efficiency faster than other M-estimators if the sample generating family $\tilde{\mathcal{P}}$ departs from the projection family $\{P_\theta : \theta \in \Theta\}$ under consideration. In addition to the minimum χ^2 and minimum Kolmogorov-Smirnov distance estimators considered by Neyman and Wolfowitz, minimum Hellinger, Anderson--Darling, and Cramér - von Mises distance estimators have been analyzed by Rao et. al (1975), Beran (1977), Parr and Schucany (1980), Millar (1981), Boos (1981). Using theoretical as well as experimental argument they have shown that, for property specified distance D, the minimum distance estimators are consistent and, as to their efficiency, they are quite competitive with the best robust M-estimators of Huber (1964,1972) and Hampel (1974).

A new general class of estimators is obtained by considering the f-divergence

$$D(P,Q) = \mathbb{E}_Q f(p/q)$$

of Csiszár (1967) as the distance on $\mathcal{P}$. Here f is continuous convex on $(0,\infty)$, strictly convex at 1 with $f(1) = 0$, and $p,q = dP/d\lambda, dQ/d\lambda$ for arbitrary $\lambda >> P,Q$. The theory of estimation obtained in this manner not only covers the above mentioned minimum distance estimators as particular cases, but also explains relation between the minimum distance estimators and the MLE or M-estimators in general.

If $\mathfrak{X}$ is finite, we generalize in this manner the minimum χ^2 estimator of Neyman (for $f(u)=(u-1)^2$), the modified χ^2 estimator and the minimum Kullback-Leibler estimator of Rao (1961) (for $f(u)=(u-1)^2/u$ and $f(u)=u\log u$), and the maximum likelihood estimator (for $f(u)=-\log u$, see also Csiszár (1978)). Using the results of Rao (1961) it can be proved (see Vošvrda (1982)) that under the hypothesis that the generating family $\tilde{\mathcal{P}}$ coincides with the projection family $\{P_\theta : \theta \in \Theta\}$ and that $\{P_\theta : \theta \in \Theta\}$ satisfies mild regularity conditions, all estimators under consideration are consistent, all estimators with

$f''(1) \neq 0$ are efficient, and all estimators with $f''(1) \neq 0$, $2f''(1) + f'''(1) = 0$ are second-order efficient in the sense of Rao.

If $\mathfrak{X}$ is neither finite nor countable then the $D(P_\theta,P_n)$ figuring in the definition of the minimum distance estimate $T(P_n)$ is a constant equal

$$\lim_{u\to 0+} f(u) + \lim_{u\to\infty} \frac{f(u)}{u}$$

on Θ as soon as all P_θ, $\theta \in \Theta$, are orthogonal to the P_n. Since the sample probability P_n is allways "discrete", this is an apparent obstacle to our f-divergence project whenever the projection family $\{P_\theta : \theta\in\Theta\}$ is "continuous". In Vajda (1982/83a), this obstacle was suggested to be overcome by replacing the f-divergence $D(P,Q)$ by the weak divergence

$$WD(P,Q) = \sup_{x\in\mathfrak{X}} \left[Q(E_x)\, f\left(\frac{P(E_x)}{Q(E_x)} \right) + (1-Q(E_x))\, f\left(\frac{1-P(E_x)}{1-Q(E_x)} \right) \right]$$

or

$$WD(P,Q) = \mathbb{E}_W \left[Q(E_x)\, f\left(\frac{P(E_x)}{Q(E_x)} \right) + (1-Q(E_x))\, f\left(\frac{1-P(E_x)}{1-Q(E_x)} \right) \right]$$

where $\{E_x : x\in\mathfrak{X}\} \subset \mathfrak{B}$ induces $\mathfrak{B}$ and where W is a σ-finite "weight" on $\mathfrak{B}$. The cited paper investigates properties of the modified concept of divergence and establishes theorems concerning consistency of the resulting general class of estimators as well (obviously, minimum Anderson-Darling or Cramér - von Mises estimators are simply particular minimum weak χ^2 estimators).

ESTIMATORS DEFINED WITH THE AID OF SAMPLE NOISE

There is another natural possibility of overcoming the above mentioned difficulty. So far we have considered a replacement of the mutually orthogonal P_θ and P_n by their non-orthogonal restrictions on subalgebras $\{E_x, \mathfrak{X}- E_x\} \subset \mathfrak{B}$ and we in fact minimized the f-divergence of the restricted variants of P_θ and P_n. Alternatively, we can replace the modifications of P_θ and P_n on subalgebras by modifications $P_\theta T^{-1}$, $P_n T^{-1}$ induced by statistics $T: (\mathfrak{X},\mathfrak{B}) \to (\mathfrak{Y},\mathfrak{F})$ or, more generally, by modifications $P_\theta * W$, $P_n * W$ induced by Markov morphisms (random statistics, observation channels, see Le Cam (1974)) $W: (\mathfrak{X},\mathfrak{B}) \to (\mathfrak{Y},\mathfrak{F})$. The final effect is the same: the modified P_θ and P_n

need not in general be orthogonal so that the minimization of the f-divergence of the modified P_θ and P_n is meaningful. A most simple approach applies the sample noise represented by a Markov morphism W only to what represents the sample evidence - it leads to a minimization of $D(P_\theta, P_n * W)$ on Θ. This problem can be further simplified by the assumption that $(\mathfrak{X}, \mathfrak{B})$ is a linear space with the Borel σ-algebra, $(\mathfrak{Y}, \mathfrak{F}) = (\mathfrak{X}, \mathfrak{B})$, and that the noise is additive. The last assumption means that the class $W = \{W_x : x \in \mathfrak{X}\}$ of probabilities is uniquely defined by a probability W and by a condition $W_x(E) = W(E-x)$. In this case the mapping $P_n * W$ reduces to the ordinary convolution

$$(P_n * W)(E) = \mathbb{E}_{P_n} W(E-x) = \frac{1}{n} \sum_{i=1}^{n} W(E-x_i).$$

For the sake of simplicity hereafter we suppose that $(\mathfrak{X}, \mathfrak{B})$ is the Borel line $(R, \mathfrak{B})$ and we restrict ourselves to a noise W uniform on the interval $(-c, c) \subset R$.
Under these assumptions

$$(P_n * W)(E) = \frac{1}{2cn} \sum_{i=1}^{n} \lambda((x_i - c, x_i + c) \cap E)$$

where λ is the Lebesque measure on $(R, \mathfrak{B})$ so that $P_n * W << \lambda$ and the $P_n * W$ is not automatically orthogonal to probabilities from absolutely continuous projection families $\{P_\theta : \theta \in \Theta\} << \lambda$.

An open problem accompanying the artifice of smoothing the sample probability P_n by an absolutely continuous additive noise W is that of specification of the "power" parameter $c > 0$ in the estimate $T(P_n) = T_c(P_n) \in \Theta$ defined by a minimization of the function $D(P_\theta, P_n * W)$. Beran (1977) who pioneered this method of estimation by considering the estimators with $f(u) = (1-u^{1/2})(1/2)$ (minimum Hellinger distance estimators) admitted the noise power $c = c_n$ tending to zero with the sample size n tending to infinity. The asymptotic considerations alone, however theoretically interesting, say virtually nothing about what c_n is most convenient when the sample size is fixed and, moreover, small. We shall look at the problem of specification of the noise power c by considering the limit

$$\lim_{c \to 0} T_c(P_n) = T_0(P_n) \tag{1}$$

for an arbitrary fixed sample $\mathbf{x} = (x_1, \ldots, x_n)$ of a fixed size n. As follows from our Theorem, for certain f (including the Hellinger distance case), this limit exists provided

(2) $$T(P_n) \text{ minimizing } \mathbb{E}_{P_n} f(p_\theta) \text{ on } \Theta$$

uniquely exists in which case both estimates (1) and (2) are identical. This result justifies a new class of estimators (2) (in fact a class of generalized M-estimators of Huber (1964) with a "goal function" $M(x) = f(p_\theta(x))$), at least for

(3) $$f(u) = (1-u^\alpha)/\alpha \quad \text{where } \alpha \in [0,1)$$

(with the limit $f(u) = -\log u$ for $\alpha = 0$). Obviously, the class of α-estimators T defined by (2), (3), contains the MLE as the particular case for $\alpha = 0$ and the minimum Hellinger distance estimator as the particular case for $\alpha = 1/2$. The consistency, equivariance, efficience, and influence curves of the α-estimators as well as of the more general estimators (2) have been studied by Vajda (1982/83b).

THE THEOREM

Let us consider an arbitrary fixed f satisfying (3), arbitrary sample $\mathbf{x} = (x_1,\dots,x_n)$, and suppose for simplicity that

$$\delta_0 = \min_{i \neq j} |x_i - x_j| > 0.$$

Suppose further that the derivative $p'_\theta(x) = dp_\theta(x)/dx$ exists for all $x \in R$, $\theta \in \Theta$, and that Θ is a compact topological space such that $\mathbb{E}_{P_n} f(p_\theta)$ is continuous and $K(\theta)$ bounded on Θ, where

$$K(\theta) = \sup_{x \in R} |f'(p_\theta(x))p'_\theta(x)| .$$

THEOREM. Let $T_c(P_n)$ exists for all $0 < c < \delta_1$ and let T_n be a non-empty set of points $T(P_n) \in \Theta$ satisfying (2). Then for each open covering $U(T_n)$ of T_n there exists $\delta > 0$ such that $T_c(P_n) \in U(T_n)$ for all $0 < c < \delta$.

Proof of this theorem is based on the fact that for f under consider consideration

(4) $$f(uv) = v^\alpha f(u) + f(v) \quad \text{for all } u,v > 0$$

and on the following Lemma.

LEMMA. It holds

$$D(P_\theta,P_n*W) = \frac{n^\alpha}{(2c)^{1-\alpha}}\left[\mathbb{E}_{P_n} f(p_\theta) + 2c\,\Psi(\theta)\right] + \frac{f(2cn)}{2c}$$

where $|\Psi(\theta)| \le K < \infty$ on Θ.

<u>Proof:</u> By the definition

$$D(P_\theta,P_n*W) = \frac{1}{n}\sum_{i=1}^{n}\int_{x_i-c}^{x_i+c}\frac{1}{2c} f(2cnp_\theta(x))dx =$$

$$= \frac{1}{2cn}\sum_{i=1}^{n}\int_{-c}^{c} f(2cnp_\theta(x_i+v))dv =$$

(cf(4))

$$= \frac{(2cn)^\alpha}{2c}\,\frac{1}{n}\sum_{i=1}^{n}\int_{-c}^{c} f(p_\theta(x_i+v))dv + \frac{f(2cn)}{2c} =$$

$$= \frac{(2cn)^\alpha}{2c}\left[\mathbb{E}_{P_n} f(p_\theta) + 2c\,\Psi(\theta)\right] + \frac{f(2cn)}{2c}$$

where

$$\Psi(\theta) = \frac{1}{n}\sum_{i=1}^{n} f'(p_\theta(x_i+v_i))\;p_\theta'(x_i+v_i) \qquad \text{for } |v_i| < c$$

and, by assumptions, $|\Psi(\theta)| \le K(\theta)$ is bounded by a constant K on Θ.

<u>Proof of the Theorem:</u> Suppose that on the contrary there exists a decreasing sequence $c_i < \delta_2 = \min\{\delta_0,\delta_1\}$ tending to zero such that

$$T_* \cap U(T_n) = \emptyset \quad \text{for} \quad T_* = \{\theta_i = T_{c_i}(P_n): i=1,2,\dots\}.$$

Let θ_* be an arbitrary accumulation point of T_*. Since $T(P_n)$ is an inner point of the open set $U(T_n)$ and T_* is disjoint with $U(T_n)$, it holds $\theta_* \neq T(P_n)$. On the other hand since $c_i < \delta_0$, the Lemma implies

$$\mathbb{E}_{P_n} f(p_{\theta_i}) + 2c_i\,\Psi(\theta_i) = \min_{\theta\in\Theta}\left[\mathbb{E}_{P_n} f(p_\theta) + 2c_i\,\Psi(\theta)\right]$$

for all i = 1,2,...

Hence

$$\mathbb{E}_{P_n} f(p_{\theta_i}) + 2c_i \Psi(\theta_i) \leq \mathbb{E}_{P_n} f(p_{T(P_n)}) + 2c_i \Psi(T(P_n))$$

for all i = 1,2,... which together with the Lemma implies

$$\mathbb{E}_{P_n} f(p_{\theta_i}) \leq \mathbb{E}_{P_n} f(p_{T(P_n)}) + 2c_i(\Psi(T(P_n))+\Psi(\theta_i))$$

$$\leq \mathbb{E}_{P_n} f(p_{T(P_n)}) + 4c_i K$$

for all i = 1,2,... Since $\mathbb{E}_{P_n} f(p_\theta)$ is supposed to be continuous on Θ, the last inequality implies that for every subsequence θ_{i_j} tending to the θ_*

$$\mathbb{E}_{P_n} f(p_{\theta_*}) = \lim_{j\to\infty} \mathbb{E}_{P_n} f(p_{\theta_{i_j}}) \leq \mathbb{E}_{P_n} f(p_{T(P_n)}) + 4K \lim_{j\to\infty} c_{i_j}$$

$$= \mathbb{E}_{P_n} f(p_{T(P_n)}).$$

This together with the condition (2) for $T(P_n)$ implies

$$\mathbb{E}_{P_n} f(p_{\theta_*}) = \mathbb{E}_{P_n} f(p_{T(P_n)})$$

so that $\theta_* = T(P_n)$ which contradicts the assumption and the Theorem is proved.

REFERENCES

Beran R. (1977): Minimum Hellinger distance estimates for parametric models. Ann. Statist. 5, 445-463.

Boss D. D. (1981): Minimum distance estimators for location and goodness of fit. JASA 76, 663-670.

Csiszár I. (1967): Information-type measures of difference of probability distributions and indirect observations. Studia Sci. Math. Hungar. 2, 209-318.

Csiszár I. (1978): Information measures: a critical survey. Trans. 7th Prague Conf. on Inform. Theory,..., Vol.B,

73-86, Praha, Academia.

Hampel F. R. (1974): The influence curve and its role in robust estimation. JASA 69, 383-393.

Huber P. I. (1964): Robust estimation of a location parameter. Ann. Math. Statist. 35, 73-101.

Huber P. I. (1972): Robust statistics: a review. Ann. Math. Statist. 43, 1041-1067.

Le Cam L. (1974): On the information contained in additional observations. Ann. Statist. 2, 630-649.

Millar P. W. (1981): Robust estimation via minimum distance methods. Zeitschr. Wahrsch. 55, 73-89.

Neyman J. (1949): Contributions to the theory of χ^2-test. Proc. 1st Berkeley Symp. on Math. Statist.,..., 239-273, Berkeley, Univ. of Calif. Press.

Parr W. C., Schucany W. R. (1980): Minimum distance and robust estimation. JASA 75, 616-624.

Rao C. R. (1961): Asymptotic efficiency and limiting information. Proc. 4th Berkeley Symp. on Math. Statist.,..., Vol. 1, 531-546, Berkeley, Univ. of Calif. Press.

Rao P. V., et al: Estimation of shift and center of symmetry based on Kolmogorov-Smirnov statistics. Ann. Statist. 3, 862-873.

Vajda I. (1982/83a): Minimum divergence principle in statistical estimation. Part I: f-estimators and weak f-estimators. Statistics and Decisions 1 (to be published).

Vajda I. (1982/83b): Minimum divergence principle in statistical estimators. Part II: Weighted f-estimators. Statistics and Decisions 1 (to be published).

Vošvrda M. (1982): On second order efficiency of minimum divergence estimators (this Transaction).

Wolfowitz J. (1957): The minimum distance method. Ann. Math. Statist. 28, 75-88.

Czechoslovak Academy of Sciences
Institute of Information Theory
and Automation
Pod vodárenskou věží 4
182 08 Praha 8 - Libeň
Czechoslovakia

ON MODELS OF COMPLICATED FUNCTIONS UNDER UNCERTAINTY

Antanas Žilinskas

Vilnius

Key words: Statistical models, stochastic functions, multimodal optimization, stochastic programming.

ABSTRACT

The problems of construction of models of complicated functions under uncertainty are discussed. Such models are important, for example, in multimodal optimization theory. The system of axioms formalizing information on a complicated function under uncertainty is suggested. It is shown that a random function is compatible with the suggested system of axioms.

DISCUSSION ON KNOWN MODELS OF COMPLICATED FUNCTONS UNDER UNCERTAINTY

Let us discuss the problem in the frames of multimodal optimization theory. The interpretation in terms of the other theories seems not difficult. In local optimization theory quadratic models are usefull: the local quadratic approximation of an objective function is used for the definition of direction of descend to the local minimum of an objective function. However a global model of an multimodal function must be adequate to represent more uncertain behaviour of the function. At the k-th minimization step the values of objective function $f(\cdot)$ are known at the points $x_i \in A \subset R^n$, $i = \overline{1,k}$: $y_i = f(x_i)$. On the base of this information the model of $f(\cdot)$ must induce some estimation of $f(x_0)$, $x_0 \in A$, $x_0 \neq x_i$, $i = \overline{1,k}$.

A model corresponding to the classical numerical analysis is a class of Lipshitian functions. It was used in multimodal optimization by Evtushenko (1974), Ivanov (1971), Nemirovskij and Judin (1979), Pijavskij (1967), Sukharev (1971), (1975), Traub and Wozniakowski (1980) and others. However, the use of Lipshitian functions is connected with some difficulties. In practical optimization problems an objective function is usually defined by complicated algorithm and analytical evaluation of Lipshitian constant of it is difficult. If this estimate is too low then the algorithm loses the theoreticalfoundation, if it is too high then the algorithm becomes inefficient. Let the Lipshitian constant of f(·) is known and y_i are marked by crosses in fig. 1. It follows from the model that $y_- \leq f(x_0) \leq y^+$, where $y^{+}_{-} = \left\{ \begin{matrix} \min \\ \max \end{matrix} \right. (y_i \pm |x - x_i| L)$, L is the Lipshitian constant. In the practical problems however the researchers and engineers never express their information on f(·) in such a way: I guarantee that $a_- \leq f(x_0) \leq a^+$.

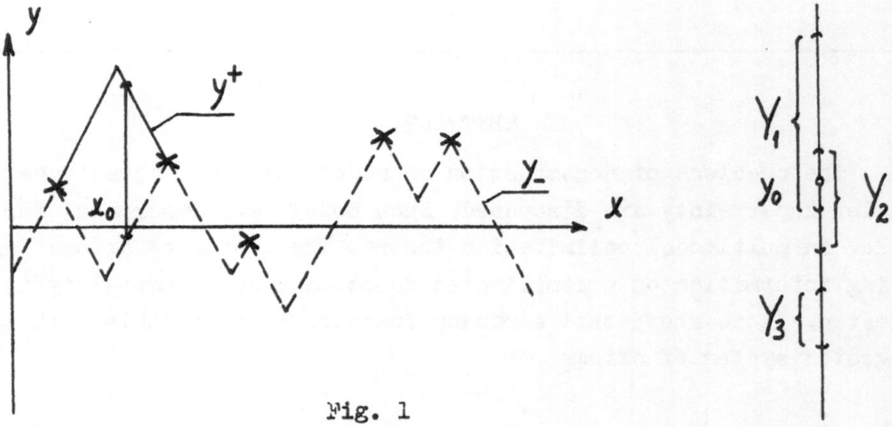

Fig. 1

They may say for example: "it seems that $f(x_0)$ will be positive", "it is more likely that $f(x_0)$ will be positive than that it will be negative", "very probable that $f(x_0)$ will be close to y_0" etc. However they feel themselves rather uncomfortable if they are asked to give some guaranteed evaluation. The impossibility of interpretation of usual assumptions on f(·) in the frame of Lipshitian model is noticeable disadvantage of it. The minimax approach for the construction of optimization algorithms is most natural in case of the use Lipshitian model. In such a case the algorithm is oriented towards the saw - like functions, example of which is showed by dotted line in fig. 1. It seems rather doubtful to use the algorithms oriented towards functions which are not common in practice.

The other class of models of complicated functions under uncertainty are stochastic functions. Such models are popular for example in automatic control theory, hydrodinamics and etc.The justification of the use of stochastic functions is based usually by the arguments on frequencies stability in the class of investigated functions. These arguments are not very convincing for the optimization problems. Besides the justification the use of stochastic functions for the construction of multimodal optimization algorithms is connected as well with the computational difficulties.

Let us discuss some models which are most popular in multimodal optimization theory. Stochastic process with discrete time $\xi(i)$, $i=\overline{0,m}$ is suggested as a one - variable model by Strongin (1969), (1978) where $\xi(i)-\xi(i-1)=\beta_i$, $i=\overline{1,m}$, $\xi(0)=\beta_0$; β_i, $i=\overline{0,m}$ are indepen - dent Gaussian random variables with expectations:

m_0, $m_i(\alpha)=\begin{cases}-m, & i\leq\alpha\\ m, & i>\alpha\end{cases}$ and with the standards $\sigma_0, \sigma_i=cm$; α is a discrete random variable with the distribution given a priori. The algorithm is justified by Strongin (1978) for the case $c\to 0$. However, in this case the sampling functions of $\xi(\cdot)$ have the form $\xi(i)=\beta_0+|i-\alpha|m-\alpha m$. For any $P\sim 1$ there exists, obviously, such a small $\varepsilon>0$ that $\xi(i)$ are unimodal for $c<\varepsilon$. Therefore the use of this model is debatable.

A Wiener process is used as one - variable model by Archetti (1980), Kushner (1964), Žilinskas (1975), (1981). It is convenient from computational point of view and justified theoretically by Žilinskas (1978).

The choice of stochastic function for a multidimensional model is more complicated from the view of numerical difficulties. Nevertheless some ideas given by Mockus (1977) and Katkauskaite (1972) may be useful.

The axiomatic approach to constructing statistical models simpler from computational point of view than stochastic functions is suggested by Žilinskas (1978). The main idea is to change the doubtful supposition of guaranteed interval for $f(x_0)$, $x_0\neq x_i$, $i=\overline{1,k}$ by more soft and human supposition on possibility to compare the likelihood of events $f(x)\in Y_j$ for any Y_j. For the example given in fig. 1 author's opinion is: $f(x_0)\in Y_2$ is no less probable than $f(x_0)\in Y_1$; both $f(x_0)\in Y_2$ and $f(x_0)\in Y_1$ are no less probable than $f(x_0)\in Y_3$. The axiomatic definition of such a comparative probability, it justification by the results of the psychological experiment and construction of statistical model are given by Žilinskas (1978). Stochastic functions are special case of the constructed models. It is interesting to examine the additional axioms specifying this important special case.

EXISTENCE OF STOXASTIC FUNCTION COMPATIBLE WITH CONDITIONAL COMPARATIVE PROBABILITY

In the paper by Žilinskas (1978) a statistical model has been constructed formalizing the supposition on comparability of likelihood of simple events, i.e. events of the type $f(x_0) \in Y$. Let us consider more special case, assuming that information on $f(\cdot)$ induces the conditional comparative probability: $Y_1/Y_2 \gtrsim Y_3/Y_4$ means that event $F_k \in Y_1$ given $F_k \in Y_2$ is no less probable than $F_k \in Y_3$ given $F_k \in Y_4$ where $F_k=(f(x_1),\ldots, f(x_k))$, Y_j are intervals in R^k. If for example $Y_1=((-\infty,y_1)\times(-\infty,\infty))$, $Y_2=((-\infty,\infty)\times(y_2,y_3))$ then Y_1/Y_2 means the event $f(x_1)\leq y_1$ given $y_2 \leq f(x_2) \leq y_3$. Let the vector $(x_1,\ldots,x_k)$ be fixed, $\tilde{\mathcal{J}}$ be the set of intervals $Y_i = \mathop{\times}\limits_{j=1}^{k} (y'_j(i), y''_j(i))$ where $\times$ denotes the cartesian product. $\bar{Y}_i$ denotes closed interval, U denotes the set of possible values of F_k. The subset of $\tilde{\mathcal{J}}$ which does not contain the impossible events is denoted by $\mathcal{J}$; if $Y \in R^k \setminus U$ then obvious $Y \notin \mathcal{J}$.

<u>Definition.</u> Intervals Y_1, Y_2 belong to j-system if $y'_i(1)= y'_i(2)$, $y''_i(1) = y''_i(2)$, $i \neq j$. Intervals Y_1, Y_2 are called j-united if they belong to a j-system and $\{y'_j(1), y''_j(1)\} \cap \{y'_j(2), y''_j(2)\} \neq \emptyset$, such intervals are denoted by $Y_1 \vee_j Y_2$.

Let us define the comparative probability (CP) formaly by means of the following axioms. For the shortness $Y_1/R^k \gtrsim Y_2/R^k$ is denoted by $Y_1 \gtrsim Y_2$.

A1. CP is week ordering on $\tilde{\mathcal{J}} \times \mathcal{J}$, i.e. it is reflexive, transitive and connected.

A2. $R^k/R^k \sim Y/Y \gtrsim Y_1/Y_2$.

A3. $Y \sim \emptyset$ if and only if $\mu(Y \cap U) = 0$, where $\mu(\cdot)$ denotes the Lebesque measure on R^k.

A4. Let $Y_1 \cap Y_2 = Y_3 \cap Y_4 = \emptyset$, $Y_1 \vee_i Y_2$, $Y_3 \vee_i Y_4$; if $Y_1/Y_5 \gtrsim Y_3/Y_6$ and $Y_2/Y_5 \gtrsim Y_4/Y_6$ then $Y_1 \cup Y_2/Y_5 \gtrsim Y_3 \cup Y_4/Y_6$: if any of the hypothesis is $>$ then the conclusion is $>$.

A5. Let $Y_1 \subseteq Y_2 \subseteq Y_3$, $Y_4 \subseteq Y_5 \subseteq Y_6$; if either $Y_1/Y_2 \gtrsim Y_4/Y_5$ and $Y_2/Y_3 \gtrsim Y_5/Y_6$ or $Y_1/Y_2 \gtrsim Y_5/Y_6$ and $Y_2/Y_3 \gtrsim Y_4/Y_5$ then $Y_1/Y_3 \gtrsim Y_4/Y_6$; if any of the hypothesis is $>$ then conclusion is $>$.

A6. Let $Y_1/Y_2 < Y_3/Y_4$; there exists such $A_j \subset Y_3$, $B_j \subset Y_3$, $a'_j(j) = y'_j(3)$, $b''_j(j) = y''_j(3)$, $j = \overline{1,k}$ that $Y_1/Y_2 \sim A_j/Y_4 \sim B_j/Y_4$, $j = \overline{1,k}$.

The meaning of axioms A1 - A4, A6 is similar to that of axioms discussed by Žilinskas (1978). The A5 is specific for conditional CP, it is similar to the rule of multiplication of conditional probabilities. The following axiom extends CP for more complicated sets.

A7. Let $X_i = \bigcup_{j=1}^{K_i} Y_{ij}$, $Y_{ij} \cap Y_{im} = \emptyset$, $j \neq m$, $Y_{1j}^* \sim Y_{3j}^*$, $j = \overline{1,k}$, $\{Y_{ij}^*$, $j = \overline{1,k}\} = \{X_i \cap X_{i+1}\}$, $i = 1,3$, $K_2 = K_4$, $Y_{2j} \sim Y_{4j}$, $j = \overline{1,K_2}$, and there exists such j that $Y_{2j} \succ \emptyset$, then $X_1/X_2 \sim X_3/X_4$.

Theorem 1. Let CP be defined on $\mathcal{J} \times \mathcal{J}$ by axioms A1 - A7. There exists a unique extension of CP on $\tilde{E} \times E$ where $\tilde{E}$ is a field of finite unions of intervals and E is subset of $\tilde{E}$ not containing impossible events. The extension of CP on $\tilde{E} \times E$ has the properties analogous to those defined by A1 - A6.

Theorem 2. Axioms A1 - A7 are necessary and sufficient for the existence of unique probability density p(·) compatible with CP, i.e. $X_1/X_2 \succsim X_3/X_4$ if and only if $\int_{X_1} p(z)dz / \int_{X_2} p(z)dz \geq \int_{X_3} p(z)dz / \int_{X_4} p(z)dz$, $p(z) > 0$ for $z \in U$, $p(z) = 0$ for $z \notin U$.

The probability density p(·) dependes on fixed vector $(x_1, \dots, x_k)$ i.e. $p(z) = p_{x_1 \dots x_k}(z_1 \dots z_k)$. Since CP of event $F_k \in \bigtimes_{i=1}^{k} (y'(i), y''(i))$ is invariant in respect of permutation of indexes, $p_{x_1 \dots x_k}(z_1 \dots z_k)$ is invariant as well. Since either denotation $F_{k-1} \in \bigtimes_{i=1}^{k-1} (y'(i), y''(i))$ and $F_k \in \bigtimes_{i=1}^{k-1} (y'(i), y''(i)) \times (-\infty, \infty)$ means the same event, there holds the eq-equality: $p_{x_1 \dots x_{k-1}}(z_1 \dots z_{k-1}) = \int p_{x_1 \dots x_k}(z_1 \dots z_k) dz_k$. Therefore the assumptions of well known theorem of A.Kolmogoroff are satisfied and there exists a stochastic function, densities of partial distributions of which equal to $p_{x_1 \dots x_k}(z_1 \dots z_k)$.

Axioms A1 - A7 are more complicated and not so intuitively acceptible as those suggested by Žilinskas (1978). Nevertheless theorem 2 justifies the use of stochastic functions as models of complicated functions under uncertainty for rather a broad class of problems. These results extend the paper of Žilinskas and Katkauskaite (1982). The proofs of the theorems are similar to those in the refered paper.

REFERENCES

Archetti F.(1980): A probabilistic algorithm for global optimization with a dimensional reduction technique. In Lecture

Notes in Cotrol and Inform. Sci., v. 23, Springer,36-42.

Evtushenko J.(1974): Methods for search of global extremum. In: Operation research, v.4, Computing Centre, Moskow, 36-68.

Ivanov V.(1971): Onoptimal algorithms for minimization in the class of functions with Lipshitz condition. In: IFIP Congress 71, 76-80.

Katkauskaitė A.(1972): Random fields with independent increments, Lith. Math. Juornal, N.4, 75-85.

Kushner H.(1964): A new method of locating the maximum point of an arbitrary multipeak curve in the presence of noise, Trans. ASME, ser. D, v. 86, 97-105.

Mockus J.(1977): On Bayesian methods of seeking the extremum and their applications. In: Information Processing 77, Amsterdam, 195-200.

Nemirovskij V., Judin D.(1979): Complexity of the problems and efficiency of optimization methods, Nauka, Moskow.

Pijavskij S.(1967): Algorithm for the search for global minimum. In: Optimal Decision Theory, v. 2, Kijev, 13-24.

Shubert B.(1972): Asequential method seeking the global maximum of a function, SIAM J. Numer. Analysis, v.9, 379-388.

Strongin R.(1969): Informational method for multiextremal minimization, Izv. AN SSSR, Technical Cybern. N.6, 118-126.
(1978): Numerical methods for multiextremal minimization, Nauka, Moskow.

Sukharev A.(1971): On optimal strategies of search for extremum, Comp. Math. and Math. Physics, N. 4, 910-924.
(1975): Optimal search for the extremum, Moskow University.

Traub J., Wozniakowski H.(1980): General theory of optimal algorithms, Academic Press, New York.

Žilinskas A.(1975): One-step Bayesian method for the search of extremum of one- variable functions, Cybernetics,N.1, 139-144.
(1978): On statistical models for multimodal optimization, Math. Operat. Stat.,ser. Statistics, v.9, 255-266.
(1981): Two algorithms for one-dimensional multimodal minimization,Math. Operat. Stat., ser. Optimization,v.12,53-63.

Žilinskas A., Katkauskaite A.(1982): On the existence of stochastic function compatible with the conditional relation of comparative probability, Cybernetics(in print).

Academy of Sciences of Lithuanian SSR
Institute of Mathematics and Cybernetics
K.Poželos 54,
Vilnius 232600, USSR

COMMUNICATIONS

AN ATTEMPT TO SOLVE APPROXIMATELY THE OPTIMAL ESTIMATION PROBLEM FOR MARKOV PROCESSES BY EXPANSION OF THE A-POSTERIORI DENSITY IN AN EDGEWORTH SERIES

Norbert Ahlbehrendt, Utz Draeger
Berlin

Key words: Markov diffusion Processes, Nonlinear estimation problems

ABSTRACT

The synthesis of optimal on-line algorithms for nonlinear estimation problems in the framework of Markov diffusion processes is an until now unsolved problem. The possibility to solve the problem by expansion of the conditional a-posteriori density in an Edgeworth serie and taking into account only the a-posteriori quasi moment-functions of lowest order is discussed.

INTRODUCTION

We discuss in our paper one aspect of the unsolved problem of the determination of the a-posteriori density function (a.p.d.) $p(z(t)/y_0^t)$ and the synthesis of nonlinear estimation algorithms for Markov diffusion processes $(z(t),y(t))$. We restrict ourselves to the following problem: for each measurement time $t > t_0$ as exact as possible informations about the state vector $z(t)$ shall be derived from the measurement ensemble y_0^t under the condition of monoton increasing measurement time t

$$y_0^t = \left\{ y(\tau) \ , \ t_0 \leq \tau \leq t \right\} .$$

That means we are looking for an estimation value $\hat{z}(t) = f\left\{ y_0^t \right\}$. as an functional of y_0^t.

The calculation of the estimation functional involves the knowledge of the a.p.d. $p(z/y_0^t)$.

We consider in the following Markov diffusion processes, which can be described by

$$\begin{aligned} dz(t) &= F(z(t),y(t),t)dt + G(z(t),y(t),t)dw \\ dy(t) &= S(z(t),y(t),t)dt + g(\ y(t),t)dw \end{aligned} \qquad (1)$$

(These equations and the following ones are written in symmetrized differentials).

w(t) is a vector Wiener process with known intensity. We assume that F,S,G,g are given functions of their arguments.
The so formulated problem is sufficiently general. Parameter identification problems are included, because a-priori unknown parameters are included in z (in this case the corresponding components of z are constant).
Assuming gg^T nonsingular, a stochastic partial differential equation for the a.p.d. $p(z/y_o^t)$ exists (after Stratonovich (196o) - Kushner (1962)):

$$L_t \, p(z(t)/y_o^t) = 0$$

$$L_t = -\frac{\partial}{\partial t} - \nabla_z^T \phi(z,y,y,t) + \frac{1}{2} \mathrm{Tr} \left[\nabla_z \nabla_z^T M(z,y,t)\right] - \frac{1}{2} \left(\Psi(z,y,y,t) - \langle \Psi(\cdot)/y_o^t \rangle\right). \quad (2)$$

$$\nabla_z^T = \left(\frac{\partial}{\partial z_1}, \ldots, \frac{\partial}{\partial z_n}\right)$$

The functions ϕ, M and Ψ are uniquely given by F,S,G and g. As known from the equation (2) the following differential equations for the a-posteriori expectation $m_z = \langle z/y_o^t \rangle$ and the a-psteriori variance $D = \langle (z-m_z)(z-m_z)^T/y_o^t \rangle$ can be derived :

$$\dot{m}_z = \langle \phi / y_o^t \rangle - \frac{1}{2} \langle (z-m_z) \cdot \Psi / y_o^t \rangle ,$$

$$\dot{D} = \langle \phi \cdot (z-m_z)^T / y_o^t \rangle + \langle (z-m_z) \phi^T / y_o^t \rangle + \langle M / y_o^t \rangle - \frac{1}{2} \langle (z-m_z)(z-m_z)^T (\Psi - \langle \Psi / y_o^t \rangle) / y_o^t \rangle , \quad (3)$$

where $\langle (\cdot)/y_o^t \rangle = \int (\cdot) \, p(z/y_o^t) \, dz$.

To calculate the expectation values in (3) we need the a.p.d. . Only for linear estimation problems (that means F,S are linear functions of z and G is independent on z) the equations (3) can be solved explicitely. In this case the equations (3) are the Kalman-Bucy equations and the solution is the Gaussian density.
As we see it, no general constructive methods for the solution of the problem of on-line estimation exists in the general nonlinear case. Therefore approximative methods are used. The almost exclusively used method is the Gauss approximation. In this case the expectations on the right side of (3) are calculated by means of a Gaussian density with expectation m_z and variance D . We call the resulting compact system for m_z and D the generalized Kalman-Bucy

equations. They are a good approximation of the optimal algorithm only if the a-posteriori cumulants of the order greater than two can be neglected. But this approximation often fails in practice.
Let us consider the system

$$\begin{aligned} \dot{x} &= A\,x + \sigma \xi_1 \\ \dot{y} &= x_1 + \sqrt{N_0}\,\xi_2 \end{aligned} \tag{4}$$

$$(z^T = (x^T, A^T)\ ;\ \dot{A}=0\ ),$$

with white independent Gaussian noises ξ_1, ξ_2. It is important for problems of optimal control of stochastic disturbed objects.
The generalized Kalman-Bucy equations result in convergent estimation algorithms for $\hat{z}$ and $\hat{A}$ in the one-dimensional case, but in divergent algorithms in the higher dimensional case.

A ESTIMATION ALGORITHM BASED ON THE EXPANSION OF THE A.P.D. IN AN FINITE EDGEWORTH SERIE

One possibility to overcome the above mentioned difficulties is to take into account in(3) a-posteriori expectations of higher order.
We think that the most suitable method of this kind is the approximative expansion of the a.p.d. in a finite Edgeworth serie:

$$p(z/y_o^t) \approx p_n(z/y_o^t) = (1 + \sum_{k=3}^{n} \frac{(-1)^k}{k!} Q_k^{(n)} \frac{\partial^k}{\partial z^k})\, p_n^{(G)}(z/y_o^t) \tag{5}$$

$Q_k^{(n)}(t)$ are the a-posteriori quasi moment-functions(a.p.q.)defined by

$$Q_k^{(n)}(t) = \frac{\partial^k}{\partial s^k} \left. \frac{\chi(s,t)}{\chi_G(s,t)} \right|_{s=0} ,$$

where $\chi(s,t)$ and $\chi_G(s,t)$ are the characteristic functions of p_n and $p_n^{(G)}$,respectively ($Q_1=Q_2=0$).
$p_n^{(G)}$ is a Gaussian density with expectation $m_z^{(n)}$ and variance D_n.
After inserting (5) in (2) we get nonlinear differential equations for the calculation of $m_z^{(n)}$ and D_n, which depend on the a.p.q.:

$$\dot{m}_z^{(n)} = \langle \phi \rangle_n^{(G)} - \frac{1}{2} \langle (z - m_z^{(n)})\, \Psi \rangle_n^{(G)}$$

$$+ \sum_{k=3}^{n} \frac{1}{k!} Q_k^{(n)} \Big[\langle \frac{\partial^k h_1}{\partial z^k} \rangle_n^{(G)} + \frac{1}{2} m_z^{(n)} \langle \frac{\partial^k \Psi}{\partial z^k} \rangle_n^{(G)} \tag{6a}$$

$$\dot{D}_n = \langle h_2^* \rangle_n^{(G)} + \sum_{k=3}^{n} \frac{1}{k!} Q_k^{(n)} \left\langle \frac{\partial^k h_2^*}{\partial z^k} \right\rangle_n^{(G)} \tag{6b}$$

$$h_2^* = h_2 - 2 m_z^{(n)} h_1 - \frac{1}{2} (D_n - m_z^{(n)2}) \Psi$$

$$h_l = l z^{l-1} \phi + \binom{l}{2} z^{l-2} M - \frac{1}{2} z^l \Psi$$

For the a.p.q. of order greater than two we get a differential equation system of the Riccati type:

$$\begin{aligned}
\dot{Q}_k^{(n)} = & -k Q_{k-1}^{(n)} \dot{m}_z^{(n)} - \binom{k}{2} Q_{k-2}^{(n)} \dot{D}_n + \sum_{l=1}^{k} \binom{k}{l} C_{k-l}^{(n)} \langle h_l \rangle_n^{(G)} \\
& + \sum_{s=3}^{n} \frac{1}{s!} Q_s^{(n)} \sum_{l=1}^{k} \binom{k}{l} C_{k-l}^{(n)} \Big[\left\langle \frac{\partial^s h_l}{\partial z^s} \right\rangle_n^{(G)} + \frac{1}{2} (D_n + m_z^{(n)2}) \cdot \\
& \cdot \left\langle \frac{\partial^s \Psi}{\partial z^s} \right\rangle_n^{(G)} \Big] + \sum_{s=3}^{n} \sum_{r=3}^{k} \frac{Q_s^{(n)}}{s!} \frac{Q_r^{(n)}}{r!} \sum_{l=s}^{k} \binom{k}{l} C_{k-l}^{(n)} \cdot \\
& \cdot \left\langle \frac{\partial^s \Psi}{\partial z^s} \right\rangle_n^{(G)} \cdot \left\langle \frac{\partial^r z^l}{\partial z^r} \right\rangle_n^{(G)} .
\end{aligned} \tag{7}$$

$$C_k^{(n)} = \frac{\partial^k}{\partial s^k} \exp(m_z^{(n)} \cdot s - \frac{1}{2} s^2 \cdot D_n)$$

$$\langle (\cdot) \rangle_n^{(G)} = \int (\cdot)\, p_n^{(G)}(z / Y_0^t)\, dz$$

The estimation algorithm (6),(7) is schematically figured in Fig.1. We have tested it for the system (4) for the one- and two-dimensional case under consideration of a.p.q. of an order of three (that are four for the one-dimensional, and ten for the two-dimensional case).

We must point out that the consideration of a.p.q. of an order of three doesn't result in an principal improvement of the convergence properties as compared with the generalized Kalman-Bucy equations. For the two-dimensional case we get only divergent algorithms, too. Simulation results for the one-dimensional case are given in Fig.2. They show only for the lowest signal to noise-ratio a little impro-

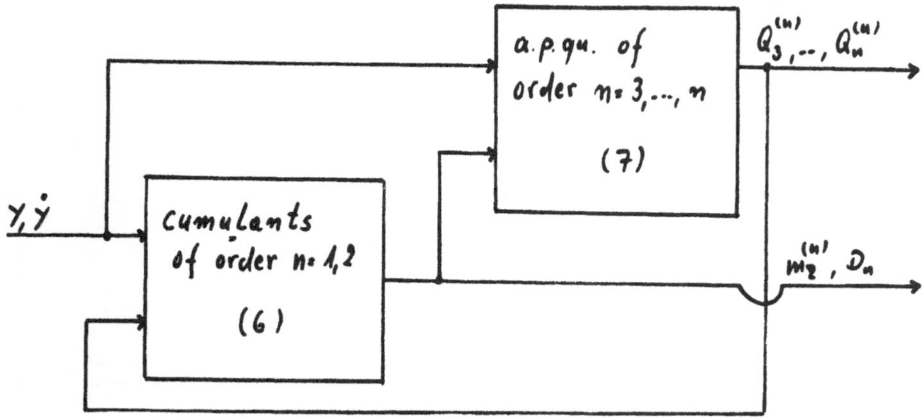

FIGURE 1 : Scheme of the estimation algorithm after (6),(7) by the help of an Edgeworth serie

vement as compared with the Gauss approximation. This doesn't justifie the higher effort, the more so as the consideration of a.p.q. of the fourth order (five for the one-dimensional case) doesn't change these results.

DISCUSSION

Since the consideration of a.p.q. of higher order ($n \geq 4$) is not practicable , especially for higher dimensional systems, we state that the synthesis of on-line estimation algorithms for Markov diffusion processes based on the Edgeworth serie as expansion of the o.p.d. is no succesful way.No better results can be expected if other a-posteriori characteristics are used (as for instance cumulants or moments).

Finally we mention that we tried to solve the nonlinear estimation problem for the system (4) in consequence of the lack of general methods for the synthesis of optimal algorithms by another way, which doesn't result from the knowledge of the a-posteriori density. It is based on a generalisation of the ideas of stochastic stability in the sense of an enlargement of the conception of the region of stability in the mean. The method is characterized by an heuristic substitute for the structur of the algorithm using measured auxiliary functions. The auxiliary functions are determined by minimization

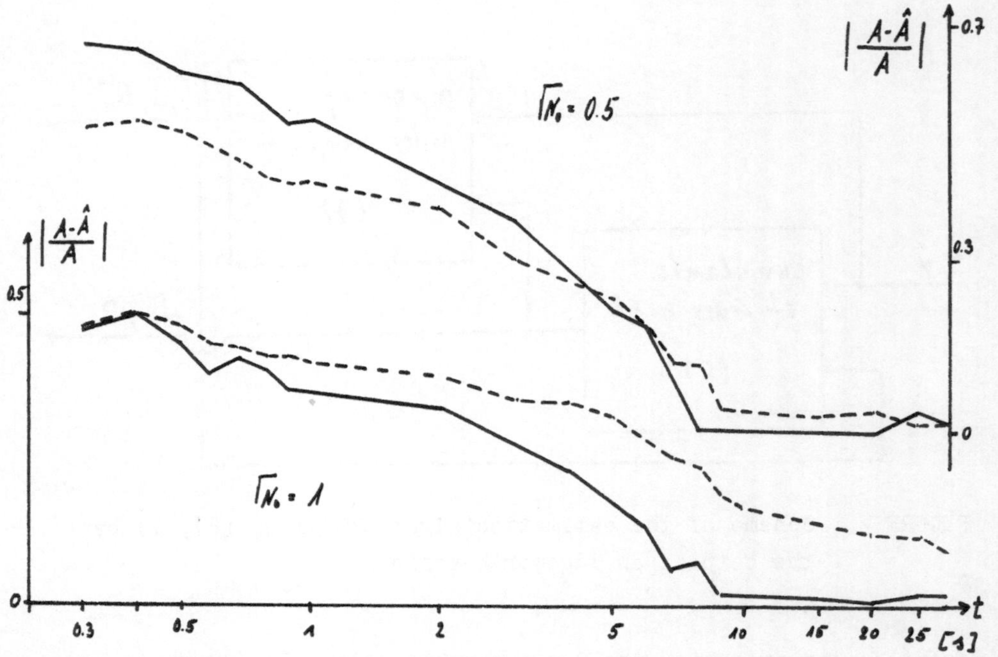

FIGURE 2: Convergence behaviour of the algorithm after (6),(7) (straight line)as compared with the Gauss approximation (broken line) for the system (4) (A=-1o ; σ = 1o)

of the region of instability of the estimation error. The minimization is performed on the basis of a Ljapunov function, dependent on the estimation error and the auxiliary functions(see Ahlbehrendt (1981)).

REFERENCES

Stratonovich R.L. (196o): Conditional Markov processes. In: Theory Prob. Appl. 5 , 156 - 178

Kushner H.J. (1962): On the differential equations satisfied by conditional probability densities of Markov processes with applications. In: J.SIAM, Control Series A,No.1, 1o6 - 119

Ahlbehrendt N. Mädiger B. (1981): A new method for synthesis of nonlinear parameter and state estimation for noisely disturbed processes. In: Proceedings of the VIII. IFAC-congress, Kyoto (Japan).

ZKI, AW-DDR, Rudower Chaussee 5, 1199 Berlin-Adlershof, GDR

MARGINAL DISTRIBUTIONS OF AUTOREGRESSIVE PROCESSES

Jiří Anděl
Prague

Key words: Autoregressive processes, marginal distribution, self-decomposable random variables

---

ABSTRACT

Consider a stationary autoregressive process $\{X_s\}$ defined by the relation $X_s=\varrho X_{s-1}+Y_s$, where $\{Y_s\}$ is a white noise. Assume that Y_s are independent and have the same distribution (normal, rectangular, Laplace or Cauchy). For these cases the distribution of X_s is calculated. Conversely, if a distribution of X_s is given (normal, exponential, gamma, Laplace, rectangular or Cauchy), the corresponding distribution of Y_s is derived. The results are applicable in Monte Carlo methods for constructing dependent random variables with a given univariate marginal distribution.

1. INTRODUCTION

Let $\{Y_s\}$ be independent identically distributed random variables such that $EY_s^2<\infty$ and Var $Y_s>0$. Let $\varrho\in(-1,1)$ be a given number. Then it is well known that there exists uniquely a process $\{X_s\}$ such that

(1.1) $$X_s=\varrho X_{s-1}+Y_s ,$$

which has the form

(1.2) $$X_s = \sum_{k=0}^{\infty} c_k Y_{s-k} , \qquad \sum_{k=0}^{\infty} c_k^2 < \infty .$$

From (1.1) and (1.2) we have

(1.3) $$X_s = \sum_{k=0}^{\infty} \varrho^k Y_{s-k} .$$

The assumption that Y_s are $N(0, \sigma^2)$ variables is made very often. Then it follows that $X_s \sim N[0, \sigma^2/(1-\varrho^2)]$ for all s. It is interesting and important to know the relation between the distribution of Y_s and the distribution of X_s also for non-normal cases. The following two problems can be considered.

Problem A. A distribution of Y_s is given and the distribution of X_s is to be calculated.

Problem B. A distribution of X_s (the same for all s) is given and the corresponding distribution of Y_s should be derived.

There were some attempts to find the solution for non-normal cases a larger time ago (e.g., see Bernier (1970)). Recently, a series of papers appeared where exponential and gamma distributions of X_s and some mixtures of them were investigated (see Gaver and Lewis (1980) and references given there). The research is connected with applications in point processes, particularly in queuing theory, where simple models giving dependent variables with exponential distribution are welcome. Gaver and Lewis observed that the mentioned problems are tightly connected with so called self-decomposable random variables (Feller (1971), p. 588). It does not seem, however, that the known results about self-decomposable random variables have contributed significantly to the solution of the problems formulated above.

Before coming to the main parts of the paper, it may be interesting to know in advance that for some distributions of X_s the Problem B has no solution. If the solution does exist, then in some simple cases it is a mixture of a continuous and a discrete distribution.

2. EVALUATION OF MARGINAL DISTRIBUTION OF X_s

Here we shall deal with the Problem A. Series (1.3) converges in the quadratic mean and, therefore, it converges also in the distribution. Denote

$$\psi(t) = Ee^{itY_s}, \quad \omega(t) = Ee^{itX_s}$$

the characteristic functions of Y_s and X_s, respectively. From (1.3) we have

$$\omega(t) = \prod_{k=0}^{\infty} \psi(\varrho^k t). \tag{2.1}$$

Because we know, that (1.3) converges in the distribution, the product

(2.1) is again a characteristic function.

a. Normal distribution. If $Y_s \sim N(0, \sigma^2)$, then $\psi(t)=\exp\{-\sigma^2t^2/2\}$ and $\omega(t) = \exp\{-\sigma^2(1-a^2)^{-1}t^2/2\}$. Therefore, $X_s \sim N[0, \sigma^2(1-a^2)^{-1}]$, which is the classical result.

b. Continuous rectangular distribution. Let Y_s have the rectangular distribution on $\langle -a,a\rangle$, where $a>0$. The density is $f(y)=(2a)^{-1}$ for $-a\le y\le a$ and $f(y)=0$ for $|y|>a$. The corresponding characteristic function is

$$\psi(t) = \frac{\sin at}{at} . \tag{2.2}$$

Then

$$\omega(t) = \prod_{k=0}^{\infty} \frac{\sin a\rho^k t}{a\rho^k t} . \tag{2.3}$$

Another form of $\omega(t)$ can be derived using the formula

$$\ln\frac{\sin x}{x} = \sum_{n=1}^{\infty} (-1)^n \frac{B_{2n}2^{2n}x^{2n}}{2n(2n)!} , \quad 0<|x|<\pi,$$

where B_{2n} are Bernoulli numbers. From here we obtain

$$\omega(t) = \exp\left\{\sum_{n=1}^{\infty} (-1)^n B_{2n} \frac{2^{2n}a^{2n}t^{2n}}{2n(2n)!(1-\rho^{2n})}\right\}, \quad |t|<\pi/a. \tag{2.4}$$

c. Laplace (double exponential) distribution. Let Y_s have the density $f(y)=(2b)^{-1}\exp\{-|y|/b\}$, $-\infty<y<\infty$, where $b>0$. Since

$$\psi(t)=(1+b^2t^2)^{-1}, \tag{2.5}$$

we get

$$\omega(t) = \prod_{k=0}^{\infty} (1+b^2\rho^{2k}t^2)^{-1}. \tag{2.6}$$

Because

$$\ln(1+x) = \sum_{n=1}^{\infty} (-1)^{n+1} \frac{x^n}{n}, \quad |x|<1,$$

we have from (2.6)

(2.7) $$\omega(t) = \exp\left\{\sum_{n=1}^{\infty} (-1)^n \frac{b^{2n}t^{2n}}{n(1-\rho^{2n})}\right\}, \quad |t| < b^{-1}.$$

d. Discrete rectangular distribution. Let $a > 0$ be a given positive number and m a positive integer. Assume that

(2.8) $$P(Y_s = \frac{j}{a}) = \frac{1}{2m+1}, \quad j=-m,\ldots,-1,0,1,\ldots,m.$$

Then

$$\psi(t) = (2m+1)^{-1} \sum_{j=-m}^{m} e^{itj/a} = 2(2m+1)^{-1}\left(\frac{1}{2} + \sum_{j=1}^{m} \cos\frac{jt}{a}\right)$$

$$= \frac{1}{2m+1} \frac{\sin\frac{(2m+1)t}{2a}}{\sin\frac{t}{a}} .$$

Consider the case

(2.9) $$\rho = (2m+1)^{-1}.$$

Since for $K \geq 0$ we have

$$\prod_{k=0}^{K} \psi(\rho^k t) = \sin\frac{t}{2\rho a} \cdot \frac{\rho^{K+1}}{\sin\frac{\rho^K t}{2a}}$$

and

$$\rho^{K+1} / \sin\frac{\rho^K t}{2a} \to 2a\rho/t$$

for $K \to \infty$, we obtain

(2.10) $$\omega(t) = \frac{\sin(t/2a\rho)}{t/2a\rho},$$

which is the characteristic function of the continuous rectangular distribution on $\langle -(2a\rho)^{-1}, (2a\rho)^{-1}\rangle$, i.e. on the interval

(2.11) $$\left\langle -\frac{2m+1}{2a}, \frac{2m+1}{2a}\right\rangle.$$

Obviously, assumption (2.9) may seem to be too strong. We shall see later that under (2.8) the continuous rectangular distribution of X_s can be achieved only if ρ is given by (2.9).

e. Cauchy distribution. Let Y_s have a Cauchy distribution $C(a,b)$, i.e.

the density

$$f(y) = \frac{1}{\pi} \frac{b}{b^2+(y-a)^2}, \quad -\infty < y < \infty ,$$

where a is a real and b a positive parameter. The corresponding characteristic function is

$$\psi(t) = \exp\{iat/b - b|t|\}.$$

Of course, in this case we cannot immediately use (1.3), because the series $\Sigma \rho^k Y_{s-k}$ does not converge in the quadratic mean. But the product of the characteristic functions of $\rho^k Y_{s-k}$ is

$$\omega(t) = \prod_{k=0}^{\infty} \psi(\rho^k t) = \exp\{iat/[b(1-\rho)] - b|t|/(1-|\rho|)\} ,$$

which is the characteristic function of

(2.12) $$C\left[\frac{a}{b(1-\rho)}, \frac{b}{1-|\rho|}\right].$$

Therefore, in this case (1.3) converges in the distribution and the limit has the Cauchy distribution (2.12).

Let us remark that in the typical cases like (2.3) and (2.6) it is difficult to derive some results about the corresponding densities, although in (2.4) and in (2.7) we have all the moments explicitly given.

3. EVALUATION OF THE DISTRIBUTION OF Y_s

This part of the paper is devoted to the solution of the Problem B. Let us recall that $\omega(t)$ and $\psi(t)$ denote the characteristic functions of X_s and Y_s, respectively. Since X_{s-1} and Y_s are independent, we have from (1.1) that

(3.1) $$\omega(t) = \omega(\rho t)\psi(t),$$

which gives

(3.2) $$\psi(t) = \omega(t)/\omega(\rho t).$$

If X is a random variable with characteristic function $\omega(t)$ such that there exists a characteristic function $\psi(t)$ satisfying (3.1) for all $\rho \in (0,1)$, we say that X belongs to the class L, or that X is a self-decomposable random variable. Some results about variables belonging to L are given in Shanbhag, Pestana and Sreehari (1977), Shanbhag and Sreehari (1977), Thorin (1977 a) and Thorin (1977 b). For completness, we shall introduce here some results from Gaver and Lewis (1980) before presenting new formulas for other distributions.

a. Normal distribution. If $X_s \sim N(0,v^2)$, then $\omega(t)=\exp\{-v^2t^2/2\}$ and

$$\psi(t) = \exp\{-v^2t^2/2\}/\exp\{-v^2\rho^2t^2/2\} = \exp\{-v^2(1-\rho^2)t^2/2\}.$$

Therefore, $Y_s \sim N[0, (1-\rho^2)v^2]$.

b. Exponential distribution [Gaver and Lewis (1980)]. Let X_s have an exponential distribution with a parameter a, i.e. the density

$$f(x)=\begin{cases} a^{-1}\exp\{-x/a\} & \text{for } x>0, \\ 0 & \text{for } x\leq 0. \end{cases} \tag{3.3}$$

Then $\omega(t) = (1-iat)^{-1}$ and from (3.2)

$$\psi(t) = \rho + (1-\rho)(1-iat)^{-1}.$$

If $0\leq\rho\leq 1$, then $\psi(t)$ corresponds to a random variable which equals to zero with the probability ρ and has the exponential distribution (3.3) with the probability $1-\rho$. If E_s is a sequence of i. i. d. variables with density (3.3), then relation (1.1) can be rewritten into form

$$X_s = \begin{cases} \rho X_{s-1} & \text{with probability } \rho, \\ \rho X_{s-1} + E_s & \text{with probability } 1-\rho. \end{cases}$$

c. Gamma distribution [Gaver and Lewis (1980)]. Let X_s have $\Gamma(a,p)$ distribution with the density

$$f(x) = \begin{cases} \dfrac{1}{a^p\Gamma(p)} e^{-x/a} x^{p-1} & \text{for } x>0, \\ 0 & \text{for } x\leq 0, \end{cases}$$

where $a>0$ and $p>0$ are given parameters. The corresponding characteristic function is $\omega(t) = (1-iat)^{-p}$. Then

$$\psi(t) = [\rho + (1-\rho)(1-iat)^{-1}]^p .$$

Let $0\leq\rho\leq 1$. The result for p=1 is given above, because $\Gamma(a,1)$ is the exponential distribution. Other simple cases are:

$$p=2, \qquad \psi(t) = \rho^2 + 2\rho(1-\rho)(1-iat)^{-1} + (1-\rho)^2(1-iat)^{-2};$$

$$p=3, \qquad \psi(t) = \rho^3 + 3\rho^2(1-\rho)(1-iat)^{-1} + 3\rho(1-\rho)^2(1-iat)^{-2}$$
$$+(1-\rho)^3(1-iat)^{-3}.$$

We can see that for p=2 the distribution corresponding to $\psi(t)$ is a mixture of degenerate random variable with mass at zero and $\Gamma(a,1)$ and $\Gamma(a,2)$ with weights ρ^2, $2\rho(1-\rho)$ and $(1-\rho)^2$, respectively. The result for p=3 is similar.

d. Laplace (double exponential) distribution. The density and the characteristic function of the Laplace distribution are given above in 2.c. The only difference is that (2.5) is denoted by $\omega(t)$ now. From (3.2) we get after elementary simplification

$$\psi(t) = \rho^2 + (1-\rho^2)(1+b^2t^2)^{-1}.$$

This characteristic function corresponds to a mixture of a degenerate random variable concentrated at zero and a random variable with Laplace distribution with the original density $f(x)=(2b)^{-1}\exp\{-|x|/b\}$. The weights are ρ^2 and $1-\rho^2$, respectively.

e. Continuous rectangular distribution. Let X_s have a continuous rectangular distribution on $\langle -a,a\rangle$, where $a>0$. We restrict ourselves to the non-trivial case $\rho \neq 0$. Because $\omega(t)=(at)^{-1}\sin at$, we have from (3.2)

$$\psi(t) = \rho \frac{\sin at}{\sin \rho at}. \tag{3.4}$$

Since $\psi(t)$ does not depend on the sign of ρ, we shall consider only the case $\rho>0$. Then it is necessary to distinguish three possibilities.

(i) $\rho = \frac{1}{2n}$, n=1,2,...

From

$$\frac{1}{2n}\frac{\sin at}{\sin(at/2n)} = \frac{1}{n}\sum_{k=1}^{n}\cos\frac{(2k-1)at}{2n}$$

$$= \frac{1}{2n}\sum_{k=1}^{n}\left[\exp\left\{\frac{i(2k-1)at}{2n}\right\} + \exp\left\{-\frac{i(2k-1)at}{2n}\right\}\right]$$

we can see that $\psi(t)$ is the characteristic function of the discrete rectangular distribution which is concentrated at the points

$$-\frac{2n-1}{2n}a,\ -\frac{2n-3}{2n}a,\ldots,\ -\frac{3}{2n}a,\ -\frac{1}{2n}a,\ \frac{1}{2n}a,\ \frac{3}{2n}a,\ldots,\ \frac{2n-3}{2n}a,\ \frac{2n-1}{2n}a.$$

Each of these points has probability 1/2n.

(ii) $\rho = \frac{1}{2n+1}$, n=1,2,...

From

$$\frac{1}{2n+1}\ \frac{\sin at}{\sin\frac{at}{2n+1}} = \frac{1}{2n+1}\left[1 + 2\sum_{k=1}^{n}\cos\frac{2kat}{2n+1}\right] = \frac{1}{2n+1}\sum_{k=-n}^{n}\exp\left\{i\frac{2kat}{2n+1}\right\}$$

it follows that $\psi(t)$ is the characteristic function of the discrete rectangular distribution which is concentrated at the points

$$-\frac{2n}{2n+1}a,\ -\frac{2n-2}{2n+1}a,\ldots,\ -\frac{2}{2n+1}a,\ 0,\ \frac{2}{2n+1}a,\ldots,\ \frac{2n-2}{2n+1}a,\ \frac{2n}{2n+1}a.$$

Each of these points has probability 1/(2n+1).

(iii) ρ is not a number of the type 1/n, n=1,2,...

If $t \to \pi/\rho a$, then $\sin\rho at \to 0$, whereas $\sin at \to \sin\pi/\rho \neq 0$. From (3.4) we have $|\psi(t)| \to \infty$ and it implies that $\psi(t)$ is not a characteristic function (the absolute value of any characteristic function cannot exceed 1). Therefore, for $\rho \neq 1/n$ there exists no distribution of Y_s which would give continuous rectangular distribution of X_s in model (1.1).

<u>f. Mixed exponential distribution.</u> If X_s has the density

$$f(x) = p_1 a_1^{-1} e^{-x/a_1} + p_2 a_2^{-1} e^{-x/a_2} \quad \text{for} \quad x > 0$$

and f(x)=0 otherwise (where $p_1 = 1-p_2 \geq 0$, $a_1 > a_2 > 0$), then the result can be found in Gaver and Lewis (1980) and in Lawrance (1980).

<u>g. Cauchy distribution.</u> If X_s has a Cauchy distribution C(a,b), then using (3.2) we have

$$\psi(t) = \exp\{ita(1-\rho)/b - b(1-|\rho|)|t|\},$$

i.e. Y_s has

$$C\left[a(1-\rho)(1-|\rho|),\ b(1-|\rho|)\right].$$

REFERENCES

Bernier J. (1970) : Inventaire des modèles et processus stochastique applicables de la description des déluts journaliers des riviers. Rev. Inst. Internat. Statist. 38(1970), 50-71.

Feller W. (1971) : An Introduction to Probability Theory and Its Applications. Wiley, New York 1971.

Gaver D. P. and Lewis P. A. W. (1980) : First-order autoregressive gamma sequences and point processes. Adv. Appl. Prob. 12(1980), 727-745.

Lawrance A. J. (1980) : The mixed exponential solution to the first-order autoregressive model. J. Appl. Prob. 17(1980), 546-552.

Shanbhag D. N., Pestana D. and Sreehari H. (1977) : Some further results in infinite divisibility. Math. Proc. Camb. Phil. Soc. 82(1977), 289-295.

Shanbhag D. N. and Sreehari M. (1977) : On certain self-decomposable distributions. Z. Wahrscheinlichkeitsth. 38(1977), 217-222.

Thorin O. (1977a): On the infinite divisibility of the Pareto distribution. Scand. Actuarial J. 4(1977), 31-40.

Thorin O. (1977b): On the infinite divisibility of the log normal distribution. Scand. Actuarial J. 4(1977), 121-148.

Charles University
Department of Statistics

Sokolovská 83
186 00 Praha 8
Czechoslovakia

ОБ ОПТИМАЛЬНЫХ АЛГОРИТМАХ ОПТИМИЗАЦИИ ФУНКЦИОНАЛОВ С БУЛЕВЫМИ ПЕРЕМЕННЫМИ

Александр Антамошкин
Кемерово

Ключевые слова: Оптимальные алгоритмы, функционалы с булевыми переменными, статистическая идентификация.

АННОТАЦИЯ

Обсуждаются: результаты построения оптимальных (по точности и скорости сходимости) детерминированных алгоритмов оптимизации функционалов с булевыми переменными на классах функционалов, вопросы идентификации функционалов реальных задач с выделенными классами по статистической информации.

До последнего времени при решении практических задач, математически сводимых к оптимизации функционалов с булевыми переменными (функционал, как правило, задаётся алгоритмически), применялись методы случайного поиска (I98Iг), так как ранее была показана несостоятельность универсальных детерминированных методов решения таких задач. К сожалению, методы случайного поиска позволяют получить только субоптимальное решение и требуют значительных затрат машинного времени.

Поэтому была предпринята попытка разработать оптимальные (очевидно, что имеет смысл говорить об оптимальности только на определённом классе) алгоритмы оптимизации функционалов с булевыми переменными.

КЛАССЫ ФУНКЦИОНАЛОВ С БУЛЕВЫМИ ПЕРЕМЕННЫМИ

Рассмотрим функционал $\mathcal{X}(X)$, заданный на векторном пространстве $Д=\{X:x_j=0 \vee 1,\ j=\overline{1,n}\}$. Очевидно, что Д является линейным пространством, как векторное пространство строк над полем $\mathbb{F}_2$.

Зададим на Д метрику: $\forall X \wedge Y \in Д\ \rho(X,Y)=\sum_{j=1}^{n}|x_j-y_j|$.

$X^* \in Д$ назовём локальным минимумом $\varkappa(X)$ относительно окрестности радиуса r - $O(X^*,r)=\{X \in Д,\ \rho(X,X^*) \leq r\}$, если $O(X^*,r) \setminus \{X^*\} \neq \emptyset \wedge \forall X \in O(X^*,r) \setminus \{X^*\}\ \varkappa(X^*) < \varkappa(X)$. Очевидно, что X^{**} будет глобальным минимумом, если $\forall X \in Д \setminus \{X^{**}\}\ \varkappa(X^{**}) < \varkappa(X)$.

Унимодальный функционал $\varkappa(X)$ монотонен на Д (монотонно убывающий), если $\forall X \in \{X: X \in Д,\ \rho(X,X^*)=к\} \wedge Y \in \{Y: Y \in Д,\ \rho(Y,X^*)=к+1\}$, $к=\overline{1,п-1}$, $\varkappa(X) < \varkappa(Y)$ и строго монотонен, если кроме указанного условия выполняется: $\varkappa(X)=\varkappa(Y)\ \forall X \wedge Y \in \{Z: Z \in Д,\ \rho(X^*,Z)=к\}$.

Полимодальный функционал, монотонный в зоне притяжения каждого локального минимума, назовём локально монотонным.

Функционал $\varkappa(X)$ называется ρ-выпуклым на ρ-выпуклом пространстве Д, если $\forall X,Y,Z \in Д \ni \rho(X,Y)+\rho(Y,Z)=\rho(X,Z)$ -

$$\varkappa(Y) \leq \frac{\rho(Y,Z)}{\rho(X,Z)}\varkappa(X)+\frac{\rho(X,Y)}{\rho(X,Z)}\varkappa(Z).$$

Функционал $\varkappa(X)$ называется однородно выпуклым над полем Д, если $\forall X \wedge Y \in Д\ \ \varkappa(X+Y) \leq \varkappa(X)+\varkappa(Y)$.

Т е о р е м а 1. ρ-выпуклый на Д функционал - унимодален.

Т е о р е м а 2. ρ-выпуклый на Д функционал - монотонен. (Из теоремы не следует строгая монотонность).

Т е о р е м а 3. Необходимым и достаточным условием ρ-выпуклости на Д монотонного функционала $\varkappa(X)$ является: $\varkappa(X)-\varkappa(Y) \geq \frac{п-1}{п} \times [\varkappa(X)-\varkappa(X^*)]\ \forall X \wedge Y \in Д \ni \rho(X,X^*) > \rho(Y,X^*)$.

Т е о р е м а 4. При введённой метрике необходимым и достаточным условием однородной выпуклости монотонного функционала $\varkappa(X)$ является: $\varkappa(X)-\varkappa(Y) \leq \varkappa(X^*)\ \forall Y \ni \rho(Y,X^*)=1 \wedge X \ni \rho(X,X^*)=п$.

Замечание 1. Однородно выпуклый унимодальный функционал может быть и не монотонным. Например, $\varkappa(X)=\sum_{j=1}^{n} j^2 x_j$.

Замечание 2. Очевидно, что строго монотонные и строго локально монотонные функционалы являются частным случаем монотонных и локально монотонных функционалов соответственно.

Таким образом, класс монотонных функционалов включает в себя классы ρ-выпуклых и строго монотонных функционалов и пересекается с классом однородно выпуклых унимодальных функционалов, причём, перечисленные классы не покрывают всего множества унимодальных функционалов с булевыми переменными, так функционал $\varkappa(X)=\prod_{j=1}^{n} j\, x_j$ -унимодален, но, как не трудно проверить, не принадлежит ни к одному из перечисленных классов.

ОПТИМАЛЬНОСТЬ АЛГОРИТМОВ

В работе (1981в) предложены оптимальные по точности алгоритмы оптимизации монотонных и локально монотонных функционалов (обозначим их ОАМ и ОАЛМ соответственно). Это регулярные алгоритмы, реализующие идею локальной дискретной оптимизации. Положение глобального минимума определяют точно. Очевидно, что по точности эти алгоритмы будут оптимальны и для классов ρ-выпуклых, строго монотонных и строго локально монотонных функционалов.

Покажем оптимальность алгоритмов ОАМ и ОАЛМ по быстродействию на множестве алгоритмов $\{A_q\}$, где A_1-ОАМ, A_2-ОАЛМ, A_3-полный перебор, A_4-случайный перебор (в пространстве Д в случайный перебор вырождается прямой случайный поиск (I982)), A_5-случайный поиск с адаптацией Лбова (I98Iа), A_6-модификация A_5 (I98Iа), A_7-случайный поиск с возвратом (I98Iа) (алгоритмы A_3, A_4 и A_5 чаще всего используются на практике для оптимизации функционалов с булевыми переменными).

Ранее было показано преимущество A_1 и A_2 перед A_3 (I98Iв) и преимущество A_7 перед A_5 и A_6 (I98Iг). Поэтому достаточно сравнить быстродействие A_1, A_2, A_4 и A_7.

В работе (I98Iв) получены аналитические выражения для верхних оценок количества вычислений функционала, необходимого для определения глобального минимума с вероятностью I алгоритмами A_1 и A_2(обозначим их N_1 и N_2 соответственно) и аналитическое выражение для верхней оценки среднего количества вычислений функционала, необходимое алгоритму A_4 для определения глобального минимума с вероятностью $P_4 = 1/2 + 1/2^n$ (обозначим его N_4).

Т е о р е м а 5. Разность $f_1(n) = N_4 - N_1 > 0$ при $n \geqslant 5$.

Из этой теоремы следует преимущество A_1 перед A_4 по быстродействию в унимодальном случае.

В полимодальном случае величина N_4 имеет тот же вид, что и в унимодальном ($N_4 = N_4(n)$), а величина N_2, кроме размерности - n, зависит и от числа экстремумов - ℓ , т.е. $N_2 = N_2(n, \ell)$.

Т е о р е м а 6. $f_2(n, \ell) = N_4 - N_2 < 0$ при $\ell = \ell^*$ и $f_2(n, \ell) > 0$ при $\ell \neq \ell^*$, где ℓ^*-максимально возможное число изолированных минимумов.

В работе (I98Iа) получено следующее выражение для определения верхней оценки среднего количества вычислений функционала, необходимое для определения X^* с вероятностью I алгоритмом A_7 при условной (на $Д' = \{X \in Д, \sum_{j=1}^{n} x_j = m, m < n$) оптимизации монотонных функционалов: $N_7 = m(n-m)\sum_{j=1}^{n} 1/j^2$. Как показано в работе (I98Iг), возможно эквивалентное преобразование задачи условной оптимизациии в безусловную, но при этом размерность задачи увеличивается вдвое, а $m = n$. После такого преобразования последнее выражение принимает вид: $N_7 = 2n^2 \sum_{j=1}^{n} 1/j^2$.

Т е о р е м а 7. Разность $f_3(n) = N_7 - N_1$ всегда положительна.

Аналитическая оценка быстродействия алгоритма A_7 в полимодальном случае не получена.

Таким образом, алгоритм A_1 оптимален как по точности, так и по быстродействию на рссмотренном множестве алгоритмов $\{A_q\}$ при оптимизации унимодальных, монотонных (соответственно ρ-выпуклых и строго монотонных) функционалов, а алгоритм A_2 оптимален по точности и быстродействию (при $\ell \neq \ell^*$) при оптимизации локально монотонных (и строго локально монотонных) функционалов на множестве алгоритмов $\{A_1, A_2, A_3, A_4\}$. С учётом того, что $\lim_{n\to\infty} P_4 = 1/2$, A_2 можно принять за оптимальный алгоритм и при $\ell = \ell^*$.

ИДЕНТИФИКАЦИЯ

Как уже указывалось, на практике вид функционала априорно не известен. В соответствии с полученными результатами, его идентификацию с одним из выделенных классов предлагается организовать по схеме, приведённой на рисунке.

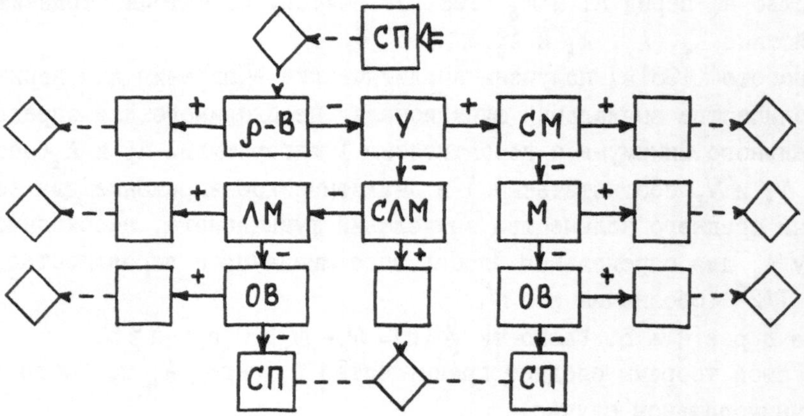

Рис.

Использованы обозначения: прямоугольники - системы операторов проверки (ρ-В - ρ-выпуклости, У - унимодальности, СМ - строгой монотонности, ЛМ - локальной монотонности, СЛМ - строгой локальной монотонности, М - монотонности, ОВ - однородной выпуклости); квадраты - алгоритмы (СП - случайного поиска, пустые - оптимальные для соответствующего класса); ромбы - память, в которой хранится накопленная статистическая информация о функционале. Циркуляция информации показана пунктирными стрелками. Сплошными стрелками показаны связи между системами операторов проверки и алгоритмами, определяющие порядок обращения к ним. Знак "+" у стрелки обозначает, что проверка дала положительный результат, знак "-" - отрицательный.

Наполнение памяти исходной статистической информацией происходит в процессе решения Т задач данного типа с помощью алгоритмов случайного поиска (1981а). Этой информацией будут множества точек поиска $\{X_\tau\}$, $\tau=\overline{1,R_t}$, $t=\overline{1,T}$, соответствующих значений функционала $\{\varkappa(X_\tau)\}$, $X_\tau \in \{X_\tau\}$ и указанных точек минимума $\{X_p^*\}$. После чего, по приведённой схеме проводится идентификация с последующим T+1 решением задачи оптимальным алгоритмом (при положительном результате идентификации) или алгоритмом случайного поиска (при отрицательном). В любом случае, информация, полученная при решении T+1-й задачи, может быть направлена в память для проверочной идентификации. Проверочная идентификация может быть осуществлена после любого решения.

В дальнейшем предполагается при отрицательном результате предусмотреть выход на блок таксономии, формирующий по свойствам новые классы функционалов.

Все рассмотренные алгоритмы и схема идентификации реализованы программно.

Пока не решён вопрос о построении оптимального алгоритма для класса однородно выпуклых функционалов.

ССЫЛКИ

Антамошкин А. (1981а): Адаптация одного класса алгоритмов случайного поиска. Автоматика и вычислительная техника, №6, 54.

(1981б): Об одной задаче оптимизации с булевыми переменными. Автоматика и вычислительная техника, №3, 54-56.

(1981в): Оптимальный метод оптимизации монотонных и локально монотонных функционалов с булевыми переменными. В кн.: Применения случайного поиска, университет, Кемерово, вып.2, 89-94.

(1981г): Оптимизация функционалов с булевыми переменными (обзор). В кн.: Теория оптимальных решений, АН ЛитССР, Вильнюс, вып.7, 9-15.

(1982): Проведение полного исследования эффективности методов оптимизации функционалов с булевыми переменными на примере прямого случайного поиска. В кн.: Применения случайного поиска, университет, Кемерово, вып.3, 61-72.

Университет. Улица Красная,6, 650043,
Кемерово, СССР.

AN EQUILIBRIUM THEORY FOR MULTI-PERSON MULTI-CRITERIA STOCHASTIC DECISION PROBLEMS WITH MULTIPLE SUBJECTIVE PROBABILITY MEASURES†

Tamer Başar
Urbana

Key Words: Stochastic multi-person decision problems, noncooperative equilibrium theory, subjective probability measures, team theory.

ABSTRACT

An equilibrium theory is developed for multi-person multi-criteria stochastic decision problems wherein the decision makers have different subjective probability measures on the uncertain quantities. Particular attention is devoted to existence and uniqueness of stable equilibria in such problems, when the loss functionals are (locally) quadratic and the subjective probability measures are Gaussian.

INTRODUCTION AND PROBLEM FORMULATION

Consider the class of two-person two-criteria stochastic decision problems with loss functionals $L_1(x, u_1, u_2)$ and $L_2(x, u_1, u_2)$ for DM1 (first decision maker) and DM2, respectively, where u_1, u_2 denote the decision variables [of DM1 and DM2, respectively] belonging to some prescribed Hilbert spaces U_1 and U_2, and $x \in X$ stands for the state of Nature. Let $y_1 \in Y_1$ and $y_2 \in Y_2$ be two stochastic variables, which are correlated with x and denote the measurements available to DM1 and DM2, respectively, so that u_i will be chosen as a measurable function of y_i, $i = 1,2$, i.e. $u_i = \gamma_i(y_i)$, where γ_i belongs to a policy space Γ_i which will be delineated in the sequel. The sets X, Y_1 and Y_2 are assumed to be structured appropriately, so that each is a well-defined Hilbert space.

So far we have adopted the standard decision-theoretic framework (see e.g. Ferguson (1967)); we depart, however, from this standard formulation in the

†This work was supported in part by the Joint Services Electronics Program under Contract N00014-79-C-0424, and in part by the Office of Naval Research.

description of the underlying probability space. Let $(\Omega,\mathfrak{F})$ be a measurable space to which the triple (x, y_1, y_2) belongs; then, we assume that the decision makers have different (not necessarily the same) subjective probability meausres $\mathcal{P}_1$ and $\mathcal{P}_2$ [for DM1 and DM2, respectively] on this measurable space $(\Omega,\mathfrak{F})$ and let the random variables (x, y_1, y_2) have finite second moments under both $\mathcal{P}_1$ and $\mathcal{P}_2$. Furthermore, we take Γ_i to be the Banach space of all measurable mappings $\gamma_i : Y_i \to U_i$, with the additional property that $\gamma_i(y_i)$, viewed as a random variable, has finite second moments.

Let $z = (x, y_1, y_2)$, and introduce, for each pair $(\gamma_1,\gamma_2) \in \Gamma_1 \times \Gamma_2$, the quantity

$$J_i(\gamma_1,\gamma_2) = \int L_i(x,\gamma_1,(y_1),\gamma_2(y_2))\, \mathcal{P}_i(dz) \tag{1}$$

as the expected cost function of DMi corresponding to the decision rules (γ_1,γ_2) and under DMi's subjective probability measure $\mathcal{P}_i$. [Here, we implicitly assume that L_i is integrable under $\mathcal{P}_i$.] We should note at this point that even in team problems (with $L_1 \equiv L_2$) the decision makers will have different expected cost functions whenever $\mathcal{P}_1$ and $\mathcal{P}_2$ do not match, since then a common probability space will not exist.

<u>Definition 1</u>

A pair of policies $(\gamma_1^o,\gamma_2^o) \in \Gamma_1 \times \Gamma_2$ constitutes an <u>equilibrium solution</u> to the decision problem formulated above if

$$J_1(\gamma_1^o,\gamma_2^o) \le J_1(\gamma_1,\gamma_2^o), \qquad \forall\, \gamma_1 \in \Gamma_1 \tag{2a}$$

$$J_2(\gamma_1^o,\gamma_2^o) \le J_2(\gamma_1^o,\gamma_2), \qquad \forall\, \gamma_2 \in \Gamma_2 \tag{2b}$$

□

<u>Definition 2</u>

An equilibrium solution (γ_1^o,γ_2^o) is a <u>locally stable equilibrium solution</u> if there exists an $\varepsilon > 0$ and an open neighborhood $N_\varepsilon(\gamma_1^o,\gamma_2^o) \subseteq \Gamma_1 \times \Gamma_2$ of (γ_1^o,γ_2^o) so that for all $(\gamma_1^{(o)}, \gamma_2^{(o)}) \in N_\varepsilon$,

$$\lim_{n\to\infty} \gamma_i^{(n)} = \gamma_i^o \;, \quad \text{in } \Gamma_i \;, \qquad i = 1,2,$$

where

$$\gamma_1^{(n)} = \arg \min_{\Gamma_1} J_1(\gamma_1,\gamma_2^{(n-1)})$$

$$\gamma_2^{(n)} = \arg \min_{\Gamma_2} J_2(\gamma_1^{(n-1)},\gamma_2) \;, \quad n=1,2,\ldots..$$

□

<u>Definition 3</u>

A locally stable equilibrium solution (γ_1^o,γ_2^o) is (globally) <u>stable</u> if $N_\varepsilon(\gamma_1^o,\gamma_2^o) = \Gamma_1 \times \Gamma_2$ in Definition 2. □

Our objective in this paper is to obtain conditions on L_1, L_2 and the probability measures $\mathcal{P}_1$, $\mathcal{P}_2$, under which the decision problem formulated above will admit a locally or globally stable equilibrium solution. We will, in particular, consider the class of problems in which L_1 and L_2 are quadratic in the decision variables u_1 and u_2, and also specialize our treatment to the case of jointly Gaussian distributions. The special case of $\mathcal{P}_1 \equiv \mathcal{P}_2$ has earlier been treated in Başar(1975) and Başar (1978), where conditions, independent of the probabilistic structure of the problem, have been obtained for stable equilibrium solutions. The present paper discusses nontrivial extensions of these results to the case $\mathcal{P}_1 \not\equiv \mathcal{P}_2$, and only outlines the method of approach and the solution because of space limitations.

QUADRATIC PROBLEMS AND GENERAL CONDITIONS FOR EXISTENCE OF A STABLE EQUILIBRIUM

Let L_1 and L_2 be defined by

$$L_1(x,u,v) = \frac{1}{2}\langle u_1,u_1\rangle + \frac{1}{2}\langle u_2,D_{22}^1u_2\rangle - \langle u_1,F_1^1x\rangle - \langle u_2,F_2^1x\rangle - \langle u_1,D_{12}^1u_2\rangle \tag{3a}$$

$$L_2(x,u,v) = \frac{1}{2}\langle u_1,D_{11}^2u_1\rangle + \frac{1}{2}\langle u_2,u_2\rangle - \langle u_2,D_{21}^2u_1\rangle - \langle u_1,F_1^2x\rangle - \langle u_2,F_2^2x\rangle \tag{3b}$$

where D_{22}^1: $U_2 \to U_2$ and D_{11}^2: $U_1 \to U_1$ are strongly positive operators, and we do not differentiate between inner products defined on different Hilbert spaces. Let $E^i[\mu(z)|y_i]$ denote the expectation of a z-measurable random variable $\mu(z)$ conditioned on the random variable y_i and under the probability measure $\mathcal{P}_i$. The following two results now follow readily from the analyses of Başar (1975) and Başar (1978).

<u>Proposition 1</u>

A pair of policies $(\gamma_1^o,\gamma_2^o) \in \Gamma_1 \times \Gamma_2$ constitutes an <u>equilibrium solution</u> to the two DM decision problem with quadratic loss functionals (3) if and only if it satisfies the pair of equations

$$\gamma_1^o(y_1) = D_{12}^1\, E^1\,[\gamma_2^o(y_2)|y_1] + F_1^1\, E^1\,[x|y_1] \tag{4a}$$

$$\gamma_2^o(y_2) = D_{21}^2\, E^2\,[\gamma_1^o(y_1)|y_2] + F_2^2\, E^2\,[x|y_2] \tag{4b}$$

□

<u>Proposition 2</u>

A pair of policies $(\gamma_1^o,\gamma_2^o) \in \Gamma_1 \times \Gamma_2$ constitutes a <u>stable</u> equilibrium solution if, for all $(\gamma_1^{(o)},\gamma_2^{(o)}) \in \Gamma_1 \times \Gamma_2$,

$$\gamma_i^o(y_i) = \lim_{n\to\infty} \gamma_i^{(n)}(y_i) \qquad \text{a.e. } \mathcal{P}_i$$

where

$$\gamma_i^{(n)}(y_i) = D_{ij}^i D_{ji}^j E^i[E^j[\gamma_i^{(n-1)}(y_i)|y_j]|y_i]$$

$$+ D_{ij}^i F_j^j E^i[E^j[x|y_j]|y_i] + F_i^i E[x|y_i], \quad (5)$$

$$i,j = 1,2;\ j\neq i;\ n = 1,2. \ . \ . \ .$$

Furthermore, such an equilibrium solution is necessarily unique. □

Let us introduce linear operators $\mathcal{L}_i:\ \Gamma_i \to \Gamma_i$ i=1,2, by

$$\mathcal{L}_i(\gamma) = D_{ij}^i D_{ji}^j E^i[E^j[\gamma(y_i)|y_j]|y_i],\ j\neq i;\ i,j=1,2. \quad (6)$$

Then, in view of Proposition 2, the quadratic decision problem will admit a unique stable equilibrium solution if, and only if, $\mathcal{L}_1$ and $\mathcal{L}_2$ are contraction mappings, i.e. there exists a constant ρ, $0 < \rho < 1$, such that

$$\|\mathcal{L}_i\| \triangleq \sup_{\gamma \in \Gamma_i} \{\langle\langle\gamma(y_i), \mathcal{L}_i\gamma(y_i)\rangle\rangle / \langle\langle\gamma(y_i), \gamma(y_i)\rangle\rangle\} < \rho,\ i=1,2, \quad (7)$$

where $\langle\langle \cdot \rangle\rangle$ denotes the inner product on Γ_1 or Γ_2. Since $\|\mathcal{L}_i\| \leq \| D_{ij}^i D_{ji}^j\|$ $\|E^i[E^j[\cdot|y_j]|y_i]\|$ by using a well-known property of linear operators defined on Banach spaces, a set of sufficient conditions for $\mathcal{L}_i$ to be a contraction mapping is existence of a pair (ρ_1,ρ_2), $0 < \rho_1,\rho_2 \leq 1$, $\min(\rho_1,\rho_2) < 1$, such that

1) $\|D_{ij}^i D_{ji}^j\| \leq \rho_1$ (8a)

2) $\|E^i[E^j[\cdot|y_j]|y_i]\| \leq \rho_2$, (8b)

which is a complete separation (in terms of sufficient conditions) of the deterministic and stochastic parts of the system.

Now, if the decision problem is a team problem with a common loss functional $L = L_1 \equiv L_2$ [which requires $D_{22}^1 = I$, $D_{12}^1 = D_{21}^{2*}$, $F_1^1 = F_1^2$, $F_2^1 = F_2^2$], and if L is strictly convex in the pair (u_1,u_2), (8a) is always satisfied with $\rho_1 < 1$. If, furthermore, the subjective probabilities $\mathcal{P}_1$ and $\mathcal{P}_2$ are the same, the second part of the linear operator $\mathcal{L}_i$ becomes a projection operator, thus leading to satisfaction of the second condition (8b) with $\rho_2 = 1$. Hence, for the strictly convex quadratic team problem with $\mathcal{P}_1 \equiv \mathcal{P}_2$, there exists a unique stable equilibrium solution (the so-called team-optimal solution), irrespective of the underlying common probability distribution — a result which is already well-established in the literature (see Radner (1962), Başar (1978)). However, for team problems with $\mathcal{P}_1 \not\equiv \mathcal{P}_2$, such a result no longer holds true, because the second part of the operator $\mathcal{L}_i$ is not necessarily a projection operator, i.e. we may not be able to find a ρ_2, $0 < \rho_2 \leq 1$, to satisfy (8b). The general condition then is (8b), which places some restrictions on the probability measures $\mathcal{P}_1$ and $\mathcal{P}_2$.

To investigate this question somewhat further, let us assume that $Y_1 = \mathbb{R}^{m_1}$, $Y_2 = \mathbb{R}^{m_2}$, and that θ_i admits a probability density function (with respect to the Lebesgue measure) denoted $p^i(x,y_1,y_2)$. By an abuse of notation, let us denote the marginal and conditional densities that involve y_1 and y_2 by $p^i(y_i)$ and $p^i(y_j|y_i)$, respectively. Then for $j\neq i=1,2$,

$$
\begin{aligned}
\|E^i[E^j[\gamma(y_i)|y_j]|y_i]\|^2 &= \| \int p^i(y_i|y_j) \int \gamma(y_i')p(y_i'|y_j)dy_i'dy_j\|^2 \\
&\leq \iint dy_j dy_i \; p^i(y_j) \; p^j(y_i|y_j) \; \langle\gamma(y_i), \; \gamma(y_i)\rangle \qquad (9a)\\
&= \int F_i(y_i) \langle\gamma(y_i), \; \gamma(y_i)\rangle \; p^i(y_i)dy_i
\end{aligned}
$$

where

$$
\begin{aligned}
F_i(y_i) &\triangleq E^i\{[p^j(y_i|y_j)/p^i(y_i|y_j)]|y_i\} \\
&= \int dy_j p^i(y_j|y_i) \; p^j(y_i|y_j)/p^i(y_i|y_j),
\end{aligned} \qquad (9b)
$$

and in arriving at the inequality we have made repeated use of the Cauchy-Buniakowski inequality. Now, under the condition

$$
F_i(y_i) \leq 1 \qquad \forall y_i \in \mathbb{R}^{m_i} \quad , \; i=1,2, \qquad (10)
$$

(9a) can be rewritten as

$$
\|E^i[E^j[\gamma(y_i)|y_j]|y_i]\|^2 \leq \quad \langle\langle\gamma(y_i), \; \gamma(y_i)\rangle\rangle
$$

thereby satisfying (8b) with $\rho_2=1$. Also note that it will be sufficient for (8a) and (8b) to be satisfied for only one i (i=1 or 2), since if, for example, $\|\mathcal{L}_1\| \leq \rho < 1$, (5) admits a well defined limit for i=1 as $n\to\infty$, which implies through (4b) that $\lim_n \gamma_2^{(n)}$ is also well-defined. This then leads to the following Corollary to Proposition 2.

<u>Corollary 1</u>

If conditions (8a) and (10) are satisfied for i=1 <u>or</u> 2, with $0 < \rho_i < 1$, the quadratic decision problem [with Y_i taken as Euclidean spaces] admits a unique stable equilibrium solution. □

<u>Remark 1</u>

If $\theta_1 = \theta_2$, $F_i(y_i) = 1 \; \forall \; y_i \in \mathbb{R}^{m_i}$, and hence (10) is always satisfied. □

JOINTLY GAUSSIAN DISTRIBUTIONS AND DERIVATION OF EXPLICIT SOLUTIONS

To explore the extent of the restrictions imposed by condition (10) on the probabilistic structure of the problem, we now further assume that the random vectors are jointly Gaussian distributed, with mean zero and covariances

$$\mathrm{cov}(y_1,y_2) = \Sigma_y^i = \begin{pmatrix} \Sigma_{y_1}^i & \Sigma_{y_1y_2}^i \\ \Sigma_{y_2y_1}^i & \Sigma_{y_2}^i \end{pmatrix} > 0, \text{ under } \theta_i. \quad (11)$$

Then, straightforward manipulations lead to

$$F_i(y_i) = \sqrt{k_i} \exp\{-\tfrac{1}{2} y_i' K_i y_i\} \quad (12)$$

where

$$k_i = \det \Sigma_{y_j}^j \det \Sigma_{y_i}^i / \{\det \Sigma_y^j \det \Sigma_{y_j}^i \det [A_i' D_i^{-1} A_i + \Sigma_{y_j}^{i^{-1}}]\} \quad (13a)$$

$$A_i = \Sigma_{y_iy_j}^j \Sigma_{y_j}^{j^{-1}} \quad (13b)$$

$$D_i = \Sigma_{y_i}^j - \Sigma_{y_iy_j}^j \Sigma_{y_j}^{j^{-1}} \Sigma_{y_jy_i}^j > 0 \quad (13c)$$

$$K_i = D_i^{-1} - \Sigma_{y_i}^{i^{-1}} - D_i^{-1} A_i [A_i' D_i^{-1} A_i + \Sigma_{y_j}^i]^{-1} A_i' D_i^{-1} \quad (13d)$$

$$j \neq i$$

For (12) to be no greater than unity uniformly in $y_i \in \mathbb{R}^{m_i}$, we will have to require K_i to be a nonnegative definite matrix, which is equivalent to the matrix inequality

$$I - D_i^{1/2} \Sigma_{y_i}^{i^{-1}} D_i^{1/2} \geq D_i^{-1/2} A_i [A_i' D_i^{-1} A_i + \Sigma_{y_j}^{i^{-1}}]^{-1} A_i' D_i^{-1/2}. \quad (14)$$

Now, in addition to (11), let us assume that x is also a zero-mean Gaussian random vector on $\mathbb{R}^n$, so that

$$\mathrm{cov}(x,y_1,y_2) = \mathrm{cov}(x,y) = \Sigma^i = \begin{pmatrix} \Sigma_x^i & \Sigma_{xy}^i \\ \Sigma_{yx}^i & \Sigma_y^i \end{pmatrix} > 0 \text{ under } \theta_i, \quad (15)$$

and that $U^i = \mathbb{R}^{r_i}$, for some integer r_i, i=1,2. Then, we have the following theorem for Gaussian decision systems.

Theorem 1

Let (14) and the strict inequality

$$k_i \| D_{ij}^i D_{ji}^j \| < 1 \quad (16)$$

be satisfied for at least one i=1,2. Then, the quadratic Gaussian decision problem formulated above admits a unique stable equilibrium solution (γ_1^o, γ_2^o) which is linear in (y_1,y_2) and is given by

$$\gamma_i^{o}(y_i) = L_i y_i \qquad , \qquad i=1,2, \tag{17}$$

where (L_1,L_2) constitutes the unique solution to the Liapunov-type matrix equations

$$L_1 - D^1_{12}D^2_{21}L_1\,\Sigma^2_{y_1y_2}\Sigma^{2^{-1}}_{y_2}\Sigma^1_{y_2y_1}\Sigma^{1^{-1}}_{y_1} - D^1_{12}F^2_2\Sigma^2_{xy_2}\Sigma^{2^{-1}}_{y_2}\Sigma^1_{y_2y_1}\Sigma^{1^{-1}}_{y_1} - F^1_1\Sigma^1_{xy_1}\Sigma^{1^{-1}}_{y_1} = 0 \tag{18a}$$

$$L_2 - D^2_{21}D^1_{12}L_2\,\Sigma^1_{y_2y_1}\Sigma^{1^{-1}}_{y_1}\Sigma^2_{y_1y_2}\Sigma^{2^{-1}}_{y_2} - D^2_{21}F^1_1\Sigma^1_{xy_1}\Sigma^{1^{-1}}_{y_1}\Sigma^2_{y_1y_2}\Sigma^{2^{-1}}_{y_2} - F^2_2\Sigma^1_{xy_2}\Sigma^{2^{-1}}_{y_2} = 0 \tag{18b}$$

<u>Proof</u>

The first part (i.e. existence and uniqueness) follows from Corollary 1 and the discussion that precedes (14), also in view of the original contraction mapping inequality (7). The second part follows by noting that if $(\gamma_1^{(o)},\gamma_2^{(o)})$ are taken to be linear in (y_1,y_2) in (5), all terms of the sequence are linear and hence the limit $(\gamma_1^{o},\gamma_2^{o})$ in Proposition 2 is linear. Denoting the coefficient gain matrices by (L_1,L_2), we readily arrive at (18a)-(18b) through straightforward manipulations. □

Conditions for existence of an <u>equilibrium</u> solution are of course less restrictive than those under which the statement of Theorem 1 is valid. In fact, the solution depicted in Theorem 1 will constitute an equilibrium solution whenever there exists a pair (L_1,L_2) satisfying (18a)-(18b). A sufficient condition for this (which is less restrictive than (14) and (16)), is provided in the following proposition, whose proof follows readily from the proof of the second part of Theorem 3 of Başar (1975).

<u>Proposition 3</u>

The quadratic Gaussian decision problem of Theorem 1 admits an equilibrium solution (not necessarily stable) given by (17)-(18), if for at least one i=1,2,

$$|\lambda_{\max}\,(D^i_{ij}\,D^j_{ji})| < 1 \tag{19a}$$

$$|\lambda_{\max}\,(\Sigma^j_{y_iy_j}\Sigma^{j^{-1}}_{y_j}\Sigma^i_{y_jy_i}\Sigma^{i^{-1}}_{y_i})| \le 1, \tag{19b}$$

where $\lambda_{\max}$ (A) denotes the eigenvalue of A which is maximum in absolute value. □

For the purpose of illustrating the various conditions of existence obtained above, we now consider a family of scalar team problems, with the decision makers having different subjective probabilities on the uncertain quantities. To be more specific, let $D^1_{22} = D^2_{11} = 1$, $D^1_{12} = D^2_{21} = d$, $|d| < 1$, $F^1_1 = F^2_1 = f_1$, $F^1_2 = F^2_2 = f_2$, and

$$\Sigma^i_y = \begin{pmatrix} \sigma^i_1 & \sigma^i_{12} \\ \sigma^i_{12} & \sigma^i_2 \end{pmatrix}$$

Then, conditions (14) and (16) are satisfied if either

$$\sigma_1^1|\sigma_1^2 \geq \sigma_2^1|\sigma_2^2 \quad ; \ |d| < 1/k_1 \tag{20}$$

or

$$\sigma_2^2|\sigma_2^1 \geq \sigma_1^2|\sigma_1^1 \quad ; \ |d| < 1/k_2 \tag{21}$$

are satisfied, where

$$k_i = \sigma_1^1\sigma_2^2|[\sigma_j^j\sigma_i^j - (\sigma_{ij}^j)^2 (1 - \sigma_j^i|\sigma_j^j)]; \ i,j=1,2; \ i\neq j,$$

which are the conditions for existence of a stable equilibrium solution.

If we are interested only in existence of equilibrium solutions (not necessarily stable), the conditions (20)-(21) can be relaxed. The conditions, in this case, follow from (19a)-(19b) to be $|d| < 1$ and either $\sigma_2^2\sigma_2^1 \leq \sigma_1^2\sigma_1^1$ or $\sigma_1^1\sigma_1^2 \leq \sigma_2^1\sigma_2^2$ which are always satisfied, provided that the loss function is strictly convex in (u_1,u_2). Hence, the conclusion is that even if the subjective probabilities are different, the scalar Gaussian team problem with strictly convex loss functional admits an equilibrium solution; this solution, however, is not necessarily stable and additional conditions (such as (20) or (21)) have to be imposed to insure stability.

CONCLUDING REMARKS

The applicability of the general approach of this paper is not restricted to the class of quadratic two-person stochastic decision problems analyzed here in considerable depth, but can readily be extended to multi-person stochastic decision problems in which the decision makers have different subjective probabilities on the uncertain quantities governing the decision process. Extensions are also possible to nonquadratic loss functionals in which case we investigate existence and uniqueness of locally stable equilibrium solutions. Because of space limitations, we have not been able to discuss such extensions in the present paper.

REFERENCES

Başar, T. (1975): Equilibrium solutions in two-person quadratic decision problems with static information structures, IEEE Trans. Automatic Control, Vol. AC-20, No. 3, 320-328.

(1978): Decentralized multicriteria optimization of linear stochastic systems, IEEE Trans. Automatic Control, Vol. AC-23, No. 2, 233-243.

Ferguson, T.S. (1967): Mathematical Statistics, New York, Academic.

Radner, R. (1962): Team-decision problems, Annals Math. Statist., Vol. 33, No. 3, 857-881.

University of Illinois
Coordinated Science Laboratory
1101 W. Springfield Avenue
Urbana, Illinois 61801 USA

PROCESSUS D'ORNSTEIN UHLENBECK GENERALISE. MESURES STATIONNAIRES DANS LE CAS GAUSSIEN.

Albert Benassi
Paris

MOTS CLES : PROCESSUS D'ORNSTEIN UHLENBECK GENERALISE , MESURE STATIONNAIRE.

RESUME

A une martingale $\{M_t \; ; \; t \geq 0\}$ à valeurs distributions et à un opérateur elliptique, nous associons un processus d'Ornstein Uhlenbeck généralisé. Dans le cas où la martingale est le processus de Siegel standart, nous décrivons l'ensemble des mesures stationnaires du processus d'Ornstein Uhlenbeck associé.

1) Soit $\{\Omega, \mathcal{A}, P\}$ un espace de probabilité, Λ un ouvert borné régulier (c.à.d. connexe de frontière C^{∞}). $\mathcal{D}(\Lambda)$ désigne l'espace de Schwartz des fonctions C^{∞} à support compact dans Λ et $\mathcal{D}'(\Lambda)$ son dual , l'espace des distributions sur Λ .

Donnons-nous une martingale $\{M_t \; ; \; t \geq 0\}$ à valeurs dans $\mathcal{D}'(\Lambda)$. Donnons-nous A un opérateur continu de $\mathcal{D}(\Lambda)$ dans $\mathcal{D}(\Lambda)$ (A se prolonge à $\mathcal{D}'(\Lambda)$ en un opérateur continu encore noté A).

Le but de cet exposé est de définir le processus d'Ornstein Uhlenbeck Généralisé associé à l'opérateur A et à la martingale M (le POUG (A;M)). Pour cela, nous étudierons l'équation de Langevin:

$$\begin{cases} dX_t = -\frac{A}{2} X_t dt + dM_t \\ X_0 = \gamma \end{cases} \qquad (*)$$

Ensuite, lorsque W_t est le processus de Siegel standart, nous étudierons les mesures stationnaires du POUG (A,W).

2) a) Soit $L^2(\Lambda)$ l'espace des fonctions de carré intégrable muni du produit scalaire usuel, $(\cdot,\cdot)_0$.

- Donnons-nous une **base** orthonormée $\{\phi_k \; ; \; k \in N\}$ de $L^2(\Lambda)$, une suite a_k de réels non nuls et positifs.

Donnons-nous une suite de martingales réelles de carré intégrable $\{(m_t^k \; ; \; t \geq 0); k \in N\}$

- Définissons les espaces de Hilbert H_+ et H_- par : $H_+ = \{$fermeture de $\mathcal{D}(\Lambda)$ pour la norme $\|\cdot\|_+ \}$, $\|f\|_+^2 = \sum_m f_m^2 a_m^{-1}$

H_- : { fermeture de $\mathcal{D}(\Lambda)$ pour la norme $\|\cdot\|_-$ } : $\|f\|_-^2 = \sum_m f_m^2 a_m$

avec si $f \in L^2(\Lambda)$; $f_m = (\phi_m, f)_0$.

Soit A l'opérateur de H_+ dans H_- défini par : $Af = \sum_m f_m a_m^{-2} \phi_m \quad \forall f \in \mathcal{D}(\,)$.

A est une isométrie positive et comme les espaces H_+ et H_- sont évidemment duaux, on a:

$\langle Af, f \rangle_{-,+} = \|f\|_+^2$.

T_t ; $t \in R$ est l'opérateur défini par :

$T_t f = \sum_m f_m e^{-\frac{t}{2a_m^2}} \phi_m$.

b) – L'équation de Langevin usuelle $dy_t = -\frac{a}{2} y_t \, dt + dm_t$; $y_0 = c$

associée au nombre réel positif a et à la martingale réelle de carré intégrable m, a pour solution :

$$y_t = c e^{-\frac{a}{2}t} + \int_0^t e^{-\frac{a}{2}(t-s)} \, dm_s .$$

Lorsque m est un élément de la suite de martingales m^k précédente et si c'est l'élément correspondant d'une suite de réels $\{c_k ; k \in N\}$, la solution de l'équation de Langevin sera dans ce cas notée y_t^k.

Pour tout nombre naturel N nous pouvons définir le processus Y_t^N par :

$$Y_t^N = \sum_{k=0}^{N} \phi_k y_t^k$$

Posons $Y_0^N = \zeta^N = \sum_{k=0}^{N} c_k \phi_k$.

Y_t^N est un processus d'Ornstein Uhlenbeck vectoriel de dimension N+1, à valeurs dans $\mathcal{D}'(\Lambda)$.

3) a) Soit H un espace de Hilbert de distributions pour lequel la suite $\{\phi_k ; k \geq 0\}$ est dense. Dans ces conditions, si $[m]_s$ désigne le processus croissant de la martingale m, l'inégalité de Doob conduit à :

$$\mathbb{E} \sup_{0 \leq t \leq T} \|Y_t^N - T_t \zeta^N\|_H^2 \leq 4 \sum_{k=0}^{N} \mathbb{E} \int_0^T e^{-\frac{T-s}{2 a_k}} \, d[m^k]_s \qquad (**)$$

Si l'on choisit judicieusement l'espace H, la connaissance du comportement de la suite a_k, celle de $\{M_t ; t \geq 0\}$ et l'inégalité (**) nous permettront de résoudre l'équation de Langevin (*).

b) Donnons-nous un opérateur différentiel A uniformément fortement elliptique, de degré 2p, formellement auto-adjoint, à coefficients bornés et C^∞.

– Soit Λ l'ouvert précédent. Puisque Λ est borné, le spectre de A est discret. Soit alors $\{\phi_k, \lambda_k, k \in N\}$ le système spectral de A dans Λ.

Posons pour tout $k \in N$ $a_k = \lambda^{-\frac{1}{2}}$; en supposant les fonctions propres de A orthonormées dans $L^2(\Lambda)$, les espaces H_+ et H_- construits avec ce système sont respectivement égaux à $H_0^p(\Lambda)$ et $H^{-p}(\Lambda)$; où $H_0^p(\Lambda)$ désigne l'espace de Sobolev usuel. Voir BENASSI [1].

Donnons-nous la suite de martingales $(m_t^k ; t \geq 0)_{k \in N}$ définie précédemment.

Nous ferons l'hypothèse suivante :

Il existe une suite de nombres réels $\{b_k ; k \geq 0\}$ et un nombre b vérifiant $0 < b_k \leq b$ $\forall k$ et tels que les mesures $d\,[m^k]_s$ vérifient :

$$d\,[m^k]_s \leq b_k^2\,ds \quad \forall k. \tag{***}$$

- Pour tout réel β définissons sur $L^2(\Lambda)$ la norme $\|\cdot\|_\beta$ par

$$\|f\|_\beta^2 = \sum_m f_m^2 \; \lambda_m^{\beta/p}$$

Définissons alors une échelle hilbertienne associée à A par :

$H^\beta(A,\Lambda)$ fermeture de $\mathcal{D}(\Lambda)$ pour la norme $\|\cdot\|_\beta$.

Dans BENASSI [2] on a montré que les échelles hilbertiennes $H^\beta(A,\Lambda)$ et $H_0(\Lambda)$ $\beta \geq 0$ sont algébriquement et topologiquement équivalentes.

PROPOSITION 1

Soient A et M l'opérateur et la martingale définis et vérifiant les hypothèses ci-dessus. Si le degré de l'opérateur A vérifie $2p > d$; alors $\forall \beta$, $0 \leq \beta < p - \frac{d}{2}$

a) Il existe un unique processus Y_t solution de l'équation de Langevin (*).

b) Y_t est à trajectoires continues (resp. c.à.d.-l.a.g.) P.p.s à valeurs dans $H^\beta(\Lambda)$ si M est à trajectoires continues (resp. c.à.d. - l.a.g.) P.p.s à valeurs dans $\mathcal{D}'(\Lambda)$.

DEMONSTRATION

D'après l'hypothèse (***) le second membre de (**) est N majoré par

$$4\,b^2 \sum_{k=0}^{N} \lambda_k^{((\beta/p)-1)}$$

mais lorsque $2p > d$ la suite ci-dessus converge [2] ; la série $Y_t = \sum_k y_t^k \,\Phi_k$ converge alors P.p.s dans $H^\beta(\Lambda)$ uniformément sur tout compact de R_+. Les résultats de régularité sont alors immédiats.

4) MESURES STATIONNAIRES DU POUG (A,W).

Dans toute la suite de ce travail, nous supposerons que la martingale M_t est le processus de Siegel standart, c.à.d. que $b_k = 1$ $\forall k \in N$, et que $(m_t^k ; t \geq 0)_{k \in N}$ est une suite de browniens indépendants.

c) Soit $\{X(t) \; ; \; t \in \Lambda\}$ le processus gaussien markovien d'ordre p associé à A sur (BENASSI [2]) par rapport au tribu $\Sigma(0)$. Les tribus $\Sigma(0)$ sont définies par $\Sigma(0) = \sigma(X(s) \; ; \; s \geq 0)$ si 0 est ouvert, $\Sigma(F) = \bigcap_{0 \supset F} \Sigma(0)$ si F est fermé. Posons $\Sigma_\infty = \Sigma(\partial 0)$ ($\partial 0$ désigne la frontière de l'ensemble borélien 0). Alors on a la décomposition $X = X_0 + X_\infty$ en une composante régulière X_0 et une composante singulière X_∞, Σ_∞ mesurable. Si μ_0 (resp. μ_∞) est la probabilité sur $H^\beta(\Lambda)$ engendrée par X_0 (resp. X_∞) on a la factorisation $\tilde{P} = \mu_0 \times \mu_\infty$ où $\tilde{P}$ est la probabilité sur $H^\beta(\Lambda)$ générée par X. Voir BENASSI [2].

Si ν est une probabilité sur $H^\beta(\Lambda)$, pour tout ouvert B, ν_B désignera la restriction de ν à $\Sigma(B)$. Définissons alors l'ensemble $\mathcal{M}_A$ des probabilités markoviennes d'ordre p de même spécifications locales que μ_0 par : $\nu \in \mathcal{M}_A$ si

(i) $\nu_B \leq \mu_{0,B}$ pour tout ouvert B ; $\bar{B} \subset \Lambda$

(ii) $\mathbb{E}^\nu[\Phi \mid \Sigma(\bar{B}^c)] = \mathbb{E}^{\mu_0}[\Phi \mid (\partial B)]$ P.p.s

où B^c désigne le complémentaire de B dans Λ, et $\bar{B}$ la fermeture de B. Φ est un élément de $L^1(H^\beta(\Lambda), \Sigma(B), \tilde{P})$.

Définissons l'espace des éléments A harmoniques $\mathcal{H}$ par :

$$\mathcal{H} = \{ H \in H^\beta(\Lambda) \mid AH = 0 \}$$

Soit $\mathcal{T}$ le groupe des translations sur $H^\beta(\Lambda)$ par les éléments de l'espace de noyaux H(A) de μ_0 cf [2]. Il est facile de vérifier que $\forall H \in \mathcal{H}$, μ_0^H est $\mathcal{T}$ ergodique. μ_0^H est la translatée de μ_0 par H. $\mathcal{M}(H)$ désigne l'ensemble des probabilités sur $\mathcal{H}$.

PROPOSITION 2

Si $\mathcal{M}_A$ est l'ensemble précédent de probabilités sur $H^\beta(\Lambda)$, alors :

$$\mathcal{M}_A = \{ \nu_m \mid \nu_m = \int_{\mathcal{H}} \mu_0^A \, m(dH) \; ; \; m \in \mathcal{M}(\mathcal{H}) \} .$$

DEMONSTRATION

$\forall H \in \mathcal{H}$ il est clair que $\mu_0^H \in \mathcal{M}_A$ car μ_0^H est $\mathcal{T}$ quasi invariante. Puisque μ_0^H est $\mathcal{T}$ ergodique, on peut appliquer le théorème 4 .17 de FÖLMER [3] de représentation intégrale par les éléments extrémaux. ■

b) Soit $\{Y_t \; ; \; t \geq 0\}$ le POUG (A,W).

LEMME 1

μ_0 est la seule mesure stationnaire régulière du POUG (A,W).

DEMONSTRATION

Montrons que μ_0 est stationnaire. Pour cela calculons la fonctionnelle caractéristique de $Y_t(\psi)$, $\psi \in H(A)$; soit :

$$\mathbb{E}^{\tilde{P}} \exp i\, Y_t(\psi) = \exp\left[-\frac{1}{2}\sum_k (\psi_k^2/\lambda_k \cdot e^{-\lambda_k t}) - \frac{1}{2}\sum_k \psi_k^2 \frac{1-e^{-\lambda_k t}}{\lambda_k}\right]$$

$$\exp\left\{-\frac{1}{2}\sum_k \frac{\psi_k^2}{\lambda_k}\right\} = \mathbb{E}^{\mu_0}(\exp i\, x(\psi)),$$

avec $\psi_k = (\phi_k, \psi)$. Comme Y_t est unique (prop. 1) et que μ_o est uniquement détermineé par sa fonction réelle caractéristique, le lemme est démontre.

PROPOSITION 3

L'ensemble des mesures stationnaires du POUG (A,W) est le simplexe des probabilités $\mathcal{M}_A$.

DEMONSTRATION

Soit $\mathcal{S}_A$ l'ensemble des mesures stationnaires du POUG (A,W).
On a alors $\mathcal{M}_A \subset \mathcal{S}_A$. En effet, puisque $H \in \mathcal{H}$ entraine TH = H on a donc $\mu_o^H \in \mathcal{M}_A$. Par linéarité on a bien $\mathcal{S}_A \in \mathcal{M}_A$. Montrons l'inclusion en sens inverse. Soit $\nu \in \mathcal{S}_A$ et $\{Y_t ; t \geq 0\}$ la solution de (*) lorsque Y_o est distribué selon ν et posons $\tilde{Y}_t = Y_t^\nu - Y_t^{\mu_o}$. Dans ce cas il est clair que $\tilde{Y}_t$ vérifie l'équation différentielle $d\tilde{Y}_t = A\tilde{Y}_t dt$.
Mais comme par construction Y_t est forcément stationnaire on en déduit que $A\tilde{Y}_t = 0 \ \forall t > 0$. Cela prouve que $\tilde{Y}_t \in \Sigma_\infty$.
Dans ces conditions, si Z est l'élément de $H^{\beta}(\Lambda)$ distribué selon ν on a la décomposition $Z = Z_o + Z_\infty$, $Z_\infty \in \Sigma_\infty$ et le théorème 4.1 de [2] permet d'écrire

$$\mathbb{E}\left(\frac{d\nu_B}{d\mu_{o,B}}\Big|\Sigma_\infty\right) = \mathbb{E}\left(\frac{d\mu_o^{Z_\infty}}{d\nu_{o,}}\Big|\Sigma_\infty\right) = 1 \ \tilde{P}.p.s$$ ce qui permet de conclure.

La proposition 2 généralise au cas non stationnaire le résultat de HOLLEY et STROOCK de [4], et la proposition [3] généralise à tous les opérateurs elliptiques de degré suffisant le résultat obtenu dans [5] pour le laplacien.

REFERENCES

[1] Benassi, A. (1980): Théorème de Traces stochastiques. Colloques internationnaux du C.N.R.S. No. 307, St Flour 22 - 29 Juin 1980 (15-25)

[2] Benassi, A. (1982): Théorème de traces stochastiques et Fonctionnelles multiplicatives pour des champs gaussien markoviens d'ordre p. Z.f.W. 59, (333-354)

[3] Föllmer, H.: Phase transition and Martin Boundary. Seminaire de Prob. IX, Lect. Notes in Math. No.465, Springer Verlag.

[4] Holley, R. A. and Stroock D. W. (198o): The D.L.R. Conditions for Translation Invariant Gaussian Measures on $\mathcal{S}'(R^d)$. Z.f.W. 53 (293-304).

[5] Holley, R. A. and Stroock D. W. (1978): Generalized Ornstein--Uhlenbeck Processes and Infinite Particle Branching Brownian Molion. Pub. R.I.M.S. Kyioto Univ. 14, 741-788.

Université Paris VI.,
4 Place Jussieu
75230 PARIS Cedex 05
France

APPLICATION OF THE STATISTICAL DECISION THEORY TO SYSTEM IDENTIFICATION

Georg Bretthauer

Dresden

Key words: Statistical decision theory, identification, model selection, multivariable systems

ABSTRACT

The paper is concerned with the application of the statistical decision theory to the model selection of linear multivariable systems. The decision algorithm represented here determines the best model from a finite number of linear models prior known in the sense of the minimal average risk. The Bayes solutions for stationary Gaussian processes and a general as well as a simple loss matrix are derived.

1. INTRODUCTION

According to the definition given by ZADEH [1], "identification is the determination, on the basis of input and output, of a system within a specified class of systems, to which the system under test is equivalent". As a result of the limitation of the measuring time, the random disturbances of the input and output signals and the incompleteness of the measuring devices no exact identification is possible. Therefore a compromise is necessary between the accuracy required and the expense of an identification experiment. An useful approach to attack this complex is the statistical decision theory. Some reasons for it are:

- statistical decision theory is a generally available and powerful approach for solving such problems
- in contrary to the methods usually used it allows to take into consideration additional information in the evaluation by means of the loss functions and

- it is a unified method applicable as well for identification as for the design of control systems.

2. IDENTIFICATION BY MEANS OF THE STATISTICAL DECISION THEORY

Up to now for single input-single output systems there have been known several papers concerning with the application of the statistical decision theory to system identification.
We shall classify them in two groups:
- application to parameter estimation [2]-[7]
- application to model selection [8]-[10]

The methods belonging to the first group can be subdivided into two subgroups. In the first subgroup the optimal estimation $\underline{\tilde{b}}$ of the unknown parameter vector $\underline{b}$ is determined without an adjustable model [2]-[6] while in the second subgroup $\underline{\tilde{b}}$ is calculated with help of an adjustable model [6]-[7].
The application of the statistical decision theory to the selection of the best model within a specified class of models is shown in [8], [9] for open loop systems and in [10] for closed loop systems. In both cases the best model is evaluated by minimizing the average risk and finding the Bayes solution. Two advantages of this decision algorithm are:
- the optimal model of the models given will be found within a relatively short measuring time and
- the expense of computation is relatively small.

Extensions of it to the model selection of multivariable systems operating in open or closed loop are not known in identification literature. Thus a derivation of an appropriate decision algorithm for the model selection of multivariable systems would be of great importance for some identification problems and will be given now.

3. DECISION ALGORITHM FOR THE MODEL SELECTION OF MULTIVARIABLE OPEN LOOP SYSTEMS

We consider the following identification problem (see Fig. 1).
Given a linear multivariable system with the transfer matrix $G = (G_{lk})$ $(l,k = 1,\dots,m)$ and a finite number M of corresponding models with the transfer matrices $G_i = (G_{ilk})$ $(i = 1,\dots,M; l,k=1,\dots,m)$. The system and the models are subjected to the random input signal vector $\underline{u}(t)$. The output signal vectors are $\underline{y}(t)$ and $\underline{y}_i^*(t)$ by which $\underline{y}(t)$ is contaminated with the random noise vector $\underline{v}(t)$.

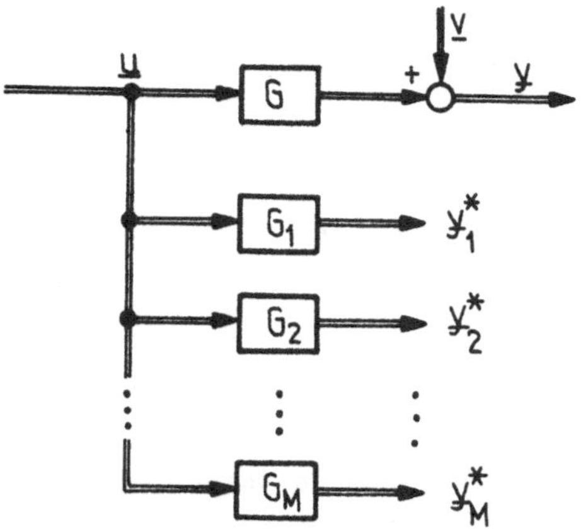

Fig. 1: Scheme of the identification problem

The problem is to find that model with the transfer matrix $G_i (i=1,...,M)$ approximating in the observation time intervall $[0,T]$ the system with the transfer matrix G in an optimal sense on the basis of the observation vector $\underline{y}$ and the prior information on the models and the random noise vector $\underline{v}(t)$.
The solution of this problem is important for the identification of multivariable systems. In these systems the main difficulty for a successful identification is to determine the structure parameters of the system to be identified. The expense of computation for it is very large. Therefore a lot of efforts is made to reduce this expense. The algorithm described now is one way for it.
To find a solution of this identification problem it will be transferred in a statistical decision problem first. Then the average risk will be evaluated and the Bayes solution will be derived.

For it we make use of the following notations:

$\underline{y}$ — n-dim. observation vector of $\underline{y}(t)$ with the elements $y_{lj} = y_l(t_j)$, $l = 1,...,m$; $j = 1,...,N$ arranged in the following way

$y_{11} \cdots y_{1N}\; y_{21} \cdots y_{2N} \cdots y_{m1} \cdots y_{mN}$, $n = mN$

$\underline{y}_i^*$ — n-dim. observation vector of $\underline{y}_i^*(t)$ with the elements $y_{ilj}^* = y_{il}^*(t_j)$, $i=1,...,M$; $l=1,...,m$; $j=1,...,N$ arranged in the following way

$y_{i11}^* \cdots y_{i1N}^*\; y_{i21}^* \cdots y_{i2N}^* \cdots y_{im1}^* \cdots y_{imN}^*$

d_i — decision for the i-th model

P_i — prior probability of the occurence of G_i

L_{ij}	element of the loss matrix L
$P(d_i/\underline{y})$	conditional probability that the decision d_i will be made for a given $\underline{y}$
$f(\underline{y}/\underline{y}_i^*)$	conditional probability density function of $\underline{y}$ for a given $\underline{y}_i^*$
$f(\underline{y}_i^*)$	prior probability density function of $\underline{y}_i^*$
Y	observation space
Y_i^*	signal space for $\underline{y}_i^*(t)$

As in the case for single input-single output systems (see [8],[10]) we get for the average risk

$$R_m = \sum_{i=1}^{M} \sum_{j=1}^{M} L_{ij} P_j \int_Y P(d_i/\underline{y}) d\underline{y} \int_{Y_j^*} f(\underline{y}/\underline{y}_j^*) f(\underline{y}_j^*) d\underline{y}_j^* \ . \tag{1}$$

A decision d_i minimizing Eq.(1) for a given $f(\underline{y}_i^*)$ is called Bayes solution.

Now we assume that

- the prior probability density functions $f(\underline{y}_i^*)$ $(i=1,\ldots,M)$ are known
- only one decision d_i and
- in any case a decision is to be found.

Then the following Bayes solution is obtained

$$P(d_i/\underline{y}) = 1 \tag{2}$$

for

$$w_i(\underline{y}) \quad = \quad \min_k w_k(\underline{y}) \tag{3}$$

with

$$w_k(\underline{y}) = \sum_{j=1}^{M} L_{kj} P_j \int_{Y_j^*} f(\underline{y}/\underline{y}_j^*) f(\underline{y}_j^*) d\underline{y}_j^* . \tag{4}$$

Eq.(4) can be simplified by limitations regarding the elements L_{ij} of the loss matrix L and the random processes used. This will be shown now.

<u>Bayes solution for a simple loss matrix</u>

In this case we assume that

$$L_{ii} = 0 \quad \text{and} \quad L_{ij} = 1. \tag{5}$$

Then from Eq. (1) it follows for the Bayes solution

$$P(d_i/\underline{y}) = 1 \tag{6}$$

for

$$\hat{w}_i(\underline{y}) \quad = \quad \max_j \hat{w}_j(\underline{y}) \tag{7}$$

with

$$\hat{w}_j(\underline{y}) = P_j \int_{Y_j^*} f(\underline{y}/\underline{y}_j^*) f(\underline{y}_j^*) d\underline{y}_j^* . \tag{8}$$

For evaluation of the Bayes solutions (Eqs.(4),(8)) it is

necessary to know $f(\underline{y}/\underline{y}_i^*)$ and $f(\underline{y}_i^*)$. But in general these probability density functions are unknown. Therefore the random processes used are now limited to the class of Gaussian processes. Then simpler Bayes solutions will be obtained which are of a great practical interest for identification

Bayes solution for Gaussian processes

Now we assume that

- $\underline{y}_i^*(t)$ and $\underline{v}(t)$ are stationary ergodic and normal distributed processes
- the expected values of $\underline{y}_i^*(t)$ and $\underline{v}(t)$ are zero and
- $\underline{y}_i^*(t)$ and $\underline{v}(t)$ are uncorrelated.

Then for $f(\underline{y}_i^*)$ holds

$$f(\underline{y}_i^*) = (2\pi)^{-\frac{n}{2}} |R_{\underline{y}_i^*\underline{y}_i^*}|^{-\frac{1}{2}} \exp(-\tfrac{1}{2}\, \underline{y}_i^{*T} (R_{\underline{y}_i^*\underline{y}_i^*})^{-1}\, \underline{y}_i^*) \tag{9}$$

with

$$R_{y_i^* y_i^*} = \begin{pmatrix} R_{y_{i1}^* y_{i1}^*} & R_{y_{i1}^* y_{i2}^*} & \cdots & R_{y_{i1}^* y_{im}^*} \\ R_{y_{i2}^* y_{i1}^*} & R_{y_{i2}^* y_{i2}^*} & \cdots & R_{y_{i2}^* y_{im}^*} \\ \vdots & \vdots & & \vdots \\ R_{y_{im}^* y_{i1}^*} & R_{y_{im}^* y_{i2}^*} & \cdots & R_{y_{im}^* y_{im}^*} \end{pmatrix} \tag{10}$$

and

$R_{\underline{y}_{il}^* \underline{y}_{il}^*}$ - (N,N) autocorrelation matrix of the l-th output signal of the i-th model

$R_{\underline{y}_{il}^* \underline{y}_{ik}^*}$ - (N,N) crosscorrelation matrix of the l-th and k-th output signals of the i-th model

In the same way for $f(\underline{v})$ holds

$$f(\underline{v}) = (2\pi)^{-\frac{n}{2}} |R_{\underline{v}\underline{v}}|^{-\frac{1}{2}} \exp(-\tfrac{1}{2}\underline{v}^T (R_{\underline{v}\underline{v}})^{-1} \underline{v}) \tag{11}$$

with

$$R_{\underline{v}\underline{v}} = \begin{pmatrix} R_{v_1 v_1} & R_{v_1 v_2} & \cdots & R_{v_1 v_m} \\ R_{v_2 v_1} & R_{v_2 v_2} & \cdots & R_{v_2 v_m} \\ \vdots & \vdots & & \vdots \\ R_{v_m v_1} & R_{v_m v_2} & \cdots & R_{v_m v_m} \end{pmatrix} \tag{12}$$

and

$R_{\underline{v}_l\underline{v}_l}$ – (N,N) autocorrelation matrix of the l-th component of $\underline{v}(t)$

$R_{\underline{v}_l\underline{v}_k}$ – (N,N) crosscorrelation matrix of the l-th and k-th components of $\underline{v}(t)$

$\underline{v}$ n-dim. vector of $\underline{v}(t)$ with the elements $v_{lj} = v_l(t_j)$, $l = 1,\dots,m$; $j = 1,\dots,N$ arranged in the following way $v_{11}\dots v_{1N}v_{21}\dots v_{2N}\dots v_{m1}\dots v_{mN}$, $n = mN$.

Applying the Eqs.(9) and (11) to the integral

$$I = \int_{Y_j^*} f(\underline{y}/\underline{y}_j^*)f(\underline{y}_j^*)d\underline{y}_j^* \tag{13}$$

yields

$$I = (2\pi)^{-\frac{n}{2}} \left|R_{\underline{y}_j^*\underline{y}_j^*}+R_{\underline{vv}}\right|^{-\frac{1}{2}}\exp(-\tfrac{1}{2}\,\underline{y}^T(R_{\underline{y}_j^*\underline{y}_j^*}+R_{\underline{vv}})^{-1}\underline{y}). \tag{14}$$

By substitution of Eq.(14) in the Eqs.(4) and (8) we finally get the Bayes solutions desired for Gaussian processes

Bayes solution (general loss matrix):

$$P(d_i/\underline{y}) = 1$$

for

$$w_i(\underline{y}) = \min_k w_k(\underline{y})$$

with

$$w_k(\underline{y}) = \sum_{j=1}^{M} L_{kj}P_j(2\pi)^{-\frac{n}{2}}\left|R_{\underline{y}_j^*\underline{y}_j^*}+R_{\underline{vv}}\right|^{-\frac{1}{2}}\exp(-\tfrac{1}{2}\,\underline{y}^T(R_{\underline{y}_j^*\underline{y}_j^*}+R_{\underline{vv}})^{-1}\underline{y}). \tag{15}$$

Bayes solution (simple loss matrix):

$$P(d_i/\underline{y}) = 1$$

for

$$\hat{w}_i(\underline{y}) = \max_j \hat{w}_j(\underline{y})$$

with

$$\hat{w}_j(\underline{y})=P_j(2\pi)^{-\frac{n}{2}}\left|R_{\underline{y}_j^*\underline{y}_j^*}+R_{\underline{vv}}\right|^{-\frac{1}{2}}\exp(-\tfrac{1}{2}\underline{y}^T(R_{\underline{y}_j^*\underline{y}_j^*}+R_{\underline{vv}})^{-1}\underline{y}) \tag{16}$$

This algorithm is an extension of an algorithm known for single input - single output systems. It is applicable for such identification problems as the determination of an actual model from a finite number of models prior known as it is the case if the parameters of the systems will be changed slowly, as the model simplification following a theretical process analysis or if as a result of an identification experiment two or more models were obtained which have only small differences in the optimization criterion and additional information is to be taken into consideration.

4. REFERENCES

[1] Zadeh, L.A.: From circuit theory to system theory
Proc. IRE, 50 (1962), 856-865

[2] Маслов, Е.П.: Применение теории статистических решений к задачам параметров объекта
Автоматика и Телемеханика, т. 24, № IO, I963 г., 1338-1350

[3] Маслов, Е.П.: Оценка параметров марковских объектов
Автоматика и Телемеханика, т. 25, № I, I964 г., 73-82

[4] Rutkowski, D.A.: Optimale Identifizierung des Leistungsverteilungskoeffizienten des Turbogenerators und ihre Anwendung an einer adaptiven Steuerung
msr 13 (1970), H. 9, 342-345

[5] Rutkowski, D.A.: Optimale Identifizierung an linearen Steuerungsobjekten
msr 14 (1971), H. 12, 468-471; msr 15 (1972), H. 5, 185-188

[6] Seidler, J.: Optimierung informationsübertragender Systeme, Bd2
VEB Verlag Technik, Berlin 1969

[7] Rutkowski, D.A.: Sonderfall der Identifizierung mit Hilfe eines einstellbaren Modells
msr 15 (1972), H. 12, 444-445

[8] Sawaragi, Y. et. al.: Statistical Decision Theory in Adaptive Control Systems
Academic Press, New York 1967

[9] Bretthauer, G.: Anwendung der statistischen Entscheidungstheorie für die Identifikation gestörter Regelstrecken
msr 22 (1979), H. 4, 202-207

[10] Bretthauer, G.: Anwendung der statistischen Entscheidungstheorie zur Systemidentifikation am geschlossenen Regelkreis
msr 23 (1980), H. 8, 439-443

Academy of Sciences of the GDR
Central Institute of Cybernetics
and Information Processes
1086 Berlin
Kurstraße 33
PSF 1298

COMPUTING FIXED POINTS FOR FUZZY MAPPINGS

Dan Butnariu

Iaşi

Key words: Fuzzy set, Fuzzy mapping, fixed point, triangulation, piecewise linear function, Eaves' fixed point method, non-cooperative n-persons fuzzy game.

ABSTRACT

In this paper we present two methods for computing fixed points for fuzzy mappings and we show that these methods can be used for solving non-cooperative n-persons fuzzy games.

1. INTRODUCTION

The problem of computing fixed points for fuzzy mappings is strongly related with that of computing "rational solutions" for non-cooperative n-persons fuzzy games seen as mathematical models for conflictual situations in which the "players" use fuzzy information in making their decisions (see Butnariu (1979)). Our aim is to present some ways for solving these problems.

A fuzzy mapping over the set X is a function R from X to $L(X)=[0,1]^X$, the class of the fuzzy subsets of X (see Zadeh (1965)). An element x of X is called fixed point of R iff $R_x(x) \geq R_x(y)$ for any y in X. D. Butnariu (1982) proves some existence criteria for fixed points of a fuzzy mapping and gives an algorithm for computing such fixed points. This algorithm can be viewed as a special case of the method B presented in this work. In the sequels we assume that the reader is familiar with the terms, notations and with the main results of D. Butnariu (1982) which will be used without special mentions. Some new terms and notations which are also needed are presented now.

Let $T=[x^0,x^1,\dots,x^n]$ be an n-simplex in E^m and let f be a linear function from the standard n-simplex $S^n=[v^0,\dots,v^n]$ to T such that $f(v^j)=x^j$, $(0 \leq j \leq n)$. This function transfers Kuhn's triangulation of order h of S^n into the triangulation $K_T(h)$ of T.

A subset C of E^m is called polyedral iff it is convex, compact and there exists a triangulation $\underline{G}$ of C such that $C=\bigcup\{T;\ T\in\underline{G}\}$. If C is a polyedral subset of E^m, then $\underline{G}_h=\bigcup\{K_T(h);\ T\in\underline{G}\}$ is a triangulation of C and mesh $\underline{G}_h \longrightarrow 0$ for $h\longrightarrow\infty$. This means that any polyedral set can be triangulated by triangulations with simplices as small as needed.

We remind that W. Zangwill (1969) describes the algorithms for optimization to be sequences $(A_k)_{k\in N}$ of point-to-set mappings from a set C to itself such that $A_o(x)=A_o\neq\emptyset$, $(x\in C)$, and for each x^o in A_o there exists a sequence $(x^k)_{k\in N}$ with $x^{k+1}\in A_k(x^k)$, $(k\in N)$ called generated sequence of $(A_k)_{k\in N}$ starting from x^o.

Let us consider the problem of optimization (P): $\max\{g(x);\ x\in C\}$, where g is a real function defined on C. The algorithm $(A_k)_{k\in N}$ solves (P) iff each generated sequence of $(A_k)_{k\in N}$ converges to a solution of the problem (P). We say that $(A_k)_{k\in N}$ solves (P) in finitely many steps iff there exists a positive integer h such that for any generated seqvence $(x^k)_{k\in N}$ of $(A_k)_{k\in N}$ is true that $x^k=x^h$ for $k\geq h$.

An important instrument for computing fixed points of fuzzy mappings is Eaves' first algorithm (cf. Eaves (1971)). In our explanation Eaves' first algorithm is considered in the form presented by Todd (1976).

2. THE METHOD (A)

In this section we assume that C denotes a polyedral subset of E^m having dim aff $(C)=n<m$ and R denotes a fuzzy mapping over C that satisfies the next conditions: (I) R is convex and closed and (II) If $x\in C$, then there exists an algorithm for optimization $(A_{x,k})_{k\in N}$ such that it solves in finitely many steps the optimization problem

(1) $$P(x):\ \max\{R_x(y);\ y\in C\}.$$

The condition (I) ensures that R has fixed points (cf. Butnariu (1982)) and the condition (II) ensures that each P(x) can be solved in finitely many computational steps. Both conditions are accomplished when R_x is a quadratic form for any x in C (see Zangwill (1969)).

The method (A) for computing fixed points for R is now:

STEP 0: Set k=1 and let $\underline{G}_k$ be as described in Section 1.

STEP 1: (a) For each x in $\underline{G}_k^o$ use the algorithm $(A_{x,k})_{k\in N}$ to compute in finitely many steps a solution of P(x). Denote $f_k(x)$ this solution (b) Define the piecewise linear function f_k w. r. t. $\underline{G}_k$ whose values for $x\in\underline{G}_k^o$ are the already computed $f_k(x)$.

STEP 2: Use Eaves' first algorithm for the computation of a fixed point x^k for the piecewise linear function f_k. Set $k:=k+1$ and $\underline{G}_k:=\underline{G}_{k+1}$. Go to the step 1(a).

The convergence of this procedure is guaranted by the next

THEOREM 1: If (I) and (II) hold, then the method (A) generates a sequence $(x^k)_{k\in N} \subseteq C$ such that $(x^k)_{k\in N}$ has cluster points in C and each cluster point of it is a fixed point of R.

Proof: It is clear that the sequence $(x^k)_{k\in N}$ described in the procedure exists because the f_k's constructed there are continuous functions and Eaves' first algorithm converges in finitely many steps. Since C is compact, it follows that $(x^k)_{k\in N}$ has a cluster point x in C. We assume (without loss of generality) that $\lim_k x^k = x$. The point-to-set mapping over C defined by $z \longrightarrow R_z^{\wedge} = \{y;\ R_z(y) = \max_w R_z(w)\}$ is compact and closed. Moreover, each $R_z^{\wedge}$ is convex (cf. (I)). It suffices to prove that $x \in R_x^{\wedge}$. To this end, let $S(k) = [u^{o,k}, \ldots, u^{n,k}]$ be the simplex of $\underline{G}_k$ which contains the point x^k. We have that $x^k = f_k(x^k) = \sum_{i=o}^{n} r_{i,k}.f^{i,k}$, where $f^{i,k} = f_k(u^{i,k})$, $r_{i,k} \geq 0$, $(o \leq i \leq n)$ and $\sum_{i=o}^{n} r_{i,k} = 1$, $(k \in N)$. Let f^i be a cluster point of the sequence $(f^{i,k})_k$ and let us denote by r^i a cluster point of $(r_{i,k})_k$ in $[0,1]$. W. l. o. g. we may assume that $\lim_k f^{i,k} = f^i$ and $\lim_k r^{i,k} = r_i$. Then $x = \sum_{i=o}^{n} r_i.f^i$, $r_i \geq 0$ for $0 \leq i \leq n$ and $\sum_{i=o}^{n} r_i = 1$. Since we have

$$\| u^{i,k} - x \| \leq \| u^{i,k} - x^k \| + \| x^k - x \| \leq \text{mesh } \underline{G}_k + \| x^k - x \| \longrightarrow 0,$$

it follows that $\lim_k u^{i,k} = x$ for $0 \leq i \leq n$. By the definition of f_k we have that $f^{i,k} \in R^{\wedge}_{u(i,k)}$ for $u(i,k) = u^{i,k}$. Hence $f^i = \lim_k f^{i,k} \in R_x^{\wedge}$ because $R^{\wedge}$ is a closed mapping. Since $R_x^{\wedge}$ is also convex, it results that $x = \sum_{i=o}^{n} r_i.f^i \in R_x^{\wedge}$ and the proof is complete.

The method (A) is an extension of Eaves' method for computing fixed points for point-to set mappings (Eaves (1971)). An essential restriction for the applicability of the method (A) is the condition (II) ensuring the existence of the values $f_k(x)$ for $x \in \underline{G}_k^o$ in each stage k. We ask when the condition (II) can be weakened. In the next section we show that the affirmative answer can be given when R is restricted to be concave.

3. THE METHOD (B)

In this section we consider C to be the set described in Section 2 and we assume that R is a fuzzy mapping over C which satisfies the next conditions: (III) R is closed; (IV) R is concave, i. e. for any $r \in [0,1]$ and for any u^1, u^2, v^1, v^2 in C we have that

(2) $$R(r.u^1 + (1-r).u^2; r.v^1 + (1-r).v^2) \geq r.R(u^1, v^1) + (1-r).R(u^2, v^2);$$

(V) For any x in C an algorithm for optimization $(A_{x,k})_{k\in N}$ on $C \times [0, \infty)$ exists such that each generated sequence (x^k, e_k), $(k\in N)$ of it has the next properties: (i) $\lim_k x^k$ exists and is a solution of P(x) and (ii) $e_k \searrow 0$ for $k \longrightarrow \infty$ and it is true that $\| x^k - \lim_h x^h \| \leq e_k$, $(k \in N)$.

The conditions (III) and (IV) ensure the existence of a fixed point of R in C (cf. Butnariu (1982)). The condition (V) shows that each problem P(x), $(x \in C)$ has a solution x^* which can be approximated by a known algorithm for optimization and in any stage of the computation of x^* by approximations the error bound e_k is known.

The method (B) is now:

STEP 0: Set k=1 and $\underline{G}_k$ the triangulation described in section 1.

STEP 1: (a) Set h=1 and choose $(y_{k,1}(v), e_{k,1}(v))$ in $A_{v,o}$ for each vertex v in $\underline{G}_k^o$. (b) For each v in $\underline{G}_k^o$ choose $(y_{k,h+1}(v), e_{k,h+1}(v))$ in the set $A_{v,h}(y_{k,h}(v), e_{k,h}(v))$. (c) If $e_{k,h+1}(v)) \leq \text{mesh } \underline{G}_k$ for all $v \in \underline{G}_k^o$, then denote $q_k(v) = y_{k,h+1}(v)$, $(v \in \underline{G}_k^o)$ and go to Step 2; if not, then set h:=h+1 and go to Step 1(b).

STEP 2: (a) Define the piecewise linear function q_k with respect to $\underline{G}_k$ whose values are $q_k(v)$ for $v \in \underline{G}_k^o$. (b) Use Eaves' first algorithm to compute a fixed point of q_k denoted by x^k. (c) Set k:=k+1 and go to Step 1(a).

The convergence of the method (B) is guaranted by the next

THEOREM 2: If the conditions (III), (IV) and (V) hold, then the method (B) converges in the following sense: (a) For any $k \in N$ the vector x^k described in the procedure exists; (b) The sequence $(x^k)_{k \in N}$ has a cluster point in C and (c) Each cluster point of $(x^k)_{k \in N}$ is a fixed point of the fuzzy mapping R.

Proof: It can be easily seen that the points x^k exist and the sequence $(x^k)_{k \in N}$ has a cluster point in the compact set C. If $v \in \underline{G}_k^o$, then $(y_{k,h}(v))_h$ converges to a solution $y_k(v)$ of P(v). Let f_k be the piecewise linear function w. r. t. $\underline{G}_k$ defined such that $f_k(v) = y_k(v)$ for any v in $\underline{G}_k^o$. Then, for each v in $\underline{G}_k^o$ we have that

(3) $\qquad \|q_k(v) - f_k(v)\| = \|y_{k,h'}(v) - y_k(v)\| \leq e_{k,h'}(v) \leq \text{mesh } \underline{G}_k$,

where h' denotes the index h attained when the procedure "Step 1" stops in the stage k. Let x be any point in C and let $(w^o, \ldots, w^j)$ be the only one open simplex in $\underline{G}_k^+$ which contains x. Let $r_o, \ldots, r_j$ be the barycentric coordinates of x in this simplex. Hence, we have

(4) $\quad \| q_k(x) - f_k(x)\| \leq \sum_{i=o}^{j} r_i \cdot \|q_k(w^i) - f_k(w^i)\| \leq \text{mesh } \underline{G}_k$, (by (3)).

In particular, for $x = x^k$ we obtain that $\|x^k - f_k(x^k)\| = \|q_k(x^k) - f_k(x^k)\| \leq \text{mesh } \underline{G}_k$. Thus, if x^o is a cluster point of R in C, then

(5) $\qquad \| x^o - f_k(x^k)\| \leq \|x^o - x^k\| + \|x^k - f_k(x^k)\| \longrightarrow 0$

(here is assumed -- without loss of generality -- that $\lim_k x^k = x^o$). Hence $x^o = \lim_k f_k(x^k)$ by (5). Let $v^{i,k}$ be the vertices of the unique open simplex in $\underline{G}_k^+$ which contains x^k. If $r_{i,k}$ are the barycentric coordinates of x^k in this simplex, then

(6) $R(x^k,f_k(x^k))=R(\sum_i r_{k,i}.x^{k,i};\ \sum_i r_{k,i}.f_k(v^{k,i}))\geq$
$$\geq \sum_i r_{k,i}.R(x^{k,i};f_k(v^{k,i})) .$$

Since each $f_k(v^{k,i})$ is a solution of $P(v^{k,i})$ it follows that for any point y in C we have $R(x^k;f_k(x^k))\geq\sum_i r_{k,i}.R(v^{k,i};y)$ by (6). If we assume (without loss of generality) that $\lim_k r_{k,i}=r_i$ and $\lim_k v^{k,i}=v^i$ for any index i, then all the v^i's are coincident to the point x^o because $\|v^{k,i}-x^k\|\leq$ mesh $\underline{G}_k\longrightarrow 0$, $(k\in N)$. Hence we have $R(x^o;x^o)\geq\sum_I r_i.R(x^o;y)$ $=R(x^o;y)$ because $\sum_i r_i=1$. The proof of the theorem is complete.

The reader may ask if fuzzy mappings which satisfy the conditions (III), (IV) and (V) exist. To answer we remind that in Butnariu (1982)it is proved that for fuzzy mappings accomplishing a "complementarity condition" Kuhn's fixed point algorithm can be used to find a simplex $S_{x,k}$ of $\underline{G}_k$ which contains a solution of P(x). Now taking $A_{x,k}(y)=$ $S_{x,k}\times\{e_{x,k}\}$, $(k\in N,\ y\in C)$ such that $e_{x,k}$ is the "error bound" determined by Bouwmann and Karamardian (1977) for Kuhn's algorithm we can conclude that each fuzzy mapping that satisfies the complementarity condition accomplishes the hypothesis of the Theorem 2. Some examples are explained in Butnariu (1982).

4. APPLICATIONS TO FUZZY GAMES

Some problems related with the problem of modelling the behaviour of the human beings in conflictual situations lead to the concept of "non-cooperative n-persons fuzzy game" introduced by Butnariu (1979).

DEFINITION 3: A non-cooperative n-persons fuzzy game with exact information (n-EFG) is a set $G=\{(Y_p,E_p)\}_{p\in P}$ where $P=\{1,2,\dots,n\}$ is a set whose elements are called "players" and for each player, Y_p denotes the set of p's alternatives in making decisions and $E_p\in L(Z)$ with $Z=\prod_{i\in P}Y_i$. For $(w^1,\dots,w^n)$ in Z, $E_p(w^1,\dots,w^n)$ is considered to be the degree of rationality of the choice w^p when the partners $i\neq p$ of p choose the alternatives w^i. It is assumed that for any player p

(7) If $w^i\in Y_i$ for $i\neq p$, then there is w^p Y_p such that $E_p(w^1,\dots,w^n)\neq 0$, i. e. any player p has a rational alternative in any state of the game.

Let $G=\{(Y_p,E_p)\}_{p\in P}$ be a game as described in Definition 3. A <u>solution</u> of G is a "play" $w=(w^1,\dots,w^n)\in Z$ such that $E_p(w)\geq E_p(w/\bar{w}^p)$ for any $\bar{w}^p\in Y_p$ and p in P, where $(w/\bar{w}^p)$ means $(w^1,\dots,w^{p-1},\bar{w}^p,w^{p+1},\dots,w^n)$.

It is interesting to observe that each normal form n-persons game Γ (see Owen (1968)) can be represented by an n-EFG denoted by $G(\Gamma)$ such that the Nash solutions of Γ are solutions of $G(\Gamma)$ and vice versa.

Analogously to the Theorem 2.4 of Butnariu (1979) it can be proved the next result:

PROPOSITION 4: Let $G=\{(Y_p,E_p)\}_{p\in P}$ and $w=(w^1,\ldots,w^n)\in Z=\prod_{i\in P} Y_i$ be a n-EFG and a play of it respectively. Then w is a solution of G if and only if w is a fixed point for the fuzzy mapping defined on Z as follows:

$$R(w;\bar{w})=\prod_{p\in P} E_p(w/\bar{w}^p), \tag{9}$$

for any $\bar{w}=(\bar{w}^1,\ldots,\bar{w}^p)$ in Z.

Now we assume that the n-EFG G is such that each Y_p is a convex and compact subset of a space $E^{m(p)}$ for $p\in P$.

PROPOSITION 5: If for any player $p\in P$ the fuzzy set E_p is a closed fuzzy subset of Z and for any w, $\bar{w}$ and $\bar{\bar{w}}$ contained in Z we have that

$$\left\{\begin{array}{c} E_k(w/r.\bar{w}^k+(1-r).\bar{\bar{w}}^k)\geqslant E_k(w/\bar{w}),\ (k\in P) \\ \text{or} \\ E_k(w/r.\bar{w}^k+(1-r).\bar{\bar{w}}^k)\geqslant E_k(w/\bar{\bar{w}}),\ (k\in P) \end{array}\right\}\ (r\in[0,1]), \tag{10}$$

then the fuzzy mapping defined by (9) has a fixed point in Z and G has a solution.

It can be easily seen that under some conditions the methods described in Section 2 and in Section 3 could be of use to approximate solution s of G viewed as fixed points of R. In practice the condition (10) means that the players of G are able to make compromistic agrements for avoiding the non-rational plays.

REFERENCES

Bowmann C. and Karamardian S. (1977): Errors bounds for approximate fixed points. In: Fixed points algorithm and applications, Karamardian S. (ed), North-Holland, Amsterdam.

Butnariu D. (1979): Solution concepts for n-persons fuzzy games. In: Advances in fuzzy sets theory and applications, M. M. Gupta et al. (eds.), North-Holland, Amsterdam.

(1982): Fixed points for fuzzy mappings. Fuzzy Sets and Systems, 7, 4, 191-207.

Eaves B. C. (1971): Computing Kakutani fixed points. SIAM J. Appl. Math. 21, No.2, 236-244.

Owen G. (1968): Game Theory. Saunders & Co., Philadelphia.

Todd M. (1976): The computation of fixed points and applications. (Lect. Notes in Math. and Economics), Springer-Verlag, Berlin.

Zadeh L. A. (1965): Fuzzy sets. Information and Control, 8, 338-352.

Zangwill W. (1969): Non-linear programming. Prentince-Hall, N. J.

Dept. of Math. Politechnic Inst.
Iaşi-6600, România.

INFORMATION SUBMARTINGALES

L. Lorne Campbell

Kingston

Key words: Entropy, information, submartingale, uncertainty

ABSTRACT

It is proposed that a measure of uncertainty be associated with each outcome of a random experiment in such a way that the uncertainties associated with a sequence of experiments form a submartingale. If some natural conditions are imposed on the submartingale, the Shannon entropy is recovered as the average uncertainty. The Doob decomposition of a submartingale is shown to be connected with well-known identities for entropy. A generalization to the entropy of degree α is also developed.

1. INTRODUCTION

An elementary, but fairly satisfactory, way of developing the notion of entropy in information theory is to begin with the notion of entropy or uncertainty of a single event. To each event of probability p is assigned a number $\phi(p)$, its uncertainty. One then postulates that a reasonable measure of uncertainty would have the property

$$\phi(pq) = \phi(p) + \phi(q) \quad . \tag{1}$$

This functional equation, together with some mild regularity condition, for example continuity or monotonicity, implies that $\phi(p) = a \log p$ for some constant a.

However, there are many situations in which successively observed events are not independent, so that the relevance of (1) is open to question. We investigate here an approach which begins in the same way, by assigning an uncertainty $\phi(p)$ to an event of probability p. Now let a sequence of experiments be performed. If the outcomes are events $E_1, E_2, \ldots,$ then after n experiments the amount of information which has been obtained is

$$X_n = \phi(P(E_1 \cap E_2 \cap \ldots \cap E_n)) \quad .$$

The sequence $\{X_n\}$ forms a random process. We use ideas from martingale theory to develop some "natural" properties for the function ϕ and then discover which functions ϕ possess these properties.

This approach leads to a different characterization of the entropy. The characterization here is similar in some respects to characterizations depending on the branching property of entropy. The Doob decomposition of submartingales is applied to $\{X_n\}$ and yields a well-known relation for entropy. Finally, some of the ideas are generalized to connect with the entropy of degree α . A related, but not identical, approach has been developed by Campbell (1982) for characterizing information measures in arbitrary spaces.

2. INFORMATION SUBMARTINGALES

Let $(\Omega, \mathcal{A}, P)$ be a probability space and let $\{\mathcal{A}_0, \mathcal{A}_1, \mathcal{A}_2, \ldots\}$ be a sequence of finite sub-σ-fields of $\mathcal{A}$. That is, each $\mathcal{A}_k$ is generated by a partition of Ω into a finite number of disjoint measurable sets. Let

$$\mathcal{B}_0 = \mathcal{A}_0 \;, \; \mathcal{B}_n = \mathcal{B}_{n-1} \vee \mathcal{A}_n \qquad (n=1,2,\ldots) \;.$$

That is, $\mathcal{B}_n$ is a finite sub-σ-field of $\mathcal{A}$ whose atoms are intersections of atoms of $\mathcal{A}_0, \mathcal{A}_1, \ldots, \mathcal{A}_n$. The sequence $\{\mathcal{B}_n\}$ is an increasing sequence. We regard $\mathcal{A}_0, \mathcal{A}_1, \ldots$ as defining successive random experiments which can be performed and $\mathcal{B}_n$ as a description of the possible sequences of observations up to time n .

Let ϕ be a real-valued function defined on $(0,1]$. Let the atoms of $\mathcal{B}_n$ be sets $B_k^{(n)}$ and let the atoms of $\mathcal{A}_n$ be $A_j^{(n)}$. Then

$$B_k^{(n)} = B_i^{(n-1)} \cap A_j^{(n)}$$

for some i and j . We define a random variable X_n by

$$X_n(\omega) = \phi(P(B_k^{(n)})) \quad \text{if} \quad \omega \in B_k^{(n)} \quad . \tag{2}$$

Thus X_n is $\mathcal{B}_n$-measurable. We will regard $X_n(\omega)$ as a measure of the amount of information acquired at time n in one particular realization of the sequence of experiments defined by $\mathcal{A}_0, \mathcal{A}_1, \mathcal{A}_2, \ldots$.

We can now compute the $\mathcal{B}_{n-1}$-measurable random variable $E(X_n | \mathcal{B}_{n-1})$ for $n>0$. Let $P(B_i^{(n-1)})>0$ and let $\omega \in B_i^{(n-1)}$. Then

$$E(X_n | \mathcal{B}_{n-1})(\omega) = \sum_j P(A_j^{(n)} | B_i^{(n-1)}) \; \phi \; (P(B_i^{(n-1)} \cap A_j^{(n)})) \quad . \tag{3}$$

In (3), terms for which $P(A_j^{(n)} | B_i^{(n-1)})=0$ are to be set equal to zero.

Now for any reasonable measure of information, it is natural to demand that, on average, the performance of another experiment should not cause information to be lost. Thus we shall require that $E(X_n|\mathcal{B}_{n-1}) \geq X_{n-1}$ for $n=1,2,\ldots$. That is, we require that $\{X_n\}$ be a submartingale (Neveu, 1975) relative to $\{\mathcal{B}_n\}$.

The difference $E(X_n|\mathcal{B}_{n-1})-X_{n-1}$ is the average gain in information obtained by doing one more experiment, given the previous outcomes. One quite "natural" restriction to place on ϕ is to require that the difference depend only on the conditional probabilities $P(A_j^{(n)}|B_i^{(n-1)})$. That is, we might assume that, for $\omega\epsilon B_i^{(n-1)}$,

(4) $$E(X_n|\mathcal{B}_{n-1})(\omega) - X_{n-1}(\omega) = H(P(A_1^{(n)}|B_i^{(n-1)}), P(A_2^{(n)}|B_i^{(n-1)}),\ldots) ,$$

for some function H . This property is much like the branching property of entropy (Aczél and Daróczy,(1975), Sec. 4.1).

If $\{X_n\}$ defined by (2) is a submartingale which satisfies (4), we shall refer to $\{X_n\}$ as an <u>information submartingale</u>. We shall call ϕ an <u>uncertainty function</u> if ϕ is a nonconstant function defined on $(0,1]$ which is such that the process $\{X_n\}$ is an information submartingale.

One example of an uncertainty function is

(5) $$\phi_0(x) = b-a \log x , \ x\epsilon(0,1] .$$

Here, b is any constant and a is a positive constant. A direct calculation shows that, if $\phi = \phi_0$ in (3) and $\omega\epsilon B_i^{(n-1)}$,

(6) $$E(X_n|\mathcal{B}_{n-1})(\omega)-X_{n-1}(\omega) = - a\sum_j P(A_j^{(n)}|B_i^{(n-1)}) \log P(A_j^{(n)}|B_i^{(n-1)}) .$$

Thus (4) is satisfied with H being the Shannon entropy of the conditional probabilities. The expression on the right of (6) is positive and thus $\{X_n\}$ is a submartingale when a is positive.

In the next section we establish a uniqueness theorem which shows that the above example is the only one which satisfies some rather modest smoothness conditions.

3. UNIQUENESS

Although we have required that (4) be satisfied for arbitrary sub-σ-fields $\mathcal{A}_n$ we now restrict our attention to $\mathcal{A}_n$ with two atoms. If we can show that only the functions ϕ_0 defined above satisfy (4) in this special case, then we need look no farther in the general case.

Let $\mathcal{A}_n$ have the atoms $A_1^{(n)}$ and $A_2^{(n)}$. Let $P(B_i^{(n-1)})>0$ and let

$$p_1 = P(A_1^{(n)}\cap B_i^{(n-1)}) , \ p_2 = P(A_2^{(n)}\cap B_i^{(n-1)}) .$$

Then

$$P(B_i^{(n-1)}) = p_1 + p_2 , \ P(A_j^{(n)}|B_i^{(n-1)}) = p_j/(p_1 + p_2) .$$

Since $p_2/(p_1 + p_2) = 1 - p_1/(p_1 + p_2)$, we can write

$$H\left(\frac{p_1}{p_1+p_2}, \frac{p_2}{p_1+p_2}\right) = h\left(\frac{p_1}{p_1+p_2}\right) .$$

Consequently, for this case (3) and (4) yield

$$\frac{p_1}{p_1+p_2}[\phi(p_1)-\phi(p_1+p_2)] + \frac{p_2}{p_1+p_2}[\phi(p_2)-\phi(p_1+p_2)] = h\left(\frac{p_1}{p_1+p_2}\right) ,$$

or with the notation $\theta(x) = x\phi(x)$,

$$\theta(p_1) + \theta(p_2) - \theta(p_1+p_2) = (p_1+p_2)\, h\left(\frac{p_1}{p_1+p_2}\right) . \tag{7}$$

Theorem 1. Let θ and h satisfy (7) on the set

$$\{(p_1,p_2) : p_1 \geq 0,\ p_2 \geq 0,\ 0 < p_1 + p_2 \leq 1\}$$

and let θ be continuous on the closed interval $[0,1]$. Then

$$\theta(x) = bx - a\, x \log x$$

for some constants a and b .

Proof. Since θ is continuous it follows from (7) that h is continuous. Next, let $p_1 = 0$ in (7). Then $\theta(0) = p_2 h(0)$ for all p_2 in $(0,1]$. Hence $\theta(0) = h(0) = 0$. Now let $p_2 = 0$ in (7) to get $p_1 h(1) = 0$ for all p_1 in $(0,1]$. Thus $h(1) = 0$.

We now obtain a functional equation for h by successively substituting $p_1 = x$, $p_2 = 1-x$ and then $p_1 = y$, $p_2 = 1-x-y$ in (7), getting

$$\theta(x) + \theta(1-x) - \theta(1) = h(x)$$

and

$$\theta(y) + \theta(1-x-y) - \theta(1-x) = (1-x)\, h\left(\frac{y}{1-x}\right) .$$

Add these and get

$$\theta(x) + \theta(y) + \theta(1-x-y) - \theta(1) = h(x) + (1-x)\, h\left(\frac{y}{1-x}\right) .$$

Since the left side of this equation is symmetric in x and y ,

$$h(x) + (1-x)\, h\left(\frac{y}{1-x}\right) = h(y) + (1-y)\, h\left(\frac{x}{1-y}\right) , \tag{8}$$

on the set

$$\{(x,y) : 0 \leq x < 1,\ 0 \leq y < 1,\ x+y \leq 1\} .$$

But the only continuous solutions of (8) which satisfy $h(0) = h(1) = 0$ are (Aczél and Daróczy, 1975, Chap. 3)

$$h(x) = -ax \log x - a(1-x) \log(1-x)$$

where a is any constant.

Finally, let

$$\psi(x) = \theta(x) + ax \log x .$$

Substitution in (7) yields

$$\psi(p_1) + \psi(p_2) - \psi(p_1+p_2) = 0 .$$

The only continuous solutions of this equation (Aczel and Daroczy, (1975), Sec. 0.3) are $\psi(x) = bx$ for arbitrary constant b . This completes the proof.

In the terminology of the previous Section we can now state:

Theorem 2. Let ϕ be an uncertainty function which is continuous on $(0,1]$ and for which $\lim_{x\to 0+} x\phi(x)$ exists. Then ϕ is the function ϕ_0 defined in (5).

4. DOOB DECOMPOSITION

If $\{X_n\}$ is a submartingale, then it has a unique Doob decomposition (Neveu, (1975), Chap. 7)

$$X_n = M_n + V_n \quad , \tag{9}$$

where M_n is a martingale and V_n is an increasing process, i.e. $V_n \le V_{n+1}$ and V_n is $\mathcal{B}_{n-1}$-measurable for $n=1,2,\ldots$. Define the random variable H_{n-1} by

$$H_{n-1} = E(X_n | \mathcal{B}_{n-1}) - X_{n-1} \quad ,$$

so that its values are given by (3). Then the increasing process V_n is given by

$$V_n = \sum_{k=0}^{n-1} H_k \quad , \tag{10}$$

so that $X_n - \sum_0^{n-1} H_k$ is a martingale.

If we form $E(X_n)$ using (9) and (10), we get

$$E(X_n) = E(X_0) + \sum_{k=0}^{n-1} E(H_k) \quad .$$

When $\phi = \phi_0$, this result in more conventional information-theoretic notation would be written (Gallager, 1968, p. 22)

$$H(\mathcal{A}_0, \mathcal{A}_1, \ldots, \mathcal{A}_n) = H(\mathcal{A}_0) + \sum_{k=1}^{n} H(\mathcal{A}_k | \mathcal{A}_0, \mathcal{A}_1, \ldots, \mathcal{A}_{k-1}) \quad .$$

5. GENERALIZATION

A generalization of (4) which may possibly be of some interest is to make the difference in (4) depend on a function of the conditional probabilities multiplied by a power of $P(B_i^{(n-1)})$, i.e. to replace (4) by

$$E(X_n | \mathcal{B}_{n-1})(\omega) - X_n(\omega) = [P(B_i^{(n-1)})]^{\alpha-1} H(P(A_1^{(n)} | B_i^{(n-1)}), \ldots) \quad ,$$

for $\alpha \neq 1$. This leads to the replacement of (7) by

$$\theta(p_1) + \theta(p_2) - \theta(p_1 + p_2) = (p_1 + p_2)^{\alpha} h\left(\frac{p_1}{p_1 + p_2}\right)$$

and the replacement of (8) by

$$h(x) + (1-x)^{\alpha} h\left(\frac{y}{1-x}\right) = h(y) + (1-y)^{\alpha} h\left(\frac{x}{1-y}\right) \quad .$$

The solution of this equation which vanishes at 0 and 1 is (Aczél and Daróczy, 1975, Theorem 6.2.8)

$$h(x) = a[(1-x)^{\alpha} + x^{\alpha} - 1]$$

for some constant a . This leads to the solution

$$\theta(x) = ax^{\alpha} + bx$$

and ultimately to the generalized uncertainty function

$$\phi_\alpha(x) = ax^{\alpha-1}+b \quad (\alpha\neq 1) \quad .$$

Then, with this ϕ, (6) is replaced by

$$\text{(11)} \qquad E(X_n \mid \mathcal{B}_{n-1})(\omega)-X_{n-1}(\omega) = a[P(B_i^{(n-1)})]^{\alpha-1}(\sum_j [P(A_j^{(n)} \mid B_i^{(n-1)})]^\alpha -1) \quad .$$

The last expression in (11) is the entropy of degree α (Aczél and Daróczy, 1975, p. 184) of the conditional probabilities. In order that this be positive, we take $a>0$ for $\alpha<1$ and $a<0$ for $\alpha>1$.

REFERENCES

ACZÉL, J. and DARÓCZY, Z. (1975): On Measures of Information and their Characterizations, Academic Press, New York.

CAMPBELL, L. L. (1982): A martingale characterization of information on arbitrary spaces, Utilitas Mathematica, to appear.

GALLAGER, R. G. (1968): Information Theory and Reliable Communication, John Wiley and Sons, New York.

NEVEU, J. (1975): Discrete Parameter Martingales, North-Holland Publishing Co., Oxford.

Department of Mathematics and Statistics
Queen's University
Kingston, Ontario
K7L 3N6
Canada

О КВАДРАТИЧЕСКОЙ МЕРЕ ОТКЛОНЕНИЯ ОЦЕНКИ ЛИНИИ РЕГРЕССИИ

Ш.А.Хашимов

Ташкент

АННОТАЦИЯ

Найдено предельное распределение квадратического отклонения оценки линии регрессии, построенной при помощи "проекционной" оценки плотности распределения.

I. Надарая Э.А. (1976) изучал предельное распределение квадратического расхождения "проекционной" оценки $\hat{h}_{n,N}(x)$ плотности распределения $h(x)$. В работах Греблицкого (1974), Хашимова Ш.А. (1977) были рассмотрены некоторые статистические свойства непараметрической оценки $\hat{r}_{n,N}(x)$ линии регрессии $r(x)$, построенной при помощи "проекционной" оценки плотности распределения.

Настоящая работа посвящена нахождению предельного распределения квадратичского расхождения

$$\int[\hat{r}_{n,N}(x)-r(x)]^2\cdot\hat{h}^2_{n,N}(x)dx$$

2. Пусть $X_1, X_2, \ldots, X_N$, $X_i=(X_{i1} X_{i2})$ — выборка из N независимых наблюдений двумерной случайной величины (ξ,η) с плотностью распределения $f(\bar{x})$, $\bar{x}=(x_1,x_2)$, обращающейся

в нуль вне множества $\mathcal{J} \subset R^2$

Предположения

Относительно $f(\bar{x})$ предположим, что существует ограниченное выпуклое открытое множество $\mathcal{J} \subset R^2$ со свойствами:

I) для любого $\varepsilon > 0$ найдется $\delta(\varepsilon) > 0$ такое, что

$f(\bar{x}) \geqslant \delta$, если $\bar{x} \in \mathcal{J}_\varepsilon = \{\bar{x} : U_\varepsilon(\bar{x}) \subset \mathcal{J}\}$,

где $U_\varepsilon(\bar{x}) = \{\bar{y} : \|\bar{x} - \bar{y}\| < \varepsilon\}$, $\bar{x}, \bar{y} \in R^2$;

2) найдется C такое, что $f(\bar{x}) \leqslant C$ для $\bar{x} \in \mathcal{J}$

3) $h(x_1), \varphi(x_1) \in L_2(a,b)$, $(a,b) \supseteq \mathcal{J}_\varepsilon^{(1)}$, где

$h(x_1) = \int f(\bar{x})\,dx_2$, $\varphi(x_1) = h(x_1) \cdot r(x_1)$, $\mathcal{J}_\varepsilon^{(1)}$ - проекция $\mathcal{J}_\varepsilon$ на ось x_1:

Пусть $\{\varphi_j(x_1)\}_1^\infty$ - полная ортонормированная, ограниченная (по x_1, j) система функций на (a,b).

В силу предположения 3)

$$h(x_1) = \sum_{j=1}^{\infty} c_j \varphi_j(x_1), \quad \varphi(x_1) = \sum_{j=1}^{\infty} a_j \varphi_j(x_1),$$

где

$$c_j = \int_a^b \varphi_j(x_1) h(x_1)\,dx_1, \quad a_j = \int_a^b \int_a^b x_2 \cdot \varphi_j(x_1) f(x_1, x_2)\,dx_1 dx_2$$

В качестве оценки кривой регрессии $r(x_1)$ по эмпирическим данным примем статистику

$$\hat{r}_{n,N}(x_1)=\frac{\hat{\varphi}_{n,N}(x_1)}{\hat{h}_{n,N}(x_1)},$$

где

$$\hat{\varphi}_{n,N}(x_1)=\sum_{j=1}^{n} a_{jN}\varphi_j(x_1),\quad \hat{h}_{n,N}(x_1)=\sum_{j=1}^{n} c_{jN}\varphi_j(x_1),$$

$$a_{jN}=\frac{1}{N}\sum_{i=1}^{N} X_{i2}\varphi_j(X_{i1}),\quad c_{jN}=\frac{1}{N}\sum_{i=1}^{N}\varphi_j(X_{i1}),\quad n=n(N)=O(N)$$

Обозначим

$$K_n(x_1,x_2)=\sum_{j=1}^{n}\varphi_j(x_1)\cdot\varphi_j(x_2),$$

$$\Delta_n=\frac{1}{n}\int_{J_\varepsilon^{(1)}}\int_a^b\int_a^b K_n^2(x_1,t)[y-r(x_1)]^2 f(t,y)\,dt\,dy\,dx_1,$$

$$\sigma_n^2=\frac{2}{n}\int_{J_\varepsilon^{(1)}}\int_{J_\varepsilon^{(1)}}\Big\{\int_a^b\int_a^b K_n(x_1,t)K_n(x_2,t)\times$$

$$[y^2-r(x_1)y-r(x_2)y+r(x_1)r(x_2)]f(t,y)\,dt\,dy\Big\}^2 dx_1dx_2,$$

$$s_k(n)=n^{-k}\int_{J^{(1)}}\cdots\int_{J_\varepsilon^{(1)}}\prod_{i=1}^{k}R_n(x_i,x_{i+1})\,dx_i,\quad x_{k+1}=x_1,$$

$$R_n(x_1,x_2)=\int_a^b\int_a^b K_n(x_1,t)K_n(x_2,t)[y^2-r(x_1)y-$$

$$-y\cdot r(x_2)+r(x_1)\cdot r(x_2)]f(t,y)\,dt\,dy-M\hat{\varphi}_{n,N}(x_1)\cdot M\hat{\varphi}_{n,N}(x_2)+$$

$$+r(x_1)M\hat{\varphi}_{n,N}(x_2)M\hat{h}_{n,N}(x_1)+r(x_2)\cdot M\hat{\varphi}_{n,N}(x_1)M\hat{h}_{n,N}(x_2)-$$

$$-r(x_1)r(x_2)M\hat{h}_{n,N}(x_1)\cdot M\hat{h}_{n,N}(x_2),$$

$$m_n = \sup_{a \le x_1 \le b} \int_a^b \left| \frac{\partial}{\partial x_2} K_n(x_1, x_2) \right| dx_2 .$$

Рассмотрим функционал

$$T_{n,N} = \frac{N}{n} \int_{J_\varepsilon^{(1)}} \left[\hat{r}_{n,N}(x_1) - r(x_1) \right]^2 \cdot \hat{h}^2_{n,N}(x_1)\, dx_1$$

Т е о р е м а I. Пусть выполнены предположения относительно $f(\bar{x}), h(x_1), \varphi(x_1)$ и при $N, n \to \infty$

$$\Delta_n = \Delta + o(n^{-1/2}),\ \sigma_n^2 = \sigma^2 + o(n^{-1/2})$$

$$n^{k-1} \cdot s_k(n) = O(1) \qquad \text{для каждого} \quad k \geq 3 .$$

Если, кроме того, $N^{-1/2} \cdot m_n \cdot \delta_n \cdot \ln^2 N \to 0$,

$$\sum_{j=n+1}^{\infty} a_j^2 = o(N^{-1} \cdot \delta_n), \quad \sum_{j=n+1}^{\infty} c_j^2 = o(N^{-1} \cdot \delta_n),$$

то при $N, n \to \infty$ случайная величина

$$\sqrt{n}\,(T_{n,N} - \Delta)$$

асимптотически нормальна со средним 0 и дисперсией σ^2.
Здесь $\delta_n \to \infty$ при $n \to \infty$ сколь угодно медленно.

Доказательство теоремы I основано на приближение нормированной и центрированной выборочной функции распределения соответствующим винеровским полем на удобном вероятностном пространстве.

Результат теоремы I позволяет построить асимптотический критерий согласия уровня λ проверки гипотезы $H_0 : r(x_1) = r_0(x_1)$.

Найдем асимптотическое распределение квадратичного функционала для последовательности "близких" альтернатив к гипотезе

$$H_0 : r(x_1) = r_0(x_1)$$

Пусть

$$H_N : r_N(x_1) = r_0(x_1) + \gamma_N \cdot \eta(x_1),$$

где $\gamma_N = N^{\frac{\alpha-2}{4}}$, $0<\alpha<\frac{1}{2}$.

Такие линии регрессии соответствуют последовательности близких к $f_0(\bar{x})$ плотностей $g_N(\bar{x}) = f_0(\bar{x}) + \gamma_N \cdot \eta_1(\bar{x})$, где $\eta_1(x_1)$ определяется по $\eta(x_1)$.

В дальнейшем индексы 0 и g_N снизу будут указывать на то, что характеристики вычислены для плотностей $f_0(\bar{x})$ и $g_N(\bar{x})$, соответственно.

Рассмотрим случайную величину

$$T_{g_N} = \frac{N}{n} \int_{J_\varepsilon^{(1)}} [\hat{r}_{g_N}(x_1) - r_0(x_1)]^2 \cdot \hat{h}^2_{g_N}(x_1)\,dx_1$$

Обозначим

$$b_j = \int_{J_\varepsilon^{(1)}} \varphi_j(x_1) h_{g_N}(x_1)\eta(x_1)\,dx_1,\quad d_j = \int_{J_\varepsilon^{(1)}} \varphi_j(x_1)\cdot\varphi_{g_N}(x_1)\cdot\eta(x_1)\,dx_1 .$$

Т е о р е м а 2. Пусть $g_N(\bar{x})$, $\varphi_{g_N}(x_1), h_{g_N}(x_1), \Delta_0, \sigma_0^2, s_k(n)$ удовлетворяют условиям теоремы I.

Если, кроме того, при $n = N^\alpha$

$$\sum_{j=1}^{\infty} |b_j| < \infty,\quad \sum_{j=1}^{\infty} |d_j| < \infty,\quad N^{-1/2}\cdot m_n\cdot\delta_n\cdot \ln^2 N \to 0,$$

$$\sum_{j=n+1}^{\infty} \tilde{a}_j^2 = o(N^{-1}\cdot\gamma_N^{-1}\cdot\delta_n^{-1}\cdot n^{1/2}),\quad \sum_{j=n+1}^{\infty} \tilde{c}_j^2 = o(N^{-1}\cdot\gamma_N^{-1}\cdot\delta_n^{-1}\cdot n^{1/2}),$$

то при $N, n \to \infty$.

$$P_{g_N}\left(T_{g_N} > \Delta + \frac{\beta_\lambda}{\sqrt{n}}\sigma\right) \to 1 - \Phi\left(\beta_\lambda - \sigma^{-1}\int_{J_\varepsilon^{(1)}} \eta^2(x_1)h_0^2(x_1)dx_1\right)$$

Здесь $\Phi(x_1)$ - функция распределения стандартного нормального закона, β_λ - квантиль уровня λ закона $\Phi(x_1)$, $\tilde{a}_j$, $\tilde{c}_j$ - коэффициенты Фурье функций $\varphi_{g_N}(x_1)$, $h_{g_N}(x_1)$ соответственно.

ЛИТЕРАТУРА

Надарая Э.А. (1976): О квадратической мере отклонения проекционной оценки плотности распределения. Теор. вер. и ее примен., XXI, 4, 864-871.

Greblicki W. (1974): Idebtyfikacja statyczna metoda szeregow orthogonalnych. Podstawy sterowanica, v.4, No 1, 3-12.

Хашимов Ш.А. (1977): Замечания о непараметрических оценках кривой регрессии. Сб.: "Предельные теоремы для случайных процессов", Изд. "Фан", Ташкент, 141-153.

Институт математики им. В.И. Романовского АН УзССР.
700143, Ташкент, 143, Академгородок, ул. Ф.Ходжаева, 29

COUPLING OF MARKOV PROCESSES AND HOLLEY'S INEQUALITIES FOR GIBBS MEASURES
STATISTICAL APPLICATION OF GIBBS MEASURES

Jacques Demongeot
Grenoble

Key words: Statistical mechanics, Gibbs measures, coupling of Markov processes, epidemiological application

---

ABSTRACT

The purpose of this paper is to generalize Holley's inequalities in the case of an attractive potential U associated with a Gibbs measure μ defined on $\mathbb{Z}^d$. The main idea is to consider U as a strongly supermodular potential. Next, using coupling techniques, we obtain new inequalities for the correlation function ρ of the Gibbs measure. Finally, we give a non-physical application of Gibbs measures by studying an epidemiological problem concerning the withering of oak trees in a French department.

INTRODUCTION

The notion of Gibbs measure can be introduced in different ways :
i) from a "static" definition, for example as Markovian measures or by means of conditional probabilities (Spitzer, Dobrushin)
ii) from a "dynamic" definition, for example as asymptotic measures of a contact process (Spitzer, Liggett).

Hence, if we have to study the asymptotical behavior of a contact process, for example in the case of an epidemiological process which has reached its "endemiological" state, we can use the notion of Gibbs measure and the related statistical results (Pickard, Demongeot(1981a)). But these last results are more or less difficult to obtain, according as the Gibbs measure associated with the studied potential is unique or not. For that, we have made precise the classical result of Holley's inequalities (see for example Preston), which permit to prove the absence or not of phase transition, i.e. the uniqueness or not for Gibbs measures. We have used the same techniques of coupling as Holley for certain set valued Markov processes. Because the potential U can be estimated (Demongeot (1981a)), remark that the generalized Holley's inequalities can be used in practice to build tests for the non-attractivity of U ; in our application, we give only a simple test for the conditional independence of certain random variables , which appear in the definition of a Markovian measure on $\mathbb{Z}^d$.

STRONGLY SUPERMODULAR POTENTIALS

If $P(\mathbb{Z}^d)$ denotes the set of all subsets of $\mathbb{Z}^d$, a potential U is a real function defined on $P(\mathbb{Z}^d)$. The interaction potential V associated with U is the real function defined on $F(\mathbb{Z}^d)$, set of all finite subsets of $\mathbb{Z}^d$, by :

$$\forall D \in F(\mathbb{Z}^d), \quad V(D) = \sum_{C \subset D} (-1)^{|D \setminus C|} U(C)$$

<u>Definitions</u>

U is called <u>strongly supermodular</u> if, for any finite family $\{A_i\}_{i=1,\dots,n}$ in $P(\mathbb{Z}^d)$, Poincaré's inequality holds :

$$U(\bigcup_{i=1}^{n} A_i) \geq \sum_{i=1}^{n} U(A_i) - \sum_{i<j} U(A_i \cap A_j) - \dots - (-1)^k \sum_{i_1<\dots<i_k} U(A_{i_1} \cap \dots \cap A_{i_k}) - \dots - (-1)^n U(\bigcap_{i=1}^{n} A_i)$$

U is called <u>attractive</u> if V satisfies:

$$\forall A \in F(\mathbb{Z}^d), \quad |A| \geq 2 \Rightarrow V(A) \geq 0$$

<u>Proposition 1</u>

U is strongly supermodular if, and only if, it is attractive

Proof :

1) If U is attractive then it is supermodular (Preston), and more, it is strongly supermodular ; denote by W the function defined by :

$$W(A_1,\dots,A_n) = \sum_{\substack{X \in T \\ X \in F(\mathbb{Z}^d)}} V(X), \text{ where } T = \{X; \exists i,j / X \subset A_i \cup A_j; X \cap (A_i \setminus A_j) \neq \emptyset, X \cap (A_j \setminus A_i) \neq \emptyset\}$$

Then W is positive, and we have :

$$W(A_1,\dots,A_n) = \sum_{Y \subset \bigcup_{i=1}^{n} A_i} V(Y) - \sum_{i=1}^{n} \sum_{Y \subset A_i} V(Y) + \sum_{i<j} \sum_{Y \subset A_i \cap A_j} V(Y) - \dots \geq 0$$

Because this last inequality is the Poincaré's inequality for U, the proof is achieved.

2) Conversely , we have, for every $D=\{x_i\}_{i=1,\dots,n}$ in $F(\mathbb{Z}^d)$:

$$V(D) = V(\bigcup_{i=1}^{n} \{x_i\}) = U(\bigcup_{i=1}^{n} \{x_i\}) - \sum_{J; |J|=n-1} U(\bigcup_{i \in J} \{x_i\}) + \dots \geq 0$$

We have expanded V with its definition formula, and this expansion is positive, because of the Poincaré's inequality ; hence V is positive, if D has more than one element.

COUPLING OF MARKOV PROCESSES

In order to generalize the coupling by Holley, and for the sake of simplicity, we are going to study the coupling between two $(P(\mathbb{Z}^d))^2$-valued Markov processes (if they are $(P(\mathbb{Z}^d))^k$-valued, the proof is similar).

Definition

A birth and death process on $(P(D))^2$, where D belongs to $F(\mathbb{Z}^d)$, is a Markov process whose infinitesimal generator G is defined by :

$$\begin{aligned}
\forall A,B \subset D,\quad & G(A,B;A\setminus\{x\},B\setminus\{x\}) = \delta_1(A,B,x), \text{ if } x\in A\cap B \\
& G(A,B;A\cup\{x\},B\cup\{x\}) = \beta_1(A,B,x), \text{ if } x\in (A\cup B)^c \\
& G(A,B;A\setminus\{x\},B) = \delta_2(A,B,x), \text{ if } x\in A\cap B \\
& G(A,B;A,B\setminus\{x\}) = \delta_3(A,B,x), \text{ if } x\in A\cap B \\
& G(A,B;A\setminus\{x\},B) = \delta_4(A,B,x), \text{ if } x\in A\setminus B \\
& G(A,B;A,B\cup\{x\}) = \beta_2(A,B,x), \text{ if } x\in A\setminus B \\
& G(A,B;A,B\setminus\{x\}) = \delta_5(A,B,x), \text{ if } x\in B\setminus A \\
& G(A,B;A\cup\{x\},B) = \beta_3(A,B,x), \text{ if } x\in B\setminus A \\
& G(A,B;A,B) = -\sum_{C,E\subset D} G(A,B;C,E) \\
& G(A,B;C,E) = 0, \text{ for all other } C,E \subset D
\end{aligned}$$

Proposition 2

If υ is a probability on $(P(D))^2$ which satisfies: $\upsilon > 0$ and

$\delta_1(A,B,x) = \upsilon(A\setminus\{x\},B\setminus\{x\})/\upsilon(A,B)$, $\delta_2(A,B,x) = \upsilon(A\setminus\{x\},B)/\upsilon(A,B)$

$\delta_3(A,B,x) = \upsilon(A,B\setminus\{x\})/\upsilon(A,B)$, $\delta_4(A,B,x) = \upsilon(A\setminus\{x\},B)/\upsilon(A,B)$

$\delta_5(A,B,x) = \upsilon(A,B\setminus\{x\})/\upsilon(A,B)$

and if we have : $\forall i=1,2,3,\ \beta_i = 1$, then υ is the unique invariant measure of the Markov process having G as generator.

Proof :

It is easy to check that : $\sum_{A,B\subset D} \upsilon(A,B)G(A,B;C,E)=0$, for all $C,E\subset D$.

Definition

The coupling of two processes having G_1 and G_2 as generators is made by considering the $(P(D))^4$-valued Markov process whose generator G' is defined by a collection of equations such that :

$G'(A,B,C,E;A\setminus\{x\},B\setminus\{x\},C\setminus\{x\},E\setminus\{x\}) = \min(\delta_{1,1}(A,B,x),$
$\delta_{1,2}(C,E,x))$, if $x\in A\cap B\cap C\cap E$, etc, ...

and where all birth coefficients are equal to 1 (in a way similar to that used by Holley (Preston).

GENERALIZATION OF HOLLEY'S INEQUALITIES

We are going to give the simplest generalization of the following Holley's inequality on the correlation functions ρ_D^Y and $\rho_D^{Y'}$ related to the conditional Gibbs measures π_D^Y and $\pi_D^{Y'}$ which express the occupation of the finite subset D of $\mathbb{Z}^d$, knowing the occupations Y and Y' of $\partial_A D = (A+D)\setminus D$, if we suppose the Gibbs measure A-Markovian (see for example Demongeot (1981a)) ; the Holley's inequality is the following : if $Y \supset Y'$, $\rho_D^Y \geq \rho_D^{Y'}$ and we have also, if the associated potential U is attractive :

<u>Proposition 3</u>

$$\forall\, Y, Y' \subset \partial_A D,\ \rho_D^{Y\cup Y'}\ \rho_D^{Y\cap Y'} \geq \rho_D^{Y}\ \rho_D^{Y'}$$

In particular, if U is symmetric with respect an axis of $\mathbb{Z}^d$, and if A admits the same symmetry, we have, if $Y \cup Y' = \partial_A D$, if $\mathrm{sym}(Y) = Y'$, $Y \cap Y' = \emptyset$ and if D tends to $\mathbb{Z}^d$:

$$\forall\, B \in F(\mathbb{Z}^d),\ \rho^{+}(B)\,\rho^{-}(B) \geq (\rho^{\pm}(B))^2$$

Proof :

It is very close to that of Holley ; we define G_1 and G_2 by considering its invariant measures :

$$\forall\, B, B' \subset D,\ \upsilon_1(B,B') = \pi_D^{Y\cup Y'}(B)\ \pi_D^{Y\cap Y'}(B')$$

$$\upsilon_2(B,B') = \pi_D^{Y}(B)\ \pi_D^{Y'}(B'),$$

where π_D^Y denotes the probability having ρ_D^Y as correlation function.

Then the invariant measure υ' of the coupled generator G' vanishes except on the elements (B_1, B_1', B_2, B_2') such that : $B_1 \supset B_2$ and $B_1' \supset B_2'$. This leads to inequalities given above, by applying the following result :

<u>Lemma</u>

If $X_1, \ldots, X_n$ are subsets of $\partial_A D$ such that : $X_1 = \bigcup_{i=2}^{n} X_i$ and if B_1, $B_2, \ldots, B_n$ are any subsets of D, we have :

$$\pi_D^{\bigcup_{i=1}^n X_i}\Big(\bigcup_{i=1}^n B_i\Big) \cdot \prod_{i<j} \pi_D^{X_i \cap X_j}(B_i \cap B_j) \ldots \geq \prod_{i=1}^n \pi_D^{X_i}(B_i) \cdot \prod_{i<j<k} \pi_D^{X_i\cap X_j \cap X_k}(B_i \cap B_j \cap B_k) \ldots$$

Proof :

It suffices to apply the Poincaré's inequality for $U((\bigcup_{i=1}^n B_i) \cup (\bigcup_{i=1}^n X_i))$, then to consider e^U, in order to have an inequality on the probabilities π_D after normalisation and by using that : $X_1 = \bigcup_{i=2}^n X_i$.

Note that, if n=3, the inequality above becomes :

$$\frac{\pi_D^{X_2\cup X_3}(B_1\cup B_2\cup B_3)\,\pi_D^{X_2\cap X_3}(B_2\cap B_3)}{\pi_D^{X_2\cup X_3}(B_1)\qquad\pi_D^{X_2\cap X_3}(B_1\cap B_2\cap B_3)} \geq \frac{\pi_D^{X_3}(B_2)\quad\pi_D^{X_2}(B_3)}{\pi_D^{X_3}(B_1\cap B_2)\,\pi_D^{X_2}(B_1\cap B_3)}$$

This last inequality gives a corresponding inequality for υ_1 and υ_2 which leads after coupling to the vanishing property of υ'.

STATISTICAL APPLICATION OF GIBBS MEASURES

It is possible, after a vertical or horizontal sampling, to estimate U and the correlation functions ρ_D (Pickard and Demongeot (1981 a & b)) and to apply these results for example to an epidemiological problem ; look at the Figure 1 (see the Appendix) : if the Gibbs measure associated to the estimated potential is S_1-Markovian, where S_1 is the set of the points of $\mathbb{Z}^d$ whose distance to the origin is equal to 1, then, for each x,y in $\mathbb{Z}^d$ such that d(x,y)=2, the random variables γ_x and γ_y are independent conditionally to a given occupation of $\partial_{S_1}\{x,y\}$ (γ_x being equal to 1, if x is occupied, here by a withering oak, and being equal to 0, if not). In these conditions, we can test for the independence of the variables whose values are encircled in the Figure 1 : conditionally to its S_2-frontier, the random variables corresponding to the surrounded variables in the others groups of trees delimited by dotted lines, are independent, if we make the S_2-Markovian assumption. In the two samples, the values taken by γ_x and γ_y have the following frequencies :

γ_y \ γ_x	0	1
0	57	7
1	5	1

The χ^2 of the contingency table above is equal to 0.18 ; hence the hypothesis of independence for γ_x and γ_y cannot be rejected, with for example a significance level equal to 0.05.

ACKNOWLEDGEMENTS

The author is indebted to D. Larroche (University of Pau) for obtaining the two samples given in the appendix.

APPENDIX

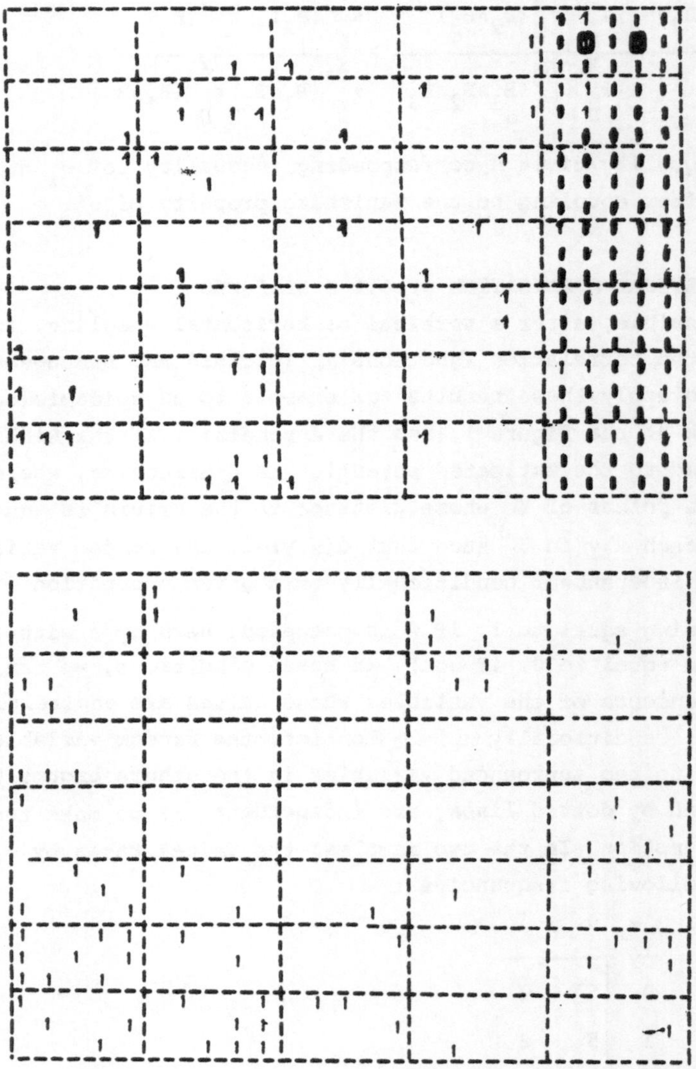

Fig. 1

Representation of two samples by using photomaps and denoting by 1 (resp. 0) each withering oak (resp. healthy oak) supposed to be setted on a regular lattice (subset of $\mathbb{Z}^d$) ; if the oak-plantation is not sufficiently regular, we can denote by 1 (resp. 0) each barycenter of a region having more (resp. less) withering oaks than healthy oaks (majority rule), the regions corresponding to a regular partition of the oak-plantation.

REFERENCES

Demongeot J. (1981a): Asymptotic inference for Markov random fields on $\mathbb{Z}^d$. Springer Series in Synergetics 9,254-267.

(1981b): Etude asymptotique d'un processus de contagion. In: Legay J.M. and Tomassone R. eds (1981) Biométrie et Epidémiologie, INRA, Paris,143-152.

(1981c): Couplage de processus de Markov. Séminaire de Statistiques de Grenoble 3, 220-248.

Dobrushin R.L. (1968) : Description of a random field by means of conditional probabilities and the conditions governing its regularity.Theory Prob.Appl. 13, 197-224.

Liggett T.M. (1977) : The stochastic evolution of infinite systems of interacting particles. Lect. Notes in Maths 598, 188-248.

Pickard D.K. (1979) : Asymptotic inference for an Ising lattice. J.Appl.Prob. 17, 583-599.

Preston C.J. (1974) : Gibbs states on countable sets. Cambridge University Press, Cambridge.

Spitzer F. (1974) : Introduction aux processus de Markov à paramètres dans $\mathbb{Z}^d$. Lect. Notes in Maths 390,114-189.

Laboratoire IMAG
BP 53 X
38 041 Grenoble Cédex
France

AN EMPIRICAL POWER OF SOME TESTS FOR LINEARITY

Czesław Domański, Andrzej Tomaszewicz
Łódź

Key words: nonparametric test, test for linearity, robust test

ABSTRACT

The paper presents an empirical power of tests for linearity based on the number of runs, modified Theil statistic and Fisher-Snedecor F-statistic. For a Monte Carlo experiment six alternative distributions were chosen. These were uniform, normal, lognormal, exponential, double exponential and Pareto distributions.

FORMULATION OF THE PROBLEM

Let a model with two explanatory variables

$$(1) \qquad y_i = g(x_{i1}, x_{i2}) + \varepsilon_i$$

be given at the usual assumptions: x_{i1}, x_{i2} - nonrandom, $[\varepsilon_1, \ldots, \varepsilon_n]^T$: : $N(0, \sigma^2 I)$ (cf. e.g. Goldberger (1964)). On the basis of a sample consisting of n independent observations

$$(2) \qquad (x_{11}, x_{12}, y_1), (x_{21}, x_{22}, y_2), \ldots, (x_{n1}, x_{n2}, y_n)$$

we verify the hypothesis H_0: $g(x_{i1}, x_{i2}) = \alpha_0 + \alpha_1 x_{i1} + \alpha_2 x_{i2}$. Without a loss in generality we assume that observations (2) are standardized.

Using the run test we base on the sign sequences of residuals

$$(3) \qquad e_i = y_i - a_0 - a_1 x_{i1} - a_2 x_{i2}$$

where a_0, a_1, a_2 are the least squares estimates of parameters $\alpha_0, \alpha_1, \alpha_2$, respectively of the model (1) under H_0. Analysing the model with two explanatory variables many criteria of ordering e_i can be used. Let f be a certain function of two variables and w the permutation ordering the values of $f_i = f(x_{i1}, x_{i2})$ such that

$$f_{w(1)} \leqslant f_{w(2)} \leqslant \dots \leqslant f_{w(n)}. \tag{4}$$

Consider such criteria of ordering the residuals e_i for which the function f is defined by one of the five formulae:

(a) $f(x_{i1}, x_{i2}) = a_0 + a_1 x_{i1} + a_2 x_{i2}$,

(b) $f(x_{i1}, x_{i2}) = x_{i1}$,

(c) $f(x_{i1}, x_{i2}) = x_{i2}$,

(d) $f(x_{i1}, x_{i2}) = x_{i1} + x_{i2}$,

(e) $f(x_{i1}, x_{i2}) = x_{i1}^2 + x_{i2}^2$.

Using a modified <u>Theil statistic</u> (cf. Theil (1950)) we find residuals in the form

$$R_{s(i)} = y_{s(i)} - a_1 x_{s(i),1} - a_2 x_{s(i),2}, \tag{5}$$

where s is the permutation ordering the values of y_i in the following way:

$$y_{s(1)} \leqslant y_{s(3)} \leqslant y_{s(5)} \leqslant \dots \leqslant y_{s(4)} \leqslant y_{s(2)} \tag{6}$$

The Theil test statistic has the form

$$T_n = \sum_{i=1}^{n} \sum_{j=i+1}^{n} \operatorname{sign}\left(R_{s(j)} - R_{s(i)}\right). \tag{7}$$

THE METHOD OF THE TEST POWER EVALUATION

The subject of our paper is the evaluation of the test power of some variants of the run test, Theil test and F test verifying the linearity hypothesis of the model with two explanatory variables. Note that both run tests and Theil test are based on test statistics with discrete distributions, while F test statistic has a continuous distribution. Therefore randomized run test and Theil test have been investigated in order to first-kind-error, probability be equal to chosen significance level. (Cf. Domański, Markowski, Tomaszewicz (1978)).

To evaluate the power of the above tests we used the Monte Carlo experiment with the following procedure. For determined sample sizes n=10,15,20,30 the values of x_{i1} were generated from the uniform distribution and the values of x_{i2} from the normal distribution in such a way that the correlation coefficient between sequences $\{x_{i1}\}$ and $\{x_{i2}\}$ be equal to the fixed number r. For each sequence of pairs $\{(x_{i1},x_{i2})\}$ generated in this way and treated as the determined ones, the sequences $\{y_i\}$ were generated, where

(8) $$y_i = g(x_{i1},x_{i2}) + \xi_i,$$
$$E\xi_i = 0, \quad E\xi_i^2 = \sigma_\xi^2, \quad E\xi_i\xi_j = 0 \quad \text{for } i,j=1,2,\dots,n \quad i \neq j$$

The variance σ_ξ^2 determines the scatter of empirical points about the surface defined by the function g. The function g has been determined in the following way:

(9) $$g(x_1,x_2) = c_0 + c_1x_1 + c_2x_2 + c_3x_1^2 + c_4x_2^2 + c_5x_1x_2.$$

Without a loss in generality of the considerations it can be assumed that $c_0 = 0$, and $c_5 = 0$. Then

(10) $$g(x_1,x_2) = c_4(2vu_1x_1 - 2u_2x_2 + v^2x_1^2 + x_2^2)$$

where $v = c_3/c_4$. In the experiment some variants of parameters r, u_1, u_2, v and the value

(11) $$\psi^2 = \frac{\sigma_\xi^2}{s_\theta^2 + \sigma_\xi^2}$$

were considered. The symbol s_θ^2 denotes variance of function (10) about its linear approximation minimizing s_θ^2 in fixed points $\{(x_{i1},x_{i2})\}$:

(12) $$s_\theta^2 = \frac{1}{n}\sum_{i=1}^{n} \left(g(x_{i1},x_{i2}) - l(x_{i1},x_{i2})\right)^2.$$

A very important problem in this experiment is the choice of alternative distributions of random terms ξ_i. Gastwirth and Selwyn (1980) give many results concerning robustness of autocorrelation tests and as alternatives of normal (NORM) distribution they consider Pareto (PAR) and the double exponential (2EXP) distributions. In our study, which is very similar, we consider besides the uniform (UNIF), lognormal (LNOR) and exponential (EXP) distributions.

Density functions of these distributions are defined as follows:

UNIF $\quad f(x) = \sqrt{12} \quad$ for $-1/2 \leq x \leq 1/2$,

NORM $f(x) = \exp(-x^2/2)/\sqrt{2\pi}$,

LNOR $f(x) = \exp(-\ln(x+\sqrt{e})^2/2)/(\sqrt{2\pi}x)$ for $x \geqslant -\sqrt{e}$,

EXP $f(x) = \exp(-x-1)$ for $x \geqslant -1$,

2EXP $f(x) = \exp(|x|/2)/\sqrt{8}$,

PAR $f(x) = 3(1+|x|)^{-4}/2$.

The range of the experiment included over 70 combinations of parameters r, u_1, u_2, v, ψ^2. The results for four of them are presented in Tables 1-4.

CONCLUSIONS

The following main conclusions can be drawn from the results of the experiments:

1. With the increase of the correlation coefficient r the power of all tests increases, except for the run test based on the ordering criterion (e) for which the power decreases.

2. The shift of (u_1, u_2) usually does not affect significantly the power of the considered tests.

3. The F test appeared to have the greatest power in almost all cases.

4. In most cases except for small values of r the power of Theil test exceeds the power of the considered run tests.

5. The F test appeared to be most robust to non-normality of distribution. This result is analogous to that obtained by Gastwirth and Selwyn (1980) who observed in this respect the predominance of a classical Durbin-Watson test over the run test based on empirical regression residuals.

6. The powers of all considered tests are similar for symmetrical distributions: NORM, UNIF and 2EXP. In the case of symmetrical distribution PAR and asymmetrical distributions LNOR and EXP significant differences in the test power are observed as compared with normal distribution.

REFERENCES

Domański C., Markowski K., Tomaszewicz A. (1978): Run Test for Linearity Hypothesis of Econometric Model with Two Exogenous Variables, Przegląd Statystyczny 25, 87-93.

Domański C.(1979): Nieparametryczne testy statystyczne (Nonparametric Statistical Tests), PWE, Warszawa.

Gastwirth J.L., Selwyn M.R. 1980 : The Robustness Properties of Two Tests for Serial Correlation, JASA 75, 138-141.

Goldberger A.S.(1964): Econometric Theory, John Wiley and Sons Inc., New York.

Theil H.(1950): A Rank-Invariant Method of Linear and Polynomial Regression Analysis, I,II and III, Proc.Kon.Ned. Akad. v. Weteruch. A 53, pp.386-392, 521-525, 1397-1412.

Institute of Econometrics and Statistics
University of Łódź
ul. Rewolucji 1905 r. 41
90-214 Łódź
Poland

TABLE 1. Percentage of rejecting H_o in the experiment with $r = 0.9$, $u_1 = u_2 = 0$, $v = 3$, $\psi^2 = 0.5$

$\alpha = 0.1$

Distribution	Run test (b)				Run test (d)				Run test (e)				Theil test				F test			
n=	10	15	20	30	10	15	20	30	10	15	20	30	10	15	20	30	10	15	20	30
UNIF	33	34	40	48	39	45	34	40	29	34	37	51	76	82	95	99	51	84	96	99
NORM	45	40	46	73	47	53	42	61	38	41	45	65	74	82	97	100	57	84	97	100
LNOR	70	80	83	96	70	86	72	84	59	81	90	90	94	97	99	100	85	89	93	96
EXP	48	60	66	82	48	64	52	67	40	58	67	78	82	90	99	100	56	87	92	98
2EXP	51	50	60	75	53	65	48	57	42	51	63	70	81	87	97	99	55	84	93	99
PAR	70	73	73	95	70	83	64	84	57	70	85	86	92	97	99	100	73	90	92	99

TABLE 2. Percentage of rejecting H_o in the experiment with $r = 0.9$, $u_1 = u_2 = 0$, $v = 3$, $\psi^2 = 0.9$

$\alpha = 0.1$

Distribution	Run test (b)				Run test (d)				Run test (e)				Theil test				F test			
n=	10	15	20	30	10	15	20	30	10	15	20	30	10	15	20	30	10	15	20	30
UNIF	15	16	15	17	13	17	15	11	15	12	13	15	24	24	29	43	13	21	21	30
NORM	14	15	18	21	14	17	17	19	17	18	12	17	21	25	38	53	12	22	30	42
LNOR	31	35	48	56	32	39	36	45	23	33	37	46	55	60	83	99	31	46	50	61
EXP	19	20	26	28	19	21	20	25	19	18	23	27	26	39	52	82	18	22	31	44
2EXP	22	21	21	24	22	27	20	22	18	16	18	23	28	26	38	55	16	22	24	30
PAR	31	35	37	47	33	39	30	38	27	33	33	45	50	49	69	90	28	29	41	53

TABLE 3. Percentage of rejecting H_o in the experiment with $r = 0.5$, $u_1 = u_2 = 0$, $v = 3$, $\psi^2 = 0.5$

$\alpha = 0.1$

Distribution	Run test (c)				Run test (d)				Run test (e)				Theil test				F test			
n=	10	15	20	30	10	15	20	30	10	15	20	30	10	15	20	30	10	15	20	30
UNIF	8	21	20	28	22	21	30	33	9	13	13	38	53	35	41	90	49	86	97	100
NORM	5	24	21	31	26	22	32	34	10	11	11	41	48	27	38	88	52	84	94	100
LNOR	1	48	44	34	49	17	40	67	6	8	16	75	77	40	67	99	81	88	91	96
EXP	5	30	40	35	35	27	33	44	7	10	15	51	61	34	48	93	62	83	92	99
2EXP	4	26	24	28	31	24	40	49	8	12	17	60	62	35	47	93	58	84	94	98
PAR	2	29	28	31	46	30	60	68	7	14	21	81	70	46	60	99	74	88	91	97

TABLE 4. Percentage of rejecting H_o in the experiment with $r = 0$, $u_1 = u_2 = 0$, $v = 1$, $\psi^2 = 0.5$

$\alpha = 0.1$

Distribution	Run test (c)				Run test (d)				Run test (e)				Theil test				F test			
n=	10	15	20	30	10	15	20	30	10	15	20	30	10	15	20	30	10	15	20	30
UNIF	31	27	35	21	3	11	20	28	17	15	22	37	3	7	34	86	46	87	99	100
NORM	35	31	35	28	2	11	23	27	10	11	20	43	1	9	31	84	55	87	96	100
LNOR	66	70	65	38	1	26	40	58	7	6	32	51	0	5	49	98	79	88	92	96
EXP	42	41	46	29	1	18	24	40	13	12	21	47	3	6	39	93	60	86	90	99
2EXP	43	44	47	23	2	13	19	44	13	14	27	58	1	7	37	90	55	86	93	98
PAR	59	65	66	24	2	15	21	58	7	8	35	74	0	9	43	98	73	88	93	97

THE OPTIMAL CONTROL OF PARTIALLY OBSERVABLE SEMI-MARKOV PROCESSES OVER THE INFINITE HORIZON: DISCOUNTED COSTS

Maria Drăguţ
Bucharest

Key words : semi-Markov decision processes, total expected discounted cost criterion, policy iteration, finite transient policy

---

ABSTRACT

This paper is devoted to the generalization of the numerical procedure for the finite-horizon partially observed semi-Markov optimization control problem due to White (1976) and also of the numerical procedure for the infinite horizon Markov optimization problem due to Sondik (1977)

INTRODUCTION

The discounted cost, optimal control problem for finite-state discrete -time semi-Markov processes with incomplete state-information is discussed. We assume that times of control reset and noise corrupted observations of the core process transition. The control employed at each time transition is allowed to be functionally dependent on the sample path of the core process only through the history of corrupted observations.

The set of values that the control vector may assume at each time of control reset is finite

PRELIMINARIES

In this section we develop the notation to be used throughout the paper following White (1976) as closely as possible.

Let the core process $[s(t),\ t=0,1,\ldots\ ]$ be a finite-state, discrete-time, controlled semi-Markov process making transitions at times $t_0=0, t_1, t_2, \dots$; $0 < t_{q+1} - t_q$ with probability one; $q = 0, 1, \dots$. Assume the core process has state space $S_s = (1, \dots, N_s)$. The controller is assumed to know the times of the core process transition; however the state of the core process is only imperfectly observed at times of core transition. We assume that the qth time of core process transition t_q , the controller is given the realization of the random variable $z(t_q)$, stochastically related to the new state of the core process $s(t_q)$. We assume that the observation process $[z(t_q), q = 1, 2, \dots]$ have a finite state space $S_z = (1, \dots, N_z)$.

The controller is allowed to select a control vector $u(t_q) \in U$ for the interval $(t_q, t_{q+1}]$, where U is a finite set. The selection of the control vector is based on the a priori density vector $p^0 = (p_1^0, \dots, p_{N_s}^0)$, $p_i^0 = P[s(0) = i]$; the set of past and present times of transition and realizations of the observation process $t_1, z(t_1), \dots, t_q, z(t_q)$; and the set of control vectors previously selected $u(t_0), u(t_1), \dots, u(t_{q-1})$.
let's notice $d_q = \{p^0, u(t_0), t_1, z(t_1), \dots, u(t_{q-1}), t_q, z(t_q)\}$ and let D_q be the set of all such data sequences.

DEFINITION. The control sequence $\{U(0), \dots, U(n-1)\}$ is said to be admissible if there exists a sequence of functions $\{\psi(0,\cdot,\cdot), \dots, \psi(n-1,\cdot,\cdot)\}$; $\psi(t, q, \cdot) : D_q \to U$ given $t = t_q$, where for $t = 0$, $q = 0$ and for $1 \le t \le n-1$, $1 \le q \le t$.

We also know time independent probability functions $c_{ij}(m, v) = P[t_{q+1} - t_q = m \ , s(t_{q+1}) = j \ / s(t_q) = i \ , u(t_q) = v]$ and $q_{jk}(v) = P[z(t_{q+1}) = j, u(t_q) = v]$ for all i, j, k and v .
We assume $c_{ij}(0, v) = 0$ for all i, j, v , so that at most one transition can occur at each time interval and $c_{ij}(m, v) = 0$ for $m > M$ for all i, j and v .
Let's define the matrices $C(m, v) = \{c_{ij}(m, v)\}$, $R_k(v) = \mathrm{diag}\{q_{jk}(v)\}$, $D_k(m, v) = C(m, v) R_k(v)$
We use the cost structure defined in Howard (1977) and White (1976).

THE INFINITE HORIZON PROBLEM WITH DISCOUNTING: CONDITIONS FOR OPTIMALITY

Be $[0, n]$ the finite control interval. Let's notice :

$\bar{g} = \max(g : t_g \leq n)$, $\bar{s}_g = \{s(t) : t_g \leq t \leq t_{g+1}\}$; $\bar{s}_{\bar{g}} = \{s(t) : t_{\bar{g}} \leq t \leq n\}$; $\gamma[\bar{s}_g, u(t_g)]$ - the cost occured during the time interval $(t_g, t_{g+1}]$ for $g < \bar{g}$; $\gamma_0[\bar{s}_{\bar{g}}, u(t_{\bar{g}})]$ -the cost occured during $(t_{\bar{g}}, n]$; $g_0^{s(n)} = \gamma_0[\bar{s}_{\bar{g}}, u(t_{\bar{g}})]$ when $t_{\bar{g}} = n$.

THEOREM 1. There exists a finite horizon completely observed semi-Markov optimization problem which is cost equivalent to the finite incompletely observed semi-Markov optimization problem.

PROOF. (see White (1976)).

Considering the optimality criterion

(1) $$E_{p^0}\left\{\sum_{g=0}^{\bar{g}-1} \gamma[\bar{s}_g, u(t_g)] + \gamma_0[\bar{s}_{\bar{g}}, u(t_{\bar{g}})]\right\}$$

the optimization problem is to find the control sequence $u(t_0), \dots, u(t_{\bar{g}})$ that minimizes (1).

Let's put: $\tilde{y}_i(g, d_g) = P[s(t) = i / d_g, t = t_g]$,

$\tilde{y}(g, d_g) = [\tilde{y}_1(g, d_g), \dots, \tilde{y}_{N_s}(g, d_g)]$, $[y(t), t = 0, 1, \dots n]$ where $y(t) = \tilde{y}(g, d_g)$ for $t_g \leq t < t_{g+1}$; $Y = (y \in R^{N_s} : y_i \geq 0, \sum_i y_i = 1)$

$$g_{n-t}^i(v) = E\{\gamma[\bar{s}_g, u(t_g)] / g < \bar{g}, t = t_g, s(t) = i, u(t) = v\} \cdot P\{g < \bar{g} / t = t_g, s(t) = i, u(t) = v\} + E\{\gamma_0[\bar{s}_g, u(t_g)] / g = \bar{g}, t = t_g, s(t) = i, u(t) = v\} \cdot P\{g = \bar{g} / t = t_g, s(t) = i, u(t) = v\}$$

$g_0^i(v) = g_0^i$; $g_{n-t}(v) = [g_{n-t}^1(v), \dots, g_{n-t}^{N_s}(v)]$,

$V(k, m, y, v) = \sum_j \sum_i q_{jk}(v) c_{ij}(m, v) y_i$;

$T_j(k, m, y, v) = \sum_i q_{jk}(v) c_{ij}(m, v) y_i / V(k, m, y, v)$

$T(k, m, y, v) = [T_1(k, m, y, v), \dots, T_{N_s}(k, m, y, v)]$

and observe that :

$$V[k, m, \tilde{y}(g, d_g), v] = P[z(t_{g+1}) = k, t_{g+1} - t_g = m / d_g, u(t_g) = v]$$

$$T_j[k, m, \tilde{y}(g, d_g), v] = P[s(t_{g+1}) = j / z(t_{g+1}) = k, t_{g+1} - t_g = m, d_g, u(t_g) = v]$$

$$T[k, m, \tilde{y}(g, d_g), v] = \tilde{y}(g+1, d_{g+1})$$

where $d_{g+1} = [d_g, (v, t_g + m, k)]$.

There exists an optimal policy u^*, $u^*(t) = \Phi^*[t, \tilde{y}(g, d_g)]$

for all $d_q \in D_q$ given $t = t_q$ and a function $C_{n-t} : Y \to R$ that satisfies :

(2) $$C_{n-t}(y) = \sum_i y_i g^i_{n-t}[\phi^*_q(t,y)] + \sum_{m=1}^{n-t} \sum_k V[k,m,y,\phi^*_q(t,y)] \cdot C_{n-t-m}[T[k,m,y,\phi^*_q(t,y)]] = \min_{v \in U} \{ \sum_i y_i g^i_{n-t}(v) + \sum_{m=1}^{n-t} \sum_k V(k,m,y,v) \cdot C_{n-t-m}[T(k,m,y,v)] \}$$

q.e.d.

Let's consider the discounted finite horizon semi-Markov control problem. The total expected discounted cost verifies the functional equation:

(3) $$C_{n-t}(y,\beta) = \min_{v \in U} \{ \sum_i y_i g^i_{n-t}(\beta,v) + \sum_{m=1}^{n-t} \beta^m \sum_k V(k,m,y,v) C_{n-m-t}[T(k,m,y,v),\beta] \}$$

THEOREM 2. If there exists :

i) a pair $m_0 \in N_+$, $\varepsilon > 0$ such that for all $v \in U$ $\sum_k V(k,m_0,y,v) < 1-\varepsilon$

ii) a uniform bound, say F on $g(y,\beta,v)$, where $g(y,\beta,v) = \lim_{n\to\infty} \sum_i y_i g^i_n(\beta,v) = \sum_i y_i g^i(\beta,v)$ then $C(y,\beta) = \lim_{n\to\infty} C_n(y,\beta)$ exists, is finite and is the unique solution for the equation :

$$C(y,\beta) = \min_{v \in U} \{ \sum_i y_i g^i(\beta,v) + \sum_{m=1}^{M} \beta^m \sum_k V(k,m,y,v) \cdot C[T(k,m,y,v)] \}$$

Furthermore there exists (see Lippman (1975)) an optimal control policy which is stationary.

PROOF. For $h : Y \to R$ let's define $\|g\| = \sup_{y \in Y} |g(y)|$ and let B be the set of all function $h : Y \to R$ for which $\|h\| < \infty$.

For each function $f : Y \to U$ and $c \in B$ let's define the usual operator

$$(T_f c)(y) = \sum_i y_i g^i(\beta, y, f(y)) + \sum_{m=1}^{M} \sum_k V(k,m,y,f(y)) c[T(k,m,y,f(y))]$$

We can show :

$$|(T_f c)(y)| \leq F + \|c\| \sum_{m=1}^{M} \beta^m (1-\varepsilon) = $$
$$= F + \|c\| \frac{\beta^{M+1}-1}{1-\beta} (1-\varepsilon)$$

and

$$|(T_f c_1 - T_f c_2)(y)| \leq \sum_{m=1}^{M} \beta^m (1-\varepsilon) \|u-v\| \leq M\beta^M (1-\varepsilon)\|u-v\|$$

Hence $T_f : B \to B$ and Denardo's N-stage contraction assumption (see Denardo (1967)) is verified.

q.e.d.

COMPUTATION OF THE OPTIMAL COST AND OF THE OPTIMAL CONTROL POLICY: POLICY ITERATION

Policy iteration for the partially observable problem consists of two steps: Value Determination and Policy Improvement. In Value Determination the cost of a stationary policy $(\delta)^\infty$ is computed and in Policy Improvement step, this cost is used to find a different stationary policy with lower cost.

Let $C(y,\beta/\delta)$ be the expected cost of following the stationary policy $(\delta)^\infty$ for all time. Based on the contraction properties of the operator

$$U_{\delta,\beta}\, c(y) = \sum_i y_i g^i(\beta, \delta(y)) + $$
$$+ \sum_m \beta^m \sum_k V(k,m,y,\delta(y)) C[T(k,m,y,\delta(y))]$$

we obtain that $C(y,\beta/\delta)$ is the unique bounded solution to the equation

$$C(y,\beta/\delta) = \sum_i y_i g^i(\beta,\delta) + \sum_m \beta^m \sum_k V(k,m,y,\delta)\cdot C[T(k,m,y,\delta),\beta/\delta] \tag{4}$$

To compute $C(y,\beta/\delta)$ the methods of Sondik (1978) and White (1976) require the piecewise linearity and concavity for $C(y,\beta/\delta)$ as function of y. Extending the notion of finitely transient policy introduced in Sondik (1978) we can show that $C(y,\beta/\delta)$ is a piecewise concave function for a finitely transient policy and that for every stationary policy $(\delta)^\infty$ we can find a finitely transient one sufficiently closed.

THEOREM 3. Let $(\delta)^\infty$ be a finitely transient policy. Then $C(y,\beta/\delta)$ is piecewise linear and concave.

PROOF. Let's denote $T_\delta(A)$ = closure$[T(k,m,y;\delta) \ : y\in A$, $(\forall) k, (\forall) m]$, $S_\delta^0 = Y$, $S_\delta^n = T_\delta(S^{n-1})$ D_δ = = closure$[y: \delta(y)$ is discountinuous at $y]$. A stationary policy$(\delta)^\infty$ is finitely transient if and only if there exists an integer $n<\infty$ such that $D_\delta \cap S_\delta^n = \emptyset$. We observe that

$C(y,\beta/\delta) = y\,\alpha(y,\beta/\delta)$ where $\alpha(y,\beta/\delta)$ is a column N_1 - vector that is the unique bounded solution to the vector equation

$$\alpha(y,\beta/\delta) = g(\beta,\delta(y)) + \sum_m \beta^m \sum_k D_k(m,\delta(y))\cdot$$
$$\cdot\alpha[T(k,m,y,\delta(y)),\beta/\delta]$$

If $(\delta)^\infty$ is a finitely transient policy there exists a finite partition of Y into sets $Y_1,\dots,Y_\ell$ with two properties : a) all points in Y_j are assigned the same control and b) under the mapping $T(k,m,\cdot,\delta)$ all points in Y_j map into the same set $Y_{\nu(j,k,m)}$. So (4) becomes the following set of vector equations

(5) $$\alpha_i = g(\beta,\delta_i) + \sum_m \beta^m \sum_k D_k(m,\delta_i)\,\alpha_{\nu(i,k,m)}$$

where $\alpha_i = \alpha(y,\beta/\delta)$ for $y\in Y_i$.

The set of linear equations (5) can be solved uniquely for $\alpha_1,\dots,\alpha_\ell$.

q.e.d.

THEOREM 4. Let $(\delta)^\infty$ be a stationary policy which is not finitely transient. There exists a finitely transient policy $(\hat{\delta})^\infty$ with the total infinite discounted expected cost as closed as desired to $C(y,\beta/\delta)$

PROOF. Let $D^n = [y : T(k,m,y,\delta)\in D^{n-1}$ for some $(k,m)]$, $n>1$, $D^0 = D_\delta$.

Let Y^k be the partition of Y defined by the sets $D_\delta, D^1,\dots,D^k$. $Y^k = [Y_j^k]$. Using Y^k we construct the k-th degree approximation. Let $y_j \in Y_j^k$ and if $T(k,m,y_j,\delta)\in Y_n^k$ then $\hat{\nu}(j,k,m) = n$. Let's consider $\hat{C}(y,\beta/\delta) = y\hat{\alpha}_j$ for $y\in Y_j^k$ where

$$\hat{\alpha}_j = g(\beta,\delta_j) + \sum_m \beta^m \sum_k D_k(m,\delta_j)\,\hat{\alpha}_{\hat{\nu}(j,k,m)}$$

Let $y\in Y_j^k$ and let's evaluate

$$C(y,\beta/\delta) - \hat{C}(y,\beta/\delta) =$$

$$= \sum_m \beta^m \sum_k V(k,m,y,\delta)\, c[T(k,m,y,\delta)/\delta] -$$

$$- \sum_m \beta^m \sum_k D_k(m,\delta_j)\, \hat{\alpha}_{\hat{\gamma}}(j,k,m)] =$$

$$= \sum_m \beta^m \sum_k \{ V(k,m,y,\delta)\, c[T(k,m,y,\delta)/\delta] - D_k(m,\delta_j) \cdot$$

$$\cdot\, \hat{\alpha}_{\hat{\gamma}}(j,k,m)]$$

By iterating n times, $n \leq k$ we obtain

$$C(y,\beta/\delta) - \hat{C}(y,\beta/\delta) =$$

$$= \sum_{m_1} \sum_{k_1} \ldots \sum_{m_n} \sum_{k_n} \beta^{m_1+m_2+\ldots+m_n} \{ V(k_1,m_1,y,\beta/\delta)\, V[k_2,m_2,$$

$$T(k_1,m_1,y,\delta)/\delta] \ldots V(k_n,m_n,T(y,(k_1,m_1,\ldots,k_n,m_n),\beta/\delta) \cdot$$

$$\cdot\, c[T(y,(k_1,m_1,\ldots,k_n,m_n),\beta/\delta] -$$

$$- D_{k_1}(m_1,\delta_1) \cdot D_{k_2}(m_2,\delta(T(k_1,m_1,y,\delta_j/\delta))) \cdot \hat{\alpha}_{\hat{\gamma}}(j,(m_1k_1,\ldots m_n k_n)]\}$$

where:

$\hat{\gamma}(j,(k_1,m_1,\ldots,k_n,m_n))$ is the compozition of $\hat{\gamma}(j,m,k)$ n times:

$$\hat{\gamma}(j,(k_1,m_1,\ldots,k_n,m_n)) = \hat{\gamma}[\ldots\hat{\gamma}[\hat{\gamma}(j;k_1,m_1),m_2,k_2],\ldots k_n,m_n]$$

$$T[y,(k_1,m_1,\ldots,k_n,m_n),\beta/\delta] =$$

$$= T[k_n,m_n,T(y,(k_1,m_1,\ldots,k_{n-1},m_{n-1}),\delta[y,(k_1,m_1,\ldots k_{n-1},m_{n-1}],\beta/\delta]$$

We observe that if $y_j \in Y_{\hat{\gamma}}(j,(k_1,m_1,\ldots,k_n,m_n))$ then $\delta[T(y,(k_1,m_1,\ldots,k_n,m_n)/\delta] = \delta(y)$ We add and abstract the quantity

$$\sum_{m_1} \ldots \sum_{m_n} V(k_1,m_1,y/\delta) \cdot V(k_2,m_2,T(y,(k_1,m_1),\beta/\delta) \ldots V(k_n,m_n,T(y,(k_1,m_1,\ldots$$

$$\ldots k_{n-1},m_{n-1}),\beta/\delta) \cdot T[y,(k_1,m_1,\ldots,k_n,m_n),\beta/\delta] \cdot \hat{\alpha}_\mu[T(y,(k_1,m_1,\ldots k_n,m_n),\beta/\delta]$$

It follows that

$$(1 - \sum_{m_1} \ldots \sum_{m_n} \beta^{m_1+\ldots+m_n}) \| C(\cdot,\beta/\delta) - \hat{C}(\cdot,\beta/\delta) \| \leq$$

$$\leq \sum_{m_1} \ldots \sum_{m_n} \beta^{m_1+\ldots+m_n} \sup_{i,j,y} |y(\hat{\alpha}_i - \hat{\alpha}_j)|$$

We observe that

$$y(\hat{\alpha}_i - \hat{\alpha}_j) \leq E + \sum_m \beta^m \sum_k [D_k(m,\delta_i)\hat{\alpha}_{\hat{\gamma}}(i,k,m) -$$

$$- D_k(m,\delta_j)\, \hat{\alpha}_{\hat{\gamma}}(j,k,m)] \leq$$

$$\leq E + 2\sum_m \beta^m \sup_{i,j,y} |y(\hat{\alpha}_i - \hat{\alpha}_j)|$$

So :

$$(1 - 2\sum_{m=1}^{M} \beta^m) \sup_{i,j,y} |y(\hat{\alpha}_i - \hat{\alpha}_j)| \leq E \qquad \text{q.e.d.}$$

REFERENCES

Denardo E.(1967) : Contraction mappings in the theory underlying dynamic programming. In SIAM Review, 9, 165-177

Howard R.(1971) : Dynamic probabilistic systems. John Wiley

Lippman S.(1975) : On dynamic programming with unbounded reward. In Management Science, vol.21, no.11, 1225-1232

Sondik E.(1978) : The optimal control of partially observable Markov processes over the infinite horizon-discounted costs. In : Operations Research, vol.26, no.2, 282-303

White C. (1976) : Procedures for the solution of a finite-horizon partially observed semi-Markov optimization problem. In : Operations Research, vol.24, no.2, 348-357

Center of Mathematical Statistics
174 Stirbei Vodă St.
77104 Bucharest
ROMANIA

INFORMATION METHODS IN IDENTIFICATION

I.S. Durgaryan, and F.F. Pashchenko

Moscow

Key words: Identification, information theory, loss function, stochasticity, entropy, correlation function, Lagrange function.

ABSTRACT

The paper concerns with analysis in terms of information theory and identification of static and dynamic systems.

An information theory approach to estimation of the control system stochasticity is considered. Identification of the control plant is done assuming that the model output contains maximal information on the plant output. Plant identification is carried out in terms of the information criterion with the model parameters taken care of.

1. INTRODUCTION

For optimal identification it is usually required that the estimated plant operator is close in terms of a certain criterion to the true operator value. For stochastic plants this closeness can be estimated in terms of the mathematical expectation of the loss function:

$$M\{\rho[y(t), y_M(t)]\} \rightarrow \underset{A \in \mathcal{A}}{extr} \quad (1)$$

where $y(t)$, $y_M(t)$ are output variables of the plant and model, respectively, A is the model operator, $\mathcal{A}$ is the class of the acceptable operators, $\rho(\cdot)$ is the loss function, and M is the mathematical expectation. Most frequently the r.m.s. mathematical expectation is used as the loss function and the problem is solved via correlation functions. The r.m.s. criterion, however, does not always represent the final goal of identification which is what the model has been developed for (Eykhoff, 1975; Durgaryan and

Pashchenko, 1980). It is desirable to obtain a criterion which would satisfy a wider range of problems including both identification proper and the destination of the model identified.

On the other hand most real systems are nonlinear. It is known that for nonlinear stochastic plants the correlation functions do not provide comprehensive communication characteristics between the variables. It can be shown that application of correlation techniques to identification of nonlinear systems leads to wrong results in quite a few cases (Durgaryan, Pashchenko, 1980). Hence, a need arises to develop more efficient methods which work even when there is no correlation.

This paper considers a new method of solving the problem of identification. The maximal amount of information on the plant output contained in the model output is chosen as the identification criterion.

2. DETERMINATION OF PLANT STOCHASTICITY

One of the problems in identification lies in determination of the stochasticity of the system studied. The solution of this problem yields the accuracy limits of both, plant and its model. The mathematical expectation of the conditional dispersion for the system output relative to input disturbance, $M\{\mathcal{D}\{y(t)/x(s)\}\}$ can be used as the stochasticity measure. For determined plants this value tends to zero $M\{\mathcal{D}\{y(t)/x(s),\ s<t\}\}=0$, whereas for stochastic plants

$$M\{\mathcal{D}\{y(t)/x(s),\ s<t\}\} = const \neq 0.$$

The response of a real system due to uncontrolled effects and noise is a sum of the determined component and noise ξ_t

$$y(t) = M\{y(t)/x(s),\ s<t\} + \xi_t \quad (2)$$

There may exist a stochastic link between the right-hand part terms.

The plant stochasticity can be defined as a difference of dispersions for the plant and model output processes

$$M\{\mathcal{D}\{y(t)/x(s),\ s<t\}\} = \mathcal{D}\{y(t)\} - \mathcal{D}\{y_M(t)\} \quad (3)$$

The following expression will henceforward be called the plant stochasticity

$$\gamma = \frac{M\{\mathcal{D}\{y(t)/x(s),\ s<t\}\}}{\mathcal{D}\{y(t)\}} \quad (4)$$

Since all terms in (3) are positive the variation limits for ν are:

$$0 \leq \nu \leq 1$$

The equality $\nu = 0$ denotes that the plant studied is a determined one. Otherwise the plant is stochastic and ν is treated as the plant stochasticity.

Similarly to the stochasticity value we can introduce the value for regularity or determinateness of the plant identified

$$q = \frac{\mathcal{D}\{M\{y(t)/x(s), s<t\}\}}{\mathcal{D}\{y(t)\}} \tag{5}$$

Note that the values introduced are of dual nature and satisfy the expression $\nu + q = 1$.

Consequently, the plant is regular (determined), if $q = 1$.

Consider the definition of the plant stochasticity from more general standpoints, i.e. in terms of the information theory. The stochasticity or randomness of a random variable can be described by entropy. On the other hand the plant stochasticity or randomness can be also described by the randomness of output variables (Petrov, 1975).

The plant stochasticity is obtained by expanding the entropy to two terms (Shannon, 1963)

$$H\{y\} = H_x\{y\} + I\{X,y\} \tag{6}$$

where $H\{y\}$ is the absolute entropy of the plant output; $H_x\{y\}$ is the conventional entropy of the output relative to the input; $I\{X,y\}$ is information contained in the plant input on its output. The plant input is assumed to be a vector. The entropy expansion formula in this case can be interpreted differently. The term $I\{X,y\}$ is the randomness of the plant output determined by the input signals observed. The terms $H_x\{y\}$ is the output randomness caused by the noise and uncontrolled disturbances.

We assume the relation

$$\nu_I = \frac{H_x\{y\}}{H\{y\}} \tag{7}$$

to be the stochasticity value in terms of the information theory.

Since $H_x\{y\} \leq H\{y\}$, and the signs of $H_x\{y\}$ and $H\{y\}$ coincide, we derive the following inequality from (7)

$$0 \leq \nu_I \leq 1$$

Let us agree on the following definitions. A plant is determined (regular) if $\nu = 0$. A plant is strictly stochastic if $\nu = 1$.

When $\nu \neq 0$, $\nu < 1$ the plant is stochastic.

Similarly to the above case we can introduce the value of the plant determinateness

$$q_I = \frac{I\{X,y\}}{H\{y\}} \tag{8}$$

The values introduced supplement each other and meet the relation

$$\nu_I + q_I = 1$$

Unlike the dispersion values (4) and (5) the information values are valid estimates of the plant stochasticity and regularity.

3. IDENTIFICATION BY THE INFORMATION CRITERION

We develop the optimal model of the plant studied assuming that the model output contains maximal information on the plant output. In this case the identification criterion can be written as

$$L = M\left\{\log \frac{\psi(y/y_M)}{\psi(y)}\right\} = \max \tag{9}$$

where $\psi(y/y_M)$ is the conventional density of distribution of y relative to y_M, $\psi(y)$ is the abosolute density of distribution of y, $M\{\log \psi(y/y_M)/\psi(y)\} = I(y,y_M)$ is information on y_M contained in y. The criterion (9) requires to maximize this information.

Assume that the joint distribution of y and y_M is normal. In this case the amount of information on the output of plant y contained in the output of model y_M can be expressed via the coefficient of correlation between these values $R(y, y_M)$:

$$I\{y,y_M\} = -\frac{1}{2}\log\left[1 - R^2(y,y_M)\right] \tag{10}$$

$$R^2(y,y_M) = \frac{[M(y\cdot y_M)]^2 - [M\{y\}\cdot M\{y_M\}]^2}{\mathcal{D}\{y\}\cdot\mathcal{D}\{y_M\}} \tag{11}$$

Note that in this case y is the output of any plant, including nonlinear, Hammerstein type etc. The output of model y_M can be written as:

$$y_M = \sum_{i=1}^{n} l_i X_i + m \tag{12}$$

where $(X_1, .., X_n)$ is the vector of inputs, m is the noise with characteristics $M\{m\} = 0$, $\mathcal{D}\{m\} = \mathcal{D}_m$. To simplify the calculations we assume that $M\{X\} = 0$, consequently, $M\{y_M\} = 0$.

Hence, a random nonlinear plant model (2) optimal in terms of (9) is being looked for.

Since logarithmic dependence is monotonous and $|R| \leq 1$, maximization of (10) requires the maximization of (11).

It is easily calculated that

$$[M(y\cdot y_M)]^2=\Big[M\Big(y\cdot\Big[\sum_{i=1}^{n} b_i X_i+m\Big]\Big)\Big]^2=\Big[\sum_{i=1}^{n} b_i K(X_i,y)+K(y,m)\Big]^2$$

$$D\{y_M\}=\sum_{i=1}^{n}\sum_{j=1}^{n} b_i b_j K(X_i,X_j)+\sum_{i=1}^{n} b_i K(X_i,m)+D_m$$

Substituting the expressions obtained in (11) we see that the problem reduces to optimization of the functional

$$V(b)=\frac{\Big[\sum_{i=1}^{n} b_i K(X_i,y)+K(y,m)\Big]^2}{\sum_{i=1}^{n}\sum_{j=1}^{n} b_i b_j K(X_i,X_j)+\sum_{i=1}^{n} b_i K(X_i,m)+D_m} \tag{13}$$

i.e. it is required to find the vector of parameters $b=(b_1,\dots,b_n)$ for which the functional (13) is maximal.

Differentiating (13) by $b_1,\dots,b_n$ and making the derivatives equal to zero we obtain a set of equations for determining the unknown coefficients

$$K(X_q,y)\cdot\mu=\sum_{i=1}^{n} b_i K(X_i,X_q)+K(X_q,m)$$

$$\mu=\frac{\sum_{i=1}^{n}\sum_{j=1}^{n} b_i b_j K(X_i,X_j)+\sum_{i=1}^{n} b_i K(X_i,m)+D_m}{\sum_{i=1}^{n} b_i K(X_i,y)+K(y,m)} \qquad q=1,\dots,n \tag{14}$$

Henceforward K denotes the correlation function. For a certain class of plants methods are developed (Durgaryan, Pashchenko, 1977) for modelling plants as

$$y_M=\sum_{i=1}^{n} b_i M\{y/x_i\}+m \tag{15}$$

The vector of parameters b_i optimal in terms of the r.m.s. criterion is obtained by solving corresponding dispersion equations of identification (Durgaryan, Pashchenko, 1977).

Identification by the information criterion for such models leads to solution of the following set of equations

$$\theta_{yx_q}\cdot\mu=\sum_{i=1}^{n} b_i\,\theta_{yyx_ix_q}+\theta_{yyx_qm}$$

$$\mu=\frac{\sum_{i=1}^{n}\sum_{j=1}^{n} b_i b_j\,\theta_{yyx_ix_j}+\sum_{i=1}^{n} b_i\,\theta_{yyx_im}+D_m}{\sum_{i=1}^{n} b_i\,\theta_{yx_i}+K_{ym}} \qquad q=1,\dots,n \tag{16}$$

θ are dispersion functions of various types (Durgaryan, Pashchenko, 1977; Rajbman, 1970).

The model output for a dynamic plant is

$$y_M(t)=\int_{-\infty}^{\infty} g(\tau)X(t-\tau)\,d\tau+m(t) \tag{17}$$

The optimal weighting function $g(\tau)$ is derived similarly, i.e. by the criterion of the maximal information on the plant output contained in the model output.

We omit the computations since they are similar to the preceding case and write down the final result which is an integral equation for determining the optimal weighting function $g(\tau)$:

$$\int_{-\infty}^{\infty} R_{xx}(\tau_2-\tau_1)g(\tau_2)d\tau_2 = \mu K_{yx}(\tau_1) + K_{mx}(\tau_1)$$
$$\mu = \frac{\int_{-\infty}^{\infty}\int_{-\infty}^{\infty} K_{xx}(\tau_2-\tau_1)\bar{g}(\tau_2)g(\tau_1)d\tau_1 + 2\int_{-\infty}^{\infty} g(\tau_1)K_{mx}(\tau_1)d\tau_1 + K_m(0)}{\int_{-\infty}^{\infty} g(\tau_1)K_{xy}(\tau_2-\tau_1)d\tau_1 + K_{my}(\tau_1)} \tag{18}$$

In obtaining equation (18) no constraints were imposed on the form of the desired weighting function. Solution of equation (18) is, however, largely dependent on the constraints. Such constraints can include, for instance, limit system characteristics and conditions for system feasibility (Pashchenko, 1980).

Note that in the absence of noise in the equations of models (12), (15), and (17) and for $\lambda = 1$, the identification results in terms of the accuracy (root-mean-square) and information criteria coincide. This follows directly from (14), (16), and (18).

4. IDENTIFICATION OF UNBIASED MODELS

A priori information is vital in mathematical simulation of real systems. A priori information characterizes design, technological and other parameters of the system, the model structure and constraints for its parameters. This type of information is usually presented as functions of the plant operators or output characteristics (Pashchenko, 1980). One of the most interesting applications of a priori information is development of unbiased models, i.e. models yielding an unbiased estimate of the value forecasted. Consider first a more general case, however. It is required to estimate parameters of model (12) by criterion (9) for the constraint

$$\sum_{i=1}^{n} b_i c_i = d \tag{19}$$

Hence, identification of a plant is presented as an optimization problem of maximizing functional (13) on a set limited by the surfaces defined by (12) and (19).

Unlike the preceding case we assume:

$$M\{X_i\} \neq 0,\ i=1,\dots,n\ ;\ K(X_i,X_j)=0,\ i\neq j\ ;\ K(X_i,m)=0$$

for any i. These assumptions are aiming only at simplification of the final results.

The Lagrange function for our case is:

$$\mathcal{L}(l) = \frac{\left[\sum_{i=1}^{n} l_i K(X_i,y) + K(y,m) - M\{y\}\left(\sum_{i=1}^{n} l_i M\{X_i\}\right)\right]^2}{\sum_{i=1}^{n} l_i^2 \mathcal{D}\{X_i\} + \mathcal{D}_m - \left[\sum_{i=1}^{n} l_i M\{X_i\}\right]^2} + \lambda\left(\sum_{i=1}^{n} c_i l_i - d\right) \tag{20}$$

The required minimal value of this functional yields the equations of identification of the model parameters (12)

$$\frac{\left[\sum_{i=1}^{n} b_i K(X_i,y) - K(y,m) - M\{y\}\left(\sum_{i=1}^{n} b_i M\{X_i\}\right)\right]\left[K(X_q,y) - M\{y\} M\{X_q\}\right]}{\sum_{i=1}^{n} b_i^2 \mathcal{D}\{X_i\} + \mathcal{D}_m - \left[\sum_{i=1}^{n} b_i M\{X_i\}\right]^2} + \lambda C_q =$$

$$= \frac{\left[\sum_{i=1}^{n} b_i K(X_i,y) + K(y,m) - M\{y\} \sum_{i=1}^{n} b_i M\{X_i\}\right]^2 \left[b_q \mathcal{D}\{X_q\} - M\{X_q\} \sum_{i=1}^{n} b_i M\{X_i\}\right]}{\left\{\sum_{i=1}^{n} b_i^2 \mathcal{D}\{X_i\} + \mathcal{D}_m - \left[\sum_{i=1}^{n} b_i M\{X_i\}\right]^2\right\}^2} ; \quad q=\overline{1,n} \tag{21}$$

Optimal estimates of the model parameters are obtained as functions of λ : $b_q^* = b_q^*(\lambda)$. The value of λ is calculated by substituting $b_q(\lambda)$ in (19). Geometry-wise the identification equations obtained show that the derivative of functional (11) on surface (19) tends to zero in point $b^* = (b_1^*, \dots, b_n^*)^T$.

The above expressions yield easily equations for estimation of parameters of the unbiased model of a system. In this case the constraint (19) and identification equations are written in the form

$$\sum_{i=1}^{n} b_i M\{X_i\} = M\{y\}$$

$$\left[\sum_{i=1}^{n} b_i K(X_i,y) - K(y,m) - M\{y\} \sum_{i=1}^{n} b_i M\{X_i\}\right]\left[K(X_q,y) - M\{y\} M\{X_q\}\right] + \lambda M\{X_q\} =$$

$$= \frac{\left[\sum_{i=1}^{n} b_i K(X_i,y) + K(y,m) - M\{y\} \sum_{i=1}^{n} b_i M\{X_i\}\right]^2 \left[b_q \mathcal{D}\{X_q\} - M\{X_q\} \sum_{i=1}^{n} b_i M\{X_i\}\right]}{\left\{\sum_{i=1}^{n} b_i^2 \mathcal{D}\{X_i\} + \mathcal{D}_m - \left[\sum_{i=1}^{n} b_i M\{X_i\}\right]^2\right\}^2} ; \quad q=\overline{1,n}$$

The mathematical model obtained contains the maximal amount of information on the plant output and provides the unbiased forecast of the output signal.

5. SUMMARY

The paper considers the information approach to solution of the identification problem. The optimality criterion is the maximal amount of data on the plant output in the model output. Estimation of stochasticity and determinateness of the systems studied is provided in terms of the information approach.

The information methods of identification can be applied to different plants: technical, economic, ecological etc. Specifically, the methods described have been applied to design of information systems and estimation of distribution of fish stock. The information methods of identification provide a more illustrative picture of processes investigated and promote optimal modelling when correlation and dispersion methods are not applicable.

REFERENCES

Durgaryan I.S., Pashchenko F.F. Mnogostupenchataya identifikatsiya i prognozirovaniye. Institute of Control Sciences, M., 1977, pp.68.

Durgaryan I.S., Pashchenko F.F. Identifikatsiya nelineynykh ob'yektov po slozhnym kriteriyam. A i T, No.7, 1980, pp.61-71.

Eykhoff P. Osnovy identifikatsii sistem upravleniya. "Mir" Publishers, 1975, pp.685.

Pashchenko F.F. Ispol'zovaniye apriornoy informatsii v zadache identifikatsii. Proc. of VIII All-Union Conference on Control Problems, vol.1, Institute of Control Problems, the ESSR Gossplan, Moskow - Tallin, 1980, pp.242-244.

Petrov V.N. et al. Obshchaya informatsionnaya teoriya proyektirovaniya dinamicheskikh sistem izmereniya, upravleniya i kontrolya. In coll. "Tekhnicheskaya kibernetika", vol.7, VINITI Publishers, 1975, pp.202-267.

Rajbman N.S. What is Identification. "Nauka" Publishers, M., 1970, pp.119.

Shannon K. Raboty po teorii i kibernetike. IL Publ., M., 1963, pp.829.

USSR Academy of Sciences
Institute of Control Sciences
Profsoyuznaya 65
117 342 Moscow
USSR

QUANTUM STOCHASTIC PROCESSES

Anatolij Dvurečenskij, Sylvia Pulmannová
Bratislava

Key words : Stochastic processes, quantum logics, joint distributions.

ABSTRACT

Stochastic processes on a quantum logic (L, M) are introduced and their properties are studied. Stochastic integral with respect to a Brownian motion process is constructed.

In the quantum logic approach, the quantum theory is introduced in terms of the set L of all yes-no measurements, which is called the logic of a physical system, and the set M of physical states of that system. For a general physical system, its logic is usually assumed to be an orthomodular σ-lattice. If the logic L is given, we can identify the states of the physical system with the probability measures on L and the observables (i.e. physical quantities) with the σ-homomorphisms from Borel subsets B^1 of the real line R^1 to L (see e.g. Mackey (1963), Varadarajan (1968)). In the following we shall suppose that the couple (L, M) is a sum quantum logic in the sense of Dvurečenskij-Pulmannová (1978, 1980). In a sum quantum logic, the sums of certain observables are defined. An important example of such a sum quantum logic is the logic of all closed subspaces of a Hilbert space.

If x is an observable, m is a state on L and $f : R^1 \to R^1$ is a Borel function, then $f(x) \to x \circ f^{-1}$ is also an observable and the expectation of $f(x)$ in the state m is defined by

$$(1) \qquad m(f(x)) = \int f(t)\, m_x(dt)$$

if the integral exists, where $m_x(E) = m(x(E))$, $E \in B^1$, is the distribution of x in m. For the details on logics, states and observables see Varadarajan (1968).

We denote by R^n the Euclidean n-dimensional space and by B^n the Borel σ-algebra of its subsets.

Definition 1. Observables $x_1, x_2, \ldots, x_n$ on (L,M) are said to have a joint distribution (j.d.) in the state m if the observables $\sum_{i=1}^n a_i x_i$ exist for any $(a_1, \ldots, a_n) \in R^n$ and if there is a measure μ on B^n such that

$$(2) \quad \mu\{(\omega_1, \ldots, \omega_n) : \sum_{i=1}^n a_i \omega_i \in E\} = m\left(\left(\sum_{i=1}^n a_i x_i\right)(E)\right)$$

for all $E \in B^1$ and all $(a_1, \ldots, a_n) \in R^n$.

By the Cramer-Wold theorem, if the j.d. exists, it is unique. Joint distributions of this type were introduced by Urbanik (1961) for the self adjoint operators on a Hilbert space and they were studied also by Varadarajan (1968) and Gudder (1968). They are called also " joint distributions of type 2 ". An example of observables having j.d. are the position and momentum operators in so-called coherent states(Urbanik (1961)).

Lemma 1. Let $x_1, \ldots, x_n$ have j.d. in m. Then the observables $y_k = \sum_{i=1}^n a_{ik} x_i$, $k = 1,2,\ldots,r$ have j.d. in m for all $(a_{1k}, a_{2k}, \ldots, a_{nk}) \in R^n$, $k = 1,2,\ldots,r$, $r = 1,2,\ldots$. In addition,

$$(3) \quad m_{y_1,\ldots,y_r}(B) = m_{x_1,\ldots,x_n} T^{-1}(B), \quad B \in B^r,$$

where $T : R^n \to R^r$, $T(\omega_1, \ldots, \omega_n) = \left(\sum_{i=1}^n a_{i1} \omega_i, \ldots, \sum_{i=1}^n a_{ir} \omega_i\right)$.

Here we denote by $m_{x_1,\ldots,x_n}$ the j.d. of $x_1, \ldots, x_n$ in m.

We shall say that the observables $\{x_\alpha : \alpha \in A\}$, where A is any set, have j.d. in m if any finite subset of A has it.

Definition 2. We shall say that $\{x_t : t \in D\}$, $D \subset R^1$, is a stochastic process on L in the state m (shortly on (L, m)) if the observables $\{x_t : t \in D\}$ have j.d. in m.

In the following, $D = [a, b)$ or (a, b) where $-\infty < a < b \leq +\infty$.

Theorem 1. Let $\{x_t : t \in D\}$ be a stochastic process on (L,m). Then there exists a stochastic process $\{p_t : t \in D\}$ on a probability space $(\Omega, \mathcal{S}, \mu)$ such that

$$(4) \quad m\left(\left(\sum_{i=1}^n a_i x_{t_i}\right)(E)\right) = \mu\left(\left(\sum_{i=1}^n a_i p_{t_i}\right)^{-1}(E)\right)$$

for any $t_1, \ldots, t_n \in D$, $(a_1, \ldots, a_n) \in R^n$ and $E \in B^1$.

Proof. It is a consequence of Kolmogoroff Extension and Existence theorems (Yeh (1973)). The probability space $(\Omega, \mathcal{S}) = (R^D, \sigma(J))$, where $\sigma(J)$ is the σ-algebra generated by the set of all cylinders with finite bases in R^D, and p_t is the projection on the coordinate t. For any $(t_1, \ldots, t_n) \in D^n$,

$$m_{x_{t_1},\dots,x_{t_n}}(B) = \mu(p_{t_1,\dots,t_n}^{-1}(B)), \quad B \in B^n ,$$

where $p_{t_1,\dots,t_n}$ is the projection on R^n indexed by $\{t_1,\dots,t_n\}$.

The observables $\{x_\alpha : \alpha \in A\}$ are independent in m if the j.d. exists and for any k-touple the j.d. equals to the direct product of the marginal probability distributions. As a cosequence of Theorem 1 we get that $\{x_t : t \in D\}$ are independent iff $\{p_t : t \in D\}$ are independent.

We say that an observable y is Gaussian in m if $m_y \in N(a, b)$, $a, b \in R^1$, $b > 0$, i.e. if

$$(5) \qquad m_y(-\infty, t) = \frac{1}{\sqrt{2\pi}\, b} \int_{-\infty}^{t} \exp\left\{\frac{-(a-z)^2}{2b^2}\right\} dz .$$

Definition 3. We shall say that the stochastic process $\{x_t : t \in D\}$ on (L, m) is Gaussian if for any $(a_1,\dots,a_n) \in R^n$ and any $(t_1,\dots,t_n) \in D^n$ the observables $y = \sum_{i=1}^n a_i x_{t_i}$ are Gaussian in m.

Definition 4. We shall say that the stochastic process $\{x_t : t \in D\}$ on (L, m) is a Brownian motion process if

(i) for $t_0, t_1, \dots, t_n \in D$, $t_0 < t_1 < \dots < t_n$, the observables $\{x_{t_j} - x_{t_{j-1}}, j = 1,\dots,n\}$ are independent,

(ii) for $t', t'' \in D$, $t' < t''$, the probability distribution of $x_{t''} - x_{t'}$ is $N(0, t'' - t')$ (i.e. Gaussian with mean zero and variance $t'' - t'$),

(iii) $m(x_a^2) = 0$.

Theorem 2. The stochastic process $\{x_t : t \in D\}$ is a Brownian motion process iff it is a centred Gaussian process with the covariance

$$(6) \quad K(s, t) = \int p_s \cdot p_t \, d\mu = \min(s-a, t-a), \quad s, t \in D ,$$

where $\{p_t : t \in D\}$ is the process corresponding to $\{x_t : t \in D\}$ by Theorem 1.

Proof. Let $\{x_t : t \in D\}$ be a centred Gaussian process with the covariance (6). By Theorem 1, $\{p_t : t \in D\}$ is centred Gaussian with covariance (6). Properties (ii) and (iii) of Definition 4 follow immediately. To show (i), let $t_0 < t_1 < \dots < t_n$. By Lemma 1, j.d. of $\{x_{t_j} - x_{t_{j-1}}, j = 1,2,\dots,n\}$ is $m_{x_{t_1},\dots,x_{t_n}} \cdot T^{-1}$, where $T : R^{n+1} \to R^n$, $T(\omega_0,\dots,\omega_n) = \{\omega_1 - \omega_0, \omega_2 - \omega_1, \dots, \omega_n - \omega_{n-1}\}$. Let us put $y_j = x_{t_j} - x_{t_{j-1}}$, $j = 1,2,\dots,n$. Then $m_{y_1,\dots,y_n}(E_1 \times \dots \times E_n) = \mu(p_{t_1,\dots,t_n}^{-1} T^{-1}(E_1 \times \dots \times E_n)) = \prod_{i=1}^n \mu((p_{t_i} - p_{t_{i-1}})^{-1}(E_i)) = \prod_{i=1}^n m_{x_{t_i} - x_{t_{i-1}}}(E_i)$, as $\{(p_{t_1} - p_{t_0}), \dots,$

$(p_{t_n} - p_{t_{n-1}})\}$ are independent by Theorem 16.13 (Yeh (1973)).

The converse statement follows from Theorem 1 and Theorem 17.1 (Yeh (1973)).

Let $S = \{x_t : t \in D\}$ be a Brownian motion process on (L, m) and let us denote by $V(S)$ the set of all finite real linear combinations of the elements of S. Let us define the map $s : V(S) \times V(S) \to R^1$ by

$$(7) \quad s(x, y) = \frac{1}{4}\left[m((x+y)^2) - m((x-y)^2)\right] = \int_{R^2} \omega_1 \omega_2 dm_{x,y}$$

where $m_{x,y}$ is the joint distribution of x, y. It can be shown that $s(x, y)$ is a real inner product. Let us write $x = y\ [m]$ if $\|x-y\|^2 = s(x-y, x-y) = m((x-y)^2) = 0$ and denote by $\mathcal{L}_2(S)$ the Hilbert space which is obtained by completing $V(S)$ with respect to s. We shall construct the stochastic integral with respect to S. To this aim, let us consider the Hilbert space $\mathcal{L}_2(D)$ of square-integrable functions on D with respect to the Lebesgue measure μ_L. Let $C(D)$ be the class of such elements of $\mathcal{L}_2(D)$ that can be written in the form of step functions with finitely many steps and with the compact supports in D. $C(D)$ is a dense linear subspace of $\mathcal{L}_2(D)$ (Yeh (1973)).

Definition 5. Let $f \in C(D)$, $f = \sum_{k=1}^n c_k \chi_{\Delta_k}$ where $\{c_k, k = 1, 2, \ldots, n\} \subset R^1$ and $\Delta_k = (t_k, t_{k+1}]$, $a \leq t_1 < t_2 < \ldots < t_{n+1} \leq b$. The stochastic integral $I(f)$ with respect to the Brownian motion process $\{x_t : t \in D\}$ is an observable on L defined by

$$(8) \qquad I(f) = \sum_{k=1}^n c_k (x_{t_{k+1}} - x_{t_k}).$$

It can be easily seen that $I(f)$ is well-defined and is not dependent on the representation of f in the form of a step function.

Proposition 1. For any $f, g \in C(D)$ and $\alpha, \beta \in R^1$ we have
(a) $s(I(f), I) = 0$, where I is the unit observable, i.e. $I(\{1\}) = 1$,
(b) $\|I(f)\|^2 = m(I(f)^2) = \|f\|_D^2$,
(c) $s(I(f), I(g)) = (f, g)_D$,
(d) $I(\alpha f + \beta g) = \alpha I(f) + \beta I(g)$.

Proof. (a) and (d) follow by the definition and the fact, that I can be added to any stochastic process.

(c) We set $f(t) = \sum_{k=1}^n c_k' \chi_{\Delta_k}(t)$, $g(t) = \sum_{k=1}^n c_k'' \chi_{\Delta_k}(t)$,

where $\Delta_k = (t_k, t_{k+1}]$, $k = 1, 2, \ldots, n$. Then

$$s(I(f), I(g)) = s\left(\sum_{k=1}^n c_k'(x_{t_{k+1}} - x_{t_k}), \sum_{k=1}^n c_k''(x_{t_{k+1}} - x_{t_k})\right) = \sum_{k=1}^n c_k' c_k''\, m((x_{t_{k+1}} - x_{t_k})^2) + \sum_{j \neq k} c_k' c_j''\, s(x_{t_{k+1}} - x_{t_k},$$

$$x_{t_{j+1}} - x_{t_j}) = \sum_{k=1}^{n} c_k' c_k'' (t_{k+1} - t_k) = \int_D f(t)g(t)\, \mu_L(dt) = (f, g)_D .$$

(b) follows from (c).

Definition 6. Let $f \in \mathcal{L}_2(D)$ and let $\{f_n\} \subset C(D)$ be such that $\lim \|f_n - f\|_D = 0$. The element $x \in \mathcal{L}_2(S)$ such that $\lim \|I(f_n) - x\| = 0$ will be called the stochastic integral of f with respect to $\{x_t : t \in D\}$ and we set $x = I(f)$.

It can be easily shown that $I(f)$ is not dependent on $\{f_n\}$. The following theorem follows from Proposition 1 by the limiting procedure.

Theorem 3. Statements (a)–(d) from Proposition 1 are fulfilled for any $f, g \in \mathcal{L}_2(D)$ and $\alpha, \beta \in R^1$.

An isomorphism theorem between $\mathcal{L}_2(D)$ and a closed subspace of $\mathcal{L}_2(S)$, analogical to the classical one, can also be proved.

In general, the stochastic integral may be only an abstract element of $\mathcal{L}_2(S)$. In the following examples, the stochastic integral is an observable.

Example 1. By Gudder (1968), the observables $x_1,\dots,x_n$ have a joint distribution of type 1 in a state m if there is a measure μ on B^n such that

$$(9) \qquad m(x_1(E_1) \wedge x_2(E_2) \wedge \dots \wedge x_n(E_n)) = \mu(E_1 \times E_2 \times \dots \times E_n)$$

for any $E_1 \times E_2 \times \dots \times E_n \in B^n$. Let $\{x_t : t \in D\}$ be observables on the Hilbert space logic $L(H)$, $3 \le \dim H \le \aleph_0$. Let m be a state on $L(H)$, defined by $m(P) = \mathrm{tr}\, \rho P$, $P \in L(H)$, where ρ is the density operator corresponding to m. In Dvurečenskij-Pulmannová (1982) there is shown, that if $\{x_t : t \in D\}$ have j.d. of type 1 in m (in the sense that any finite subset has one), then they have also a j.d. of type 2 and, moreover, there is a subspace $H_0 \subset H$, which reduces all x_t, $t \in D$. The operators x_t/H_0, $t \in D$, are mutually compatible and $\rho H_0 \subset H_0$. From this it follows that to any stochastic integral with respect to $\{x_t : t \in D\}$ there is a self-adjoint operator x, reduced by H_0, which is in the same equivalence class in $\mathcal{L}_2(S)$. The observable x is Gaussian in m.

Example 2. In some analogy with Cocroft and Hudson (1977), let us consider a two dimensional Brownian motion process $\{(x_t, y_t) : t \in D\}$ on (L, m). It is defined by following properties:

(i) $\{x_t, y_t : t \in D\}$ have j.d. of type 2 in m,

(ii) for any $t_0, t_1, \dots, t_n \in D$, $t_0 < t_1 < \dots < t_n$, the set $\{x_{t_i} - x_{t_{i-1}}, y_{t_i} - y_{t_{i-1}}, i = 1,\dots,n\}$ is independent,

(iii) for $t' < t''$, $t', t'' \in D$, both observables $x_{t''} - x_{t'}$ and $y_{t''} - y_{t'}$

are $N(0, t'' - t')$,

(iv) $m(x_a^2) = m(y_a^2) = 0$,

(v) $\{x_t : t \in D\}$ and $\{y_t : t \in D\}$ are two compatible sets of observables.

Then for any $f \in \mathcal{L}_2(D)$, the stochastic integral $(I_1(f), I_2(f))$ is a pair of Gaussian observables.

In both examples we use the fact, that to any sequence of mutually compatible observables $\{x_n\}$, which is Cauchy in $\mathcal{L}_2(S)$, there is a limit which is an observable compatible with $\{x_n\}$. Indeed, by Varadarajan (1968) there is an observable y and Borel functions f_n, $n = 1,2,\ldots$, such that $x_n = f_n(y)$. We have

$$m((x_{n+p} - x_n)^2) = \int (f_{n+p} - f_n)^2 \, dm_y \longrightarrow 0,$$

so that the sequence $\{f_n\}$ is Cauchy in $\mathcal{L}_2(R^1, B^1, m_y)$. Let f be such that $f_n \to f$. Put $x = f(y)$, then

$$\int (f_n - f)^2 \, dm_y = m((x_n - x)^2) \to 0.$$

REFERENCES

Cocroft, A.M., Hudson, R.L. (1977): Quantum mechanical Wiener processes. Journ. Multivariate Analysis 7, No.1, 107-124.

Dvurečenskij, A., Pulmannová, S. (1978): Stochastic processes on quantum logics. Rep. Math. Phys., to be published.

Dvurečenskij, A., Pulmannová, S. (1980): On the sum of observables in a logic. Math. Slovaca 30, No. 4, 393-399.

Dvurečenskij, A., Pulmannová, S. (1982): A note on joint distributions of observables. Math. Slovaca 32, No. 2, 155-166.

Gudder, S.P. (1968): Joint distributions of observables. J.Math.Mech. 18, No.4, 326-335.

Urbanik, K. (1961): Joint probability distributions of observables in quantum mechanics. Studia Math. 21, 118-133.

Varadarajan, V.S. (1968): Geometry of Quantum Theory I. Van Nostrand, Princeton.

Yeh, J. (1973): Stochastic Processes and the Wiener Integral. Marcel Dekker, New York.

A. Dvurečenskij
Institute of Measurement
and Measuring Technique
Slovak Academy of Sciences
842 19 Bratislava
Czechoslovakia

S. Pulmannová
Mathematical Institute
Slovak Academy of Sciences
814 73 Bratislava
Czechoslovakia

SYMBOL ERROR RATE OF BINARY BLOCK CODES

Michele Elia

Turin

Key words: perfect binary block code
symbol error rate

ABSTRACT

Symbol error rate versus error probability for the binary symmetric channel is evaluated for perfect binary codes, i.e. repetition codes (2m+1, 1, 2m+1), Hamming codes (2^m-1, 2^m-1-m, 3) and Golay code (23, 12, 7), and for the parity check codes (n+1, n, 2).

INTRODUCTION

The symbol error rate, henceforth shortened to SER, of a block code is defined as the average that an information symbol is in error after complete decoding. Complete decoding means that every received word is assigned to a definite codeword; in the case of linear codes of block length n complete decoding according to a maximum likelihood criterion is therefore obtained from a coset decomposition of the n-dimensional space with respect to the code. In particular for binary block code (n, k, d) on the binary symmetric channel with error probability p, the set of coset leaders, i.e. the set of correctable error patterns, is made up of minimum weight coset vectors. The symbol error rate P_s, for group codes, may assume (see Macwilliams and Sloane(1977)) the following form

$$(1) \qquad P_s = \frac{1}{k} \sum_{\underline{e}} f(\underline{e})\, p^{wt(\underline{e})} (1-p)^{n-wt(\underline{e})}$$

where $f(\underline{e})$ is the number of incorrect information symbols after decoding, if the all zero codeword is transmitted, the error pattern is $\underline{e}$ and $wt(\underline{e})$ denotes its Hamming weight.

For computational purposes (1) will be written as follows

(2) $$P_s = \sum_{i=0}^{n} B_i \, p^i (1-p)^{n-i} = \sum_{i=0}^{n} E_i \, p^i$$

(3) $$kB_i = \sum_{wt(\underline{e})=i} f(\underline{e})$$

where kB_i is the number of incorrect information symbols, after decoding pertaining to the error patterns of weight i.

In this paper SER will be evaluated for the BCH code (15, 7, 5), the Golay code (23, 12, 7) and the parity check codes (n, n-1, 2). Moreover, the expressions of SER for repetition codes (2m+1, 1, 2m+1) and Hamming codes (2^m-1, 2^m-1-m, 3) will be recalled to collect the symbol error rate of all binary perfect block codes.
All the computations refer to systematic codes.

COUNTING CRITERIA

The computation of B_i, i=0,1, ... ,n , by means of equation (3) requires counting the number of information symbols equal to one, associated to codewords of given weight; this counting is based on the cyclic property of all codes here considered.

Let C be a systematic cyclic code (n, k, d) having generator polynomial g(x) and let A_i be its weight distribution, i.e. A_i is the number of codewords of weight i, therefore it follows from a result of Peterson and Weldon (1972) that k iA_i/n counts the number of information symbols equal to one, pertaining to the set of codewords of weight i . Moreover if n and i are relatively prime, then n divides A_i, and the set of codewords of weight i can be partitioned into A_i/n disjoint subsets which are generated by cyclic shifts of A_i/n distinguished codewords. A similar result holds for the set of coset leaders. Let l(x) be the polynomial representation of a coset leader $\underline{l}$ of weight i prime with n, let us suppose g(x) has at least a primitive factor which does not divide l(x), therefore the n distinct vectors obtained by cyclic shifting $\underline{l}$ can be taken as coset leaders as well, otherwise $x^j l(x)$, for some j<n, belongs to the lateral having leader l(x), that is

$$l(x)+g(x)f(x)=x^j l(x)$$

where g(x)f(x) is a codeword; rewriting the above equation we obtain

$$(1+x^j)\,l(x)=g(x)f(x)$$

which is impossible because the primitive factor of g(x) can divide neither l(x) nor $1+x^j$ for j<n.

SYMBOL ERROR RATE

Let us consider first the Golay code (23, 12, 7) which is a triple error correcting perfect code, with weight distribution $A_0=A_{23}=1$, $A_7=A_{16}=253$, $A_8=A_{15}=506$, $A_{11}=A_{12}=1288$ and the remaining A_i are zero. To get B_i's the following argument applies. The code is perfect and corrects three or less errors, therefore received vectors of weight i may be divided for their decoding, into seven exhaustive and exclusive sets D_j, j=-3,-2,-1,0,1,2,3 , of

$\binom{26-i}{3}A_{i-3}$, $\binom{25-i}{2}A_{i-2}$, $(\binom{24-i}{1}+\binom{24-i}{2}\binom{i-1}{1})A_{i-1}$, $(\binom{i}{0}+\binom{i}{1}\binom{23-i}{1})A_i$, $(\binom{i+1}{1}+\binom{i+1}{2}\binom{22-i}{1})A_{i+1}$, $\binom{i+2}{2}A_{i+2}$, $\binom{i+3}{3}A_{i+3}$ cardinality, according to their vectors respectively fill in codewords of weights i-3, i-2, i-1, i, i+1, i+2 and i+3. As the codewords length is a prime number every group of A_h codewords contributes to B_i with $12A_h\ h/23$, h=i-3,..., i+3, so that using (3) we get

$$B_i=\frac{1}{23}\{\binom{26-i}{3}(i-3)A_{i-3}+\binom{25-i}{2}(i-2)A_{i-2}+[\binom{24-i}{1}+\binom{24-i}{2}\binom{i-1}{1}](i-1)A_{i-1}+$$
$$+[\binom{i}{0}+\binom{i}{1}\binom{23-i}{1}]\,iA_i+[\binom{i+1}{1}+\binom{i+1}{2}\binom{22-i}{1}](i+1)A_{i+1}+\binom{i+2}{2}(i+2)A_{i+2}+$$
$$+\binom{i+3}{3}(i+3)A_{i+3}\}$$

The resulting B_i and E_i are summarized in Table 1.

Considering now the BCH code (15, 7, 5), matter is not so simple since the code length is not a prime. Its weight distribution is $A_0=A_{15}=1$, $A_5=A_{10}=18$, $A_6=A_9=30$, $A_7=A_8=15$ and the other A_i's are zero. Codewords of weight 6,7,8 and 9 are respectively generated by 15 cyclic shifts of convenient codewords, while codewords of weight 5 and 10 are respectively generated by 15 cyclic shifts of one codeword and by only 3 cyclic shifts of another codeword. This code corrects all the error patterns of weight 0, 1 and 2 and Berlekamp (1968) has shown that the remaining correctable error patterns can be chosen of weight 3. Therefore we have 1 leader of weight 0, 15 leaders of weight 1, 105 leaders of weight 2 and 135 leaders of weight 3. Coset leaders of weights 1 and 2 are generated by 15 cyclic shifts respectively of 1 and 7 leaders, while the leaders of weight 3 are respectively generated by 15 cyclic shifts of 8 leaders and 5 cyclic shifts of 3 leaders. Summarizing, the 256 leaders are single out by means of 20 leaders. Therefore, received vectors of weight i may be divided for their decoding, into seven exhaustive and exclusive sets D_h, h=-3,-2,-1,0,1,2,3 , and the vectors of D_h respectively fill in codewords of weights i-3, i-2, i-1, i, i+1, i+2 and i+3. Each set D_h contributes to B_i with $7(i+h)|D_h|/15$.

The results of rather length computations are summarized in Table 1.

TABLE 1
Coefficient for symbol error rate

i	GOLAY B_i	GOLAY E_i	BCH B_i	BCH E_i
0	0	0	0	0
1	0	0	0	0
2	0	0	0	0
3	0	0	116	116
4	2695	2695	507	-885
5	11473	-39732	1193	3272
6	31339	285670	2191	-7374
7	84029	-1304820	3105	10836
8	188848	4190340	3330	-10356
9	341440	-9948400	2814	5880
10	519288	17918824	1810	-1232
11	657944	-24751440	858	-608
12	694134	26186160	339	416
13	624778	-20931680	105	-64
14	475750	12270720	15	0
15	301466	-4987136	1	0
16	161128	1256640	0	0
17	69608	-147840	0	0
18	22176	0	0	0
19	6160	0	0	0
20	1771	0	0	0
21	253	0	0	0
22	23	0	0	0
23	1	0	0	0

Perfect repetition codes (2m+1, 1, 2m+1), with a single information symbol per word, are m errors correcting, that is a codeword is in error if and only if m+1 or more symbols are in error, thus SER results

$$P_s = \sum_{h=m+1}^{2m+1} \binom{2m+1}{h} p^h (1-p)^{n-h}$$

Perfect Hamming codes $(2^m-1,\ 2^m-1-m,\ 3)$, $m>2$, correct one or less errors, therefore getting B_i the same argument used for Golay code applies. In decoding weight i received vectors, three exhaustive and exclusive sets D_{-1}, D_0 and D_1, respectively, of cardinality $(n-i+1)A_{i-1}$, A_i and $(i+1)A_{i+1}$, may be considered according to their vectors respectively filling in codewords of weight i-1, i and i+1. Recalling that the contribution to B_i of each group of A_h codewords, is $(2^m-1-m)hA_h/(2^m-1)$, we can write

$$B_i = \frac{1}{k}\left[(i+1)^2 A_{i+1} k/n + iA_i k/n + (i-1)(n-i+1)A_{i-1} k/n\right] =$$

$$= \frac{1}{2^m-1}\left[(i+1)^2 A_{i+1} + iA_i + (i-1)(2^m-i)A_{i-1}\right]$$

from this equation, using a generating function technique, we obtain SER in the closed form as reported in van Lint (1973)

$$P_s=\frac{1}{2^m}[1+(2^m-2)p-(1-2p)^{2^m-1}(1+2(2^m-2)(p-p^2))]$$

Finally we consider the parity check codes (n, n-1, 2), n>1, which used with complete decoding, do not improve SER. Optimal complete decoding requires two coset leaders: one is the all zero word, the other one is a word of weight 1, which can be chosen either to affect an information symbol or the parity check symbol, and the first choice leads to the following expression for SER

$$P_s= p[1+\frac{1}{n-1}\sum_{h=0}^{\lfloor (n-1)/2 \rfloor}\binom{n-1}{2h}\frac{n-(4h+2)}{2h+1}p^{2h}(1-p)^{n-2h-1}]$$

while the second leader choice gives $P_s=p$, therefore proving that SER depends on the leader choice even for codes having an automorphism group of high order.

REFERENCES

Berlekamp E.R. (1968): Algebraic Coding Theory, McGraw Hill, New York.

Peterson W.W. and Weldon E.J. (1972): Error-Correcting Codes, MIT Press, Boston.

van Lint J.H. (1973): Coding Theory, Lecture Notes in Mathematics n.201, Springer Verlag, Berlin.

MacWilliams F.J. and Sloane N.J.A. (1977): The Theory of Error Correcting Codes, vol.XX, North-Holland, Amsterdam.

Politecnico di Torino
Istituto Matematico
corso Duca d'Abruzzi 24
10129 Torino
Italy

BINARY COMMUNICATION OVER A CHANNEL SUBJECT TO ACTIVE INTERFERENCE

Thomas Ericson

Linköping

Keywords: Binary communication, active interference, game theory, jamming, error probability

---

ABSTRACT

Coding for a channel with active interference (jamming) is viewed as a game theory problem. The legal user controls the encoder and the decoder by the choice of a key, while the jammer controls the channel by the choice of an interference signal. For any given pair of encoder and decoder this leads to a two persons zero sum game. We study this game and solve it in one simple case.

THE PROBLEM

Let $M = \{0,1\}$ and let R^n be the n-dimensional Eucledian space. Suppose $c:R^n \times R^n \to R^n$ is a deterministic two-input channel which for any pair $(x,s) \in$ $\in R^n \times R^n$ of inputs produces an output $y = c(x,s) \in R^n$. A legal user has access to the x-input and tries to use it for transmission of a message $m \in M$. A jammer controls the s-input. The legal user has also at his disposal a key variable $z \in R^n$ with the aid of which he is able to simultaneously control the encoder and the decoder.

We will consider the simple case $c(x,s) = x + s$. The encoder will be assumed to be

$$f(m,z) = \begin{cases} z & m = 0 \\ -z & m = 1 \end{cases}$$

and the decoder will be

$$g(y,z) = \begin{cases} 0 & (y,z) \geq 0 \\ 1 & (y,z) < 0 \end{cases}$$

where (y,z) denotes the inner product:

$$(y,z) \triangleq \sum_{i=1}^{n} y_i z_i \, .$$

Let $\hat{m} \triangleq g[f(m,z) + s,z]$. For $m \in M$ we define

$$e_m(z,s) = \begin{cases} 0 & \hat{m} = m \\ 1 & \hat{m} \neq m \end{cases}$$

and notice that

$$e_0(z,s) = 0 \leftrightarrow \quad ||z||^2 + (s,z) \geq 0$$
$$e_1(z,s) = 0 \leftrightarrow - ||z||^2 + (s,z) < 0 \quad .$$

where $||z||^2 \triangleq (z,z)$.

We will consider the two persons zero sum game defined by the control variables z and s and the objective function

$$e(z,s) \triangleq \frac{1}{2} e_0(z,s) + \frac{1}{2} e_1(z,s)$$

which the legal user tries to minimize and the jammer tries to maximize.

The control variables will be restricted so that

(1) $$||z||^2 \leq E_x; \quad ||s||^2 \leq E_s \ .$$

The quantities E_x and E_s can be interpreted as transmitted energy.

We will allow mixed strategies - i.e. random choices of the control variables - and our problem is to solve the game, i.e. to find a pair (Z_0,S_0) of independent random variables such that

$$E\ e(Z_0,S_0) = \min_Z \max_S E\ e(Z,S) = \max_S \min_Z E\ e(Z,S)$$

where the min and max are taken over all pairs (Z,S) of independent random variables with support in the regions defined by (1).

We emphasize the dependence on n, E_x and E_s by writing

$$P_n(E_x/E_s) \triangleq E\ e(Z_0,S_0) \ .$$

The quantity P_n will be called the <u>error probability</u> of the system.

BASIC RESULTS

It is obvious that e(z,s) depends on (z,s) only through the two scalar quantities $||z||^2$ and (s,z). From (1) and Schwarz' inequality we have

$$(s,z)^2 \leq ||s||^2 \cdot ||z||^2 \leq E_s \cdot ||z||^2 \ .$$

By this observation it is easy to see that any pair (z,s) of pure strategies

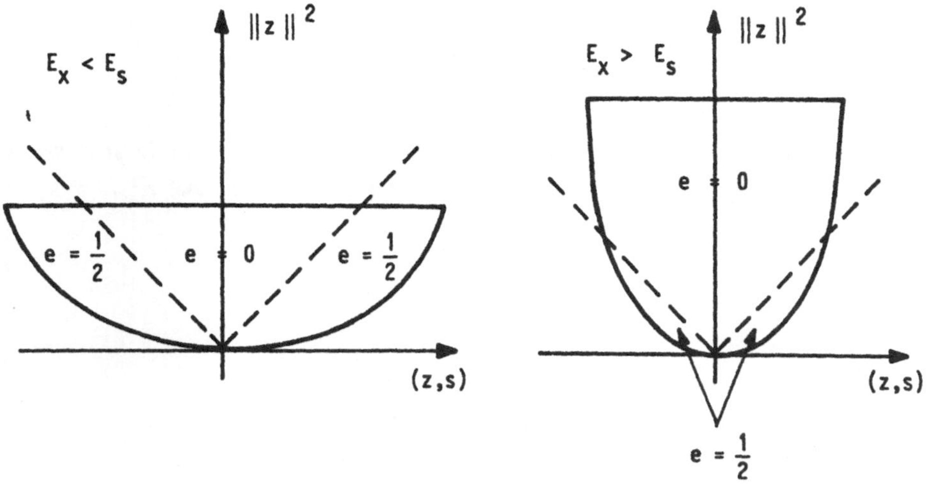

Fig. 1. Possible outcomes in the $[(z,s),||z||^2]$ - plane.

generates a point in the region

$$0 \leq ||z||^2 \leq E_x$$
$$0 \leq |(z,s)| \leq \sqrt{E_s}\, ||z|| \quad .$$

Conversel y it is also easy to see that any point in this region is achieved by some pair of strategies. The two cases $E_x < E_s$ and $E_x > E_s$ are illustrated in Fig. 1.

It seems reasonable that both parties should utilize all their available energy. This is indeed the case that follows from

Proposition 1: for any $\lambda > 1$

$$e(\lambda z,s) \leq e(z,s) \leq e(z,\lambda s)$$

The proof is simple and will be omitted.

Let Q_Z be the probability distribution of the random variable Z over R^n and let U be an arbitrary orthogonal n x n matrix. We say that Z is isotropic if $Q_{UZ} = Q_Z$ for all orthogonal matrices U. We have

Proposition 2: if Z is isotropic then

$$E\, e(Z,Us) = E\, e(Z,s)$$

for any $s \in R^n$ and any orthogonal matrix U.

The proof follows by a simple change of variables argument. The same result also holds with a fixed z and an isotropic S.

THE MIN-MAX SOLUTION

Combining propositions 1 and 2 we find that the min-max point is achieved for $n \geq 2$ if Z_0 and S_0 are uniform over the n-spheres of radii $\sqrt{E_x}$ and $\sqrt{E_s}$, respectively. For n = 0,1,... let

$$T_n(a) \triangleq \int_0^a \sin^n x \, dx .$$

Straightforward calculations show that P_n is given by

$$P_n(E_x/E_s) = \frac{T_{n-2}(a)}{T_{n-2}(\pi)}$$

$$a = \arccos \sqrt{\frac{E_x}{E_s}}$$

$$n = 2,3,...$$

In Fig. 2 this function has been plotted for n = 2,3,...,10.

COMMENTS

Mc Eliece and Stark have investigated the information theoretic capacity of the present channel. In their case the saddle point was found to be Gaussian. This seems to be in agreement with our result, as with increasing dimensionality n the Gaussian distribution approaches a uniform distribution on an n-sphere with radius $\sqrt{n}\,\sigma$, where σ is the standard deviation.

REFERENCES

Blackwell D. and Girshick M.A. (1954): Theory of games and statistical behaviour. 5th printing. Wiley.

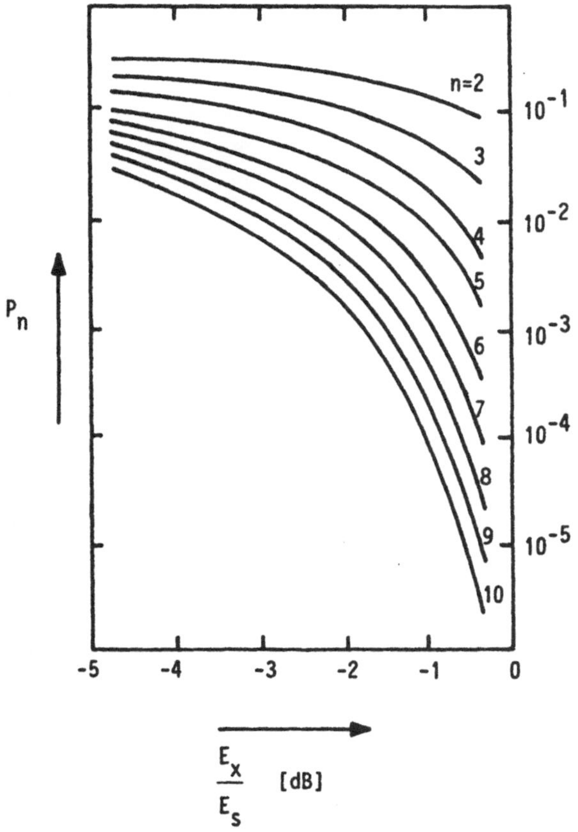

Fig. 2. Error Probability P_n as a function of Signal to Jamming Energies E_x/E_s; $n = 2,3,\ldots,10$.

Mc Eliece R.J. and Stark W.E. (1981): An information theoretic study of communication in the presence of jamming. In: IEEE International Conference on Communications, Denver, 45.3.1-45.3.5.

Linköping Institute of Technology
Department of Electrical Engineering
S-581 83 LINKÖPING
Sweden

GENERALIZATIONS OF THE MAXIMUM ENTROPY PRINCIPLE AND THEIR APPLICATIONS

Thomas Fischer
Dresden

Key words: Maximum entropy principle, universal encoding and decoding

---

ABSTRACT

Three inequalities closely related to the principle of maximum entropy and analogous principles are presented. They are applied to various problems of universal encoding and decoding.

PRINCIPLE OF MAXIMUM ENTROPY

The first inequality concerns the inaccuracy function,

$$H(p,q) = -\sum_{i=1}^{n} p_i \log q_i ,$$

and the Shannon entropy, $H(p) = H(p,p)$.

Theorem 1. Let P be a convex and compact class of finite probability distributions and let p_o be such that $H(p_o) = \max_{p \in P} H(p)$. Then the inequality

$$H(p,p_o) \leq \max_{p \in P} H(p)$$

is satisfied for any $p \in P$.

The estimation of Theorem 1 applies to the problem of universal source coding. Let C_o denote a block code designed for p_o and let R_o be the rate of C_o. Then, for any discrete memoryless source having a distribution $p \in P$, the error probability vanishes with increasing block length, if $R_o > H(p,p_o)$, cf. with Fischer (1977). From Theorem 1 we obtain that the latter inequality is certainly satisfied, if $R_o > \max_{p \in P} H(p)$. Thus C_o is universal for P.

Note that codes designed according to an entropy maximizing distribution are universal for any finite block length. This and the simple construction are the main advantages of maximum entropy codes. Universal variable length codes and universal search trees can be constructed in a very similar way. For large block length, however, known asymptotic methods have better performance, cf. with Davisson (1973), and Krichevsky and Trofimov (1981). A more detailed survey of the applications of Theorem 1 together with a generalization to entropy of order α can be found in a forthcoming paper.

PRINCIPLE OF MINIMUM MUTUAL INFORMATION

Let $k = (k_{ij})$ and $w = (w_{ij})$ be two stochastic matrices, $i = 1,\dots,m$, $j = 1,\dots,n$, and let

$$I(p;k,w) = \sum_{i,j} p_j k_{ij} \log \frac{w_{ij}}{\sum_j p_j w_{ij}}$$

denote the generalization of Shannon's mutual information function introduced in Fischer (1978). Define $I(p;k) = I(p;k,k)$.

Theorem 2. Let K be a convex and compact set of stochastic matrices and let p_o and k_o be chosen such that $I(p_o;k_o) = \min_{k \in K} \max_{p} I(p;k)$. Then the inequality

$$I(p_o;k,k_o) \geq I(p_o;k_o)$$

is satisfied for any $k \in K$.

Proof. For any $k \in K$ and $\lambda \in [0,1]$ define

$$f(\lambda) = I(p_o;(1-\lambda)\, k_o + \lambda k).$$

From the convexity of K we have $(1-\lambda)\, k_o + \lambda k \in K$, which implies $f(\lambda) \geq f(0) = I(p_o;k_o)$ for any $\lambda \in [0,1]$. Note that f must be also convex, since $I(p;k)$ is a convex function of k. Thus f is monotonically increasing, which implies $f'(\lambda) \geq 0$ for $\lambda \in [0,1]$ whenever the first derivation exists. Now a rather straightforward calculation yields

$$f'(0) = I(p_o;k,k_o) - I(p_o;k_o),$$

which proves the inequality of Theorem 2. It remains to show that the first derivation for $\lambda = 0$ exists. From the definition of the generalized mutual information function it is easy to see that the only critical case would be the occurence of a pair i,j with $k_{o\,ij} = 0$ while $p_{o\,j}\, k_{ij} > 0$. However, this would imply $f'(0) < 0$ also in a neighborhood of $\lambda = 0$, which contradicts to the fact that f increases.

Since the generalized mutual information function has a similar importance for the channel decoding problem as inaccuracy in source coding theory, cf. with Fischer (1978), Theorem 2 may be used for the evaluation of the performance of a maximum likelihood decoding rule based on k_o. Let C be a channel code of block length n and let $P(C,k,k_o)$ denote the resulting error probability when using C and maximum likelihood decoding based on k_o for any discrete memoryless channel having channel matrix k. Then the expectation of $P(C,k,k_o)$ with respect to p_o can be upperbounded by

$$2^{-n\,E(R,p_o,k,k_o)}$$

where R denotes the rate of C and the exponent is a function that is positive, if $R < I(p_o;k,k_o)$. Theorem 2 implies that the exponent is still positive, if the latter condition is replaced by $R < I(p_o;k_o)$. Since $I(p_o;k_o)$ is the capacity of the class K, the maximum likelihood decoding rule used above has been shown to be universal for K.

A similar approach was investigated in a paper of Stiglitz (1966) where a channel matrix was determined that optimizes the error exponent instead of mutual information. For a derivation of the above exponent we refer to the very similar investigations of Fischer (1978).

PRINCIPLE OF MAXIMUM CONDITIONAL ENTROPY

Let p, q, and k be chosen as before and define the a posteriori probabilities

$$k'_{ji} = p_j\,k_{ij}/p'_i\,,$$
$$w'_{ji} = q_j\,k_{ij}/q'_i$$

where

$$p'_i = \sum_j p_j\,k_{ij}\,,$$
$$q'_i = \sum_j q_j\,k_{ij}$$

for $i = 1,\dots,m$, $j = 1,\dots,n$. The third inequality deals with conditional inaccuracy,

$$H(k',w'/p') = -\sum_i p'_i \sum_j k'_{ji} \log w'_{ji}\,,$$

a generalization of conditional Shannon entropy, $H(k'/p') = H(k',k'/p')$.

<u>Theorem 3</u>. Let P be a convex and compact set of probability vectors and let p_o be such that $H(k_o'/p_o') = \max H(k'/p')$ where k_o' and p_o' are defined as k' and p' with p_o instead of p and the maximization is extended for P. Then the inequality

$$H(k',k_o'/p') \leq \max_{p\in P} H(k'/p')$$

is satisfied for any $p \in P$.

The idea of the proof is similar to that of Theorem 2. First observe that $H(k'/p') = H(p) - I(p;k)$. Then define for any $p\in P$ the function $g(\lambda) = H((1-\lambda)p_o + \lambda p) - I((1-\lambda)p_o + \lambda p;\, k)$. Since $H(k'/p')$ can be shown to be a concave function of the a priori distribution p, g must be also concave. Moreover, since $g(\lambda) \leq g(0)$ for $\lambda \in [0,1]$, g decreases. Then another straightforward calculation gives $g'(0) = H(k',k_o'/p') - H(k_o'/p_o') \leq 0$, which proves Theorem 3.

If we choose k such that $I(p;k) = 0$, we have $H(k'/p') = H(p)$. Hence, it is easy to see that Theorem 1 is only a special case of Theorem 3.

The inequality of Theorem 3 can be used for upperbounding the error probability of Bayes decisions based on an entropy maximizing a priori distribution instead of the true one. Let $H_1,\ldots,H_n$ be a given set of hypotheses and suppose that the associated a priori distribution p is only known to be an element of some convex and compact class P. Then a possible choice of a decision rule consists in deciding for that hypothesis H_j which has greatest a posteriori probability $k'_{o\,ji}$, i. e. the decision rule is based on p_o instead of the unknown p. Here the index i stands for a result of an observation as usual. Then the error probability of this decision, which depends on p_o, k, and the actual distribution p, can be upperbounded by

$$P(p,p_o,k) \leq H(k',k_o'/p').$$

This can be shown in the same way as in the usual case $P = \{p\}$, cf. Rényi (1966). Then Theorem 3 yields

$$P(p,p_o;k) \leq \max_{p\in P} H(k'/p')$$

for $p \in P$. Consequently, if the decision is based on a sample of independent identical distributed observations, $P(p,p_o,k)$ vanishes with increasing sample size. Thus the decision rule based on p_o has been shown to be universal for P in similar sense as above.

REFERENCES

Davisson, L.D. (1973): Universal noiseless coding. IEEE Trans. Inform. Theory IT-19, 783 - 795

Fischer, Th. (1977): Über Verallgemeinerungen der Bongard-Entropie und Codierungssätze für Quellen mit unbekanntem statistischen Verhalten (On generalizations of the Bongard entropy and coding theorems for unknown source statistics). EIK 13, 125 - 135

(1978): Some remarks on the role of inaccuracy in Shannon's theory of information transmission. In: Trans. of the Eighth Prague Conference, Prague 1978. Academia, Prague, Vol. A, 211 - 226

(1981): Das Prinzip der maximalen Entropie und seine Anwendung in der Codierungs- und Suchtheorie (On the principle of maximum entropy and its applications to coding and search theory). to appear

Krichevsky, R.E.; Trofimov, V.K. (1981): The performance of universal encoding. IEEE Trans. Inform. Theory IT-27, 199 - 207

Rényi, A. (1966): On the amount of missing information and the Neyman - Pearson lemma. In: Festschrift für J. Neyman. Wiley, London, 281 - 288

Stiglitz, I.G. (1967): A coding theorem for a class of unknown channels. IEEE Trans. Inform. Theory IT- 13, 217 - 220

Department of Mathematics
Technical University of Dresden
8027 Dresden, Mommsenstraße 13
German Democratic Republic

A MODIFICATION OF THE EXTENDED KALMAN FILTER ALGORITHM WITH APPLICATION IN HYDROLOGY

László Gerencsér
Budapest

Key words: Extended Kalman filter, implicit state space equations, convergence of recursive stochastic algorithms, wave propagation.

ABSTRACT

In this paper we propose a method for the solution of a hydrological problem, which is equivalent to the problem of estimating the parameters of a linear parabolic differential equation. The same problem was formulated and solved in Bagchi et al. (1980), where the authors used an explicit discrete scheme and a usual extended Kalman filter /EKF/ method. In this paper we propose to use an implicit discrete scheme, which has better stability properties. As a result we have to modify the EKF algorithm for the case, when the state space equation is implicit.

We shall describe and explain the algorithm in a detailed form, but for the convergence proof we refer to Gerencsér (1982).

THE HYDROLOGICAL PROBLEM

We shall investigate the motion of a waterwave in a section $[x_0, x_L]$ of the river. The water level at time t at point x will be denoted by $f(x,t)$. This function satisfies the following diffusion equation

$$(1)\quad \frac{\partial f(x,t)}{\partial t} = -C\,\frac{\partial f(x,t)}{\partial x} + D\,\frac{\partial^2 f(x,t)}{\partial x^2} \qquad C,D > 0.$$

Here C is the translation velocity, D is the diffusion coefficient.

The solution of this equation is uniquely determined by the initial condition $f(x,0) = f_0(x)$ $x \in [x_0, x_L]$ and the boundary conditions $f(x_0,t)$,

$f(x_L,t)$, $0 \leq t \leq T$. The boundary values are measured at discrete times t_i where $0=t_0<t_1<\ldots t_{m+1}=T$ is an equidistant subdivision of the time interval $[0,T]$ with stepsize $\Delta t=t_i-t_{i-1}$. The measurements are noisy, say we have

$$(2) \qquad g_0(t_k) = f(x_0,t_k) + v_0(t_k)$$

$$(3) \qquad g_L(t_k) = f(x_L,t_k) + v_L(t_k).$$

The coefficient vector $\theta=(C,D)$ is unknown, and we want to estimate it on the basis of measured values of $f(x,t)$. These measurements are of the form

$$(4) \qquad y(t_k) = f(x_c,t_k) + \varepsilon(t_k)$$

where x_c is an intermediate point of the interval $[x_0,x_L]$ and $\varepsilon(t_k)$ is a white noise process.

This problem was formulated also by Bhagchi et al. (1980) and these authors propose the application of an explicit discrete scheme of the partialdifferential equation and estimate c,d by an extended Kalman filter method.

Our approach is slightly different. From the numerical point of view it is useful to use an implicit scheme. This is obtained as follows: $x_0<x_1\ldots<x_{N+1}=x_L$ is an equidistant subdivision of the interval $[x_0,x_L]$ with stepsize Δx. Let $f_{i,j}$ be an approximation of $f(x_i,t_j)$, and let us assume that the vector

$$(5) \qquad \phi(j) = (f_{1,j},\ldots,f_{N,j})^T$$

has already been computed. To compute the values $f_{i,j+1}$ we approximate the derivatives using 6 points. The derivatives in (1) are approximated by the following formulae:

$$(6) \qquad \frac{\partial f(x,t)}{\partial t} \approx (f_{i,j+1}-f_{i,j})/\Delta t$$

$$\frac{\partial f(x,t)}{\partial x} \approx \frac{1}{2}(f_{i+1,j+1}-f_{i-1,j+1})/\Delta x + \frac{1}{2}(f_{i+1,j}-f_{i-1,j})/2\Delta x$$

$$\frac{\partial f(x,t)}{\partial x^2} \approx \frac{1}{2}(f_{i+1,j+1}-2f_{i,j+1}+f_{i-1,j+1})/(\Delta x)^2 + \frac{1}{2}(f_{i+1,j}-2f_{i,j}+f_{i-1,j})(\Delta x)^2$$

THE FILTERING PROBLEM

Thus we have to solve a special case of the following estimation problem: find θ^{*} from a linear disrete time stochastic system:

$$(7)\quad F(\theta)\phi(t,\theta) = A(\theta)\phi(t-1,\theta) + B(\theta)u(t) + C(\theta)e(t) \qquad \phi(t,\theta)\in R^{n}$$

$$(8)\quad y(t,\theta) = D(\theta)\phi(t,\theta) + E(\theta)v(t) \qquad y(t,\theta)\in R^{m}$$

where the matrices A,B,C,D,E,F are known, continuously differentiable functions of the unknown parameter θ^{*} in some neighbourhood of θ^{*}. The noise process /e(t),v(t)/ is an n+m dimensional white noise process, which is also independent of the initial value $\phi(0,\theta)$.

The traditional EKF method in its simplest form can be obtained by extending the state vector $\phi(t)$ by the unknown parameter θ^{*}. Thus we get a bilinear model, from which $\phi(t)$ and θ^{*} can be estimated by linearization and by application of the linear Kalman filter. Thus for $F(\theta)\equiv 1$ we get the following

Conceptual algorithm 1 /explicit EKF method/:

$$(9)\quad \hat{\phi}(t+1)=A(\hat{\theta}(t))\hat{\phi}(t)+B(\hat{\theta}(t))u(t+1)+K(t)(y(t)-D(\hat{\theta}(t)\hat{\phi}(t))$$

$$(10)\quad \hat{\theta}(t+1)=\hat{\theta}(t)+L(t)(y(t)-D(\hat{\theta}(t))\hat{\phi}(t)).$$

This algorithm is conceptual in the sense that the computation of the gain matrices has to be specified /see Ljung (1979)/.

It can be shown that the EKF algorithm can be analyzed by Ljung's scheme and thus convergence can be proved for the parameter vector $(\hat{\theta}(t), R(t))$ where $R(t) = (tE\tilde{\theta}(t)\tilde{\theta}(t)^{T})^{-1}$ and $\tilde{\theta}(t)=\theta^{*}-\hat{\theta}(t)$.

For implicit state space equation we proceed as follows: first we introduce a new variable G(t) which is an approximation of $F^{-1}(\hat{\theta}(t))$. To update G(t) we select a fast iterative procedure. For fixed θ we shall write it in the form

$$(11)\quad G(t+1) = G(t) + \psi(G(t), F(\theta))$$

In any case the mapping $G\to\psi(G,F(\theta))$ should be a contraction in some neighbourhood of $G^{*}=F^{-1}(\theta^{*})$. We shall use the same iterative method when θ^{*} is changing. Thus we get the following

Conceptual algorithm 2 /implicit EKF method/:

$$(12)\quad \hat{\phi}(t+1)=G(t)A(\hat{\theta}(t))\hat{\phi}(t)+ G(t)B(\hat{\theta}(t))u(t+1)+K(t)(y(t)-D(\hat{\theta}(t))\hat{\phi}(t))$$

(13) $\hat{\theta}(t+1) = \hat{\theta}(t) + L(t)(y(t) - D(\hat{O}(t))\,\hat{\phi}(t))$

(14) $G(t+1) = G(t) + \psi(G(t), F(\hat{\theta}(t))$

where in the computation of the gain matrices $K(t)$, $L(t)$ we have to use an explicit state space equation with matrices $G(t)A(\theta)$, $G(t)\,B(\theta)$, $G(t)G(\theta)$.

It can be shown that this algorithm is in the class of algorithms which can be analyzed with a slight generalization on Ljung's theorem /see Gerencsér (1981), (1982), Lipcsey (1982), Ljung (1977)/.

In the preparation of this work we were helped by Dr. Szőllössy--Nagy and Dr. R. Farzan, whom we express our gratitude.

APPENDIX

We shall write out the details of the second algorithm. Let us introduce the notations $\tilde{\phi}(t)= \phi(t)- \hat{\phi}(t)$, $\tilde{\theta}(t)=\theta-\hat{\theta}(t)$. The estimaed error covariance matrix will be denoted by $P(t)$, it is split into four blocks: $P_{\phi\phi}$, $P_{\phi\theta}$, $P_{\theta\phi}$, $P_{\theta\theta}$. The estimated covariance matrix of the innovation is

(15) $S(t) = D(\hat{\theta}(t))\,P_{\phi\phi}(t)D(\hat{\theta}(t)) + E(\hat{\theta}(t))\,E^T(\hat{\theta}(t))$.

The transition matrices of the linearized model are

(16) $\bar{A}(t)= G(t)A(\hat{\theta}(t))$ $\bar{M}(t)=G(t)A_\theta(\hat{\theta}(t))\hat{\phi}(t)$, $\bar{B}(t)=G(t)B(\hat{\theta}(t))$,
$\bar{C}(t)= G(t)C(\hat{\theta}(t))$

and in the observation model we have

(17) $\bar{D}(t) = D(\hat{\theta}(t))$, $\bar{N}(t) = D_\theta(\hat{\theta}(t))\hat{\phi}(t)$.

With these notations the propagation of the covariance matrices are described by

(18) $P_{\phi\phi}(t+1) = \bar{A}(t)P_{\phi\phi}(t)\bar{A}^T(t) + \bar{C}(t)\bar{C}(t)^T - K(t)S(t)K^T(t)$

(19) $P_{\phi\theta}(t+1) = \bar{A}(t)P_{\phi\theta}(t) + \bar{M}(t)P_{\theta\theta}(t) - K(t)S(t)L^T(t)$

(20) $P_{\theta\theta}(t+1) = P_{\theta\theta}(t) - L(t)S(t)L^T(t)$.

The gain matrices have to be computed in the following way:

(21) $K(t) = \bar{A}(t)\, P_{\phi\phi}\; \bar{D}(t)\, S^{-1}(t)$

(22) $L(t) = (P_{\theta\phi}(t)\, \bar{D}(t) + P_{\theta\theta}\; \bar{N}(t))S^{-1}(t).$

We obtain an alternative form of Eq /19/, if we substitute L(t) by the expression (22):

(22) $P_{\phi\theta}(t+1)=(\bar{A}(t)-K(t)\bar{D}(t))P_{\phi\theta}(t)+(\bar{M}(t)-K(t)\bar{N}(t))P_{\theta\theta}(t).$

The above equations are slightly simpler than those in Ljung (1979) because we neglected small terms which do not influence asymptotic behaviour.

REFERENCES

Bhagchi, A., Strijbos, R.C.W., Thé, G. (1980): Identification of a distributed-parameter system with boundary noise. Int. J. System Sciences 1, No. 1, 49-56.

Gerencsér, L. (1981): Stability theorems for 2x2 hypermatrices. J. Optimization Theory and Applications 35, No. 1, 1-7.

Gerencsér, L. (1982): Implicit stochastic recursive algorithms. Computer and Automation Inst. Hung. Acad. Sci., Budapest , Working Paper.

Lipcsey, Zs. (1982):A generalization of Ljung's theorem on the convergence of stochastic recursive algorithms. Computer and Automation Inst., Hung. Acad. Sci., Budapest , Working Paper.

Ljung, L. (1977): Analysis of recursive stochastic algorithms. IEEE Trans. Automatic Control 22, No. 4, 551-575.

Ljung, L. (1979): Aşymptotic behaviour of the extended Kalman-filter as a parameter estimator for linear systems. IEEE Trans. Automatic Control 24, No. 1, 36-50.

Szőllösy-Nagy, A. (1981): A prediction method for wave propagation. Water Management Research Institute Budapest , Technical Report.

Computer and Automation Res. Inst., Hung. Acad. Sci.
Kende u.13-17, Budapest,1111, Hungary.

RELATIONS BETWEEN THE CRUDE, FACTORIAL AND INVERSE FACTORIAL MOMENTS

Tadeusz Gerstenkorn

Łódź

Key words: crude moments, factorial moments

ABSTRACT

In the paper there have been given the relations between the crude, factorial and inverse factorial moments based on Stirling numbers of different kinds. Appropriate formulae and tables are made up for the proposed Stirling numbers of the third and fourth kinds.

One knows that, in the case of some distributions, it is easier and more convenient to calculate factorial moments than crude ones. This can be exemplified by the Pólya distribution, for which factorial moments are calculated in a quite simple manner (cf. Eggenberger and Pólya (1923), p. 281, Dyczka (1969), p. 130), whereas crude ones - - after rather complicated endeavours (Gerstenkorn (1971), p. 24). In the problems of the so-called compound distributions the use, in some cases, of inverse factorial moments allows one to express certain formulae in a simple way. However, since in statistical practice we make use of crude (or central) moments because of their simple interpretation, it is needful to point out the formulae enabling us to pass from one type of moments to another.

Many interesting relations between various kinds of moments are shown in Gerstenkorn's paper (1969).

Bohlmann (1914), p. 398, (54) was the first, as it seems, to point to the possibility of expressing crude moments by means of factorial ones on the basis of Stirling numbers. (He was not using this denomination, but employing finite differences divided by factorials).

The Stirling numbers of the second kind S_i^r $(i = 0,1,\ldots)$ are defined as coefficients of descending factorial polynomials in the identity

(1) $$x^r = \sum_{i=0}^{r} S_i^r x^{[i]}$$

where $x^{[i]} = x(x-1)\dots(x-i+1)$ and $x^{[o]} = 1$.

Taking (1) into consideration, one easily obtains (by affixing the operator of the expected value) for the crude moments

$$\alpha_r = E(X^r) ,$$

the relation

(2) $$\alpha_r = \sum_{i=0}^{r} S_i^r \alpha_{[i]} ,$$

where

(3) $$\alpha_{[i]} = E(X^{[i]})$$

are factorial moments.

The Stirling numbers of the second kind S_i^r are effectively expressed by the formula

$$S_i^r = \frac{1}{i!} \sum_{j=0}^{i} (-1)^{i-j} \binom{i}{j} j^r ,$$

with that the recurrence relation

$$S_i^{r+1} = S_{i-1}^r + iS_i^r , \quad i = 1,2,\dots,r ,$$

holds for them, which enables one to make up easily a table of these numbers (e.g. Łukaszewicz and Warmus (1956), p.409, Kaufmann (1975), p. 48). Such a table is employed while using relation (2).

There is also a possibility of passing from crude moments to factorial ones (cf., e.g., Gerstenkorn (1969), p. 217, (7.7) or (1971), p. 29, (2.29)) when use is made of the formula

(4) $$\alpha_{[r]} = \sum_{i=0}^{r} s_i^r \alpha_i , \quad r = 0,1,2,\dots,$$

in which s_i^r stand for the Stirling numbers of the first kind, defined as coefficients of a polynomial in the identity

$$x^{[r]} = \sum_{i=0}^{r} s_i^r x^i .$$

The table of those numbers is made up on the basis of the recurrence relation

$$s_i^{r+1} = s_{i-1}^r - rs_i^r \quad , \quad i = 1,2,\ldots,r \; .$$

This table can be found, for instance, in Łukaszewicz and Warmus (1956) p. 409 or in Kaufmann (1975), p. 49.

In turn, we proceed to inverse factorial moments.

By the inverse factorial moment of order r of a random variable X we mean the expected value of an ascending factorial polynomial, i.e.

(5) $$\alpha_{[r,-1]} = \alpha_{[-r]} = E(X^{[r,-1]}) \quad , \quad r = 0,1,2,\ldots,$$

where

$$x^{[r,-1]} = x^{[-r]} = x(x+1)(x+2)\ldots(x+r-1) \quad , \quad x^{[0]} = 1$$

We shall give a connection of these moments with crude moments and factorial ones by introducing Stirling numbers of the third and of the fourth kind.

The Stirling numbers of the third kind $\bar{s}_i^r$ $(i = 0,1,2,\ldots,r)$ are defined as coefficients of a polynomial in the relation

(6) $$x^{[-r]} = \sum_{i=0}^{r} \bar{s}_i^r \, x^i \quad .$$

It can be seen that $\bar{s}_o^r = 0$ for $r > 0$; $\bar{s}_i^r = 0$ for $i < 0$ or $i > r$; $\bar{s}_r^r = 1$ for $r = 0,1,2,\ldots$.

The recurrence relation

(7) $$\bar{s}_i^{r+1} = \bar{s}_{i-1}^r + r\bar{s}_i^r$$

holds. We justify it by paying regard to the equality

$$x^{[-(r+1)]} = x^{[-r]}(x+r) \quad , \quad r = 0,1,2,\ldots \; .$$

In conformity with (6), we have

$$\sum_{i=0}^{r+1} \bar{s}_i^{r+1} \, x^i = \sum_{i=0}^{r} \bar{s}_i \, x^{i+1} + r \sum_{i=0}^{r} \bar{s}_i^r \, x^i$$

After having transformed the above sums, we get

$$\sum_{i=1}^{r} \bar{s}_i^{r+1} \, x^i = \sum_{i=1}^{r} x^i (\bar{s}_{i-1}^r + r\bar{s}_i^r)$$

whence one obtains relation (7).

On the ground of (7), we make up a table for the first eight terms.

Table of Stirling numbers of the third kind

r	$\bar{s}_1^r$	$\bar{s}_2^r$	$\bar{s}_3^r$	$\bar{s}_4^r$	$\bar{s}_5^r$	$\bar{s}_6^r$	$\bar{s}_7^r$	$\bar{s}_8^r$
1	1							
2	1	1						
3	2	3	1					
4	6	11	6	1				
5	24	50	35	10	1			
6	120	274	225	85	15	1		
7	720	1764	1624	735	175	21	1	
8	5040	13068	13132	6769	1960	322	28	1

From (6) we obtain a relation between inverse factorial moments and crude ones

(8) $$\alpha_{[-r]} = \sum_{i=0}^{r} \bar{s}_i^r \alpha_i \quad , \quad r = 0,1,2,\ldots .$$

Making use of (2), from (8) we obtain a relationship expressing an inverse factorial moment by factorial ones

(9) $$\alpha_{[-r]} = \sum_{i=1}^{r} \bar{s}_i^r \sum_{j=1}^{i} S_j^i \alpha_{[j]} \, , \quad r = 0,1,2,\ldots .$$

In order to derive a relation between crude moments and inverse factorial ones, we introduce Stirling numbers of the fourth kind.

The Stirling numbers of the fourth kind $\bar{S}_i^r$ $(i = 0,1,\ldots,r)$ are defined as coefficients of ascending factorial polynomials in the identity

(10) $$x^r = \sum_{i=0}^{r} \bar{S}_i^r x^{[-i]} .$$

It is easy to notice that $\bar{S}_o^r = 0$ for $r > 0$ and for $i < 0$ or $i > r$; $\bar{S}_r^r = 1$ for $r = 0,1,2,\ldots$.

For Stirling numbers of the fourth kind, the recurrence relation

(11) $$\bar{S}_i^{r+1} = \bar{S}_{i-1}^r - i\bar{S}_i^r \, , \quad i = 1,2,\ldots,r,$$

is true. Its justification is based on (10) and on the equality

$$\sum_{i=0}^{r+1} \bar{S}_i^{r+1} x^{[-i]} = x^{r+1} = \sum_{i=0}^{r} \bar{S}_i^r x x^{[-i]} =$$

$$= \sum_{i=0}^{r} \bar{S}_i^r x^{[-(r+1)]} - \sum_{i=0}^{r} i\bar{S}_i^r x^{[-i]} .$$

But

$$\sum_{i=0}^{r+1} \bar{\bar{S}}_i^{r+1} x^{[-i]} = \sum_{i=1}^{r} \bar{\bar{S}}_i^{r+1} x^{[-i]} + x^{[-(r+1)]} ,$$

$$\sum_{i=0}^{r} \bar{\bar{S}}_i^{r} x^{[-(r+1)]} = \sum_{i=1}^{r+1} \bar{\bar{S}}_{i-1}^{r} x^{[-i]} = \sum_{i=1}^{r} \bar{\bar{S}}_{i-1}^{r} x^{[-i]} + s^{[-(r+1)]},$$

and consequently

$$\sum_{i=1}^{r} \bar{\bar{S}}_i^{r+1} x^{[-i]} = \sum_{i=1}^{r} (\bar{\bar{S}}_{i-1}^{r} - i\bar{\bar{S}}_i^{r}) \quad x^{[-i]}$$

whence we get (11).

Making use of relation (11), we present a table for the first eight terms.

Table of Stirling numbers of the fourth kind

r	$\bar{\bar{S}}_1^r$	$\bar{\bar{S}}_2^r$	$\bar{\bar{S}}_3^r$	$\bar{\bar{S}}_4^r$	$\bar{\bar{S}}_5^r$	$\bar{\bar{S}}_6^r$	$\bar{\bar{S}}_7^r$	$\bar{\bar{S}}_8^r$
1	1							
2	-1	1						
3	1	-3	1					
4	-1	7	-6	1				
5	1	-15	25	-10	1			
6	-1	31	-90	65	-15	1		
7	1	-63	301	-350	140	-21	1	
8	-1	127	-966	1701	-1050	266	-28	1

From (10), by affixing the operator of the expected value, we obtain

$$\alpha_r = \sum_{i=0}^{r} \bar{\bar{S}}_i^r \alpha_{[-i]} \quad , \quad r = 0,1,2,\dots . \tag{12}$$

REFERENCES

Bohlmann G. (1913): Formulierung und Begründung zweier Hilfssätze der mathematischen Statistik. Math. Ann. 74, 341-409.

Dyczka W. (1969): The moments of Pólya distribution special case. Ann. Soc. Math. Polonae, Ser.I: Comment. Math. (Prace Matemat.) 13, No 1, 129-139.

Eggenberger F. and Pólya G. (1923): Über die Statistik verketteter Vorgänge. Z. Angew. Math. Mech. 3, 279-289.

Gerstenkorn T. (1969): Numerische Methoden zur Andwendung der Formeln für die Momente der Wahrscheinlichkeitsverteilungen. Wissensch. Zeitschrift der Tech. Hochschule Otto von Guericke Magdeburg 13, No 3/4, 213-219.

(1971): The recurrence relations for the moments of the discrete probability distributions. Dissert. Math. (Rozprawy Matemat.) 83, PWN, Warszawa.

Kaufmann A. (1968): Introduction à la combinatorique en vue des applications. Dunod, Paris.

Кофман А. (1975): Введение в прикладную комбинаторику. Наука, Москва.

Łukaszewicz J. and Warmus M. (1956): Metody numeryczne i graficzne. PWN, Warszawa.

Łódź University
Institute of Mathematics
ul. Stefana Banacha 22
90-238 Łódź
Poland

THE BERNSTEIN-VON MISES THEOREM FOR
NON-STATIONARY MARKOV PROCESSES

Heinz Gillert
Dresden

Key words: Markov process, arbitrary initial distribution, bayesian estimate, a posteriori density, strong consistency

---

ABSTRACT

The Bernstein-von Mises theorem is shown to be valid under any initial distribution for a class of Markov processes containing also periodical Markov processes. As a corollary it follows that the bayesian estimate of the unknown parameter is strong consistent and has asymptotically normal distribution under any initial distribution.

INTRODUCTION

The Bernstein-von Mises theorem is proved by Borwanker et al. (1971) for stationary ergodic Markov processes under some regularity conditions which ensure strong consistency and asymptotically normal distribution of the maximum likelihood estimate. From the view point of applications stationarity is a strong condition since it is difficult to test this hypothesis; on the other hand, often it is known that the Markov process under consideration is not stationary. Therefore, the results of Borwanker et al. (1971) will be extended to the case of arbitrary initial distributions for the class of Markov processes satisfying condition Z given below (see Gillert (1978)).

NOTATIONS AND CONDITIONS

Let $[X_n, n = 0,1,\ldots]$ be a homogeneous Markov process with state space $[\mathfrak{X},\mathfrak{B}]$ and the family of transition probabilities $\{P(x,A;\theta),\ x \in \mathfrak{X},\ A \in \mathfrak{B},\ \theta \in \Theta \subset R^1\}$.

Condition Z: There exists a partition $\mathfrak{X} = \bigcup_{i=0}^{d} S_i$ with $S_1 \neq \emptyset$ such that

(i) $P(x,S_{i+1}) = 1$ for $x \in S_i$ $(i = 1,\ldots,d)$, $(S_{d+1} = S_1)$

(ii) $\lim_{n\to\infty} P^n(y, \bigcup_{i=1}^{d} S_i) = 1$ for $y \in S_0$ (if $S_0 \neq \emptyset$)

(iii) the d-step Markov process $[X_{nd}, n = 0,1,\ldots]$ on $[S_1,\mathfrak{B}_1)$ is ergodic.

Condition $\overline{Z}$: Condition Z holds for all $\theta \in \Theta$ with the same partition.

Condition B:

(a) There exists a σ-finite measure μ which dominates all $P(x,.;\theta)$; all densities $f(x,y;\theta)$ are $\mathfrak{B}\times\mathfrak{B}$-measurable and for any $x \in \mathfrak{X}$ the set $\{y \in \mathfrak{X}: f(x,y;\theta) > 0\}$ is independent from θ.

(b) there exist first and second derivatives of the densities in θ which are continuous in θ for $(\mu \times \mu)$ - almost all (x,y) and

$$\frac{\partial}{\partial\theta}\int f(x,y;\theta)\,\mu(dy) = \int \frac{\partial}{\partial\theta} f(x,y;\theta)\,\mu(dy)$$

$$\frac{\partial^2}{\partial\theta^2}\int f(x,y;\theta)\mu(dy) = \int \frac{\partial^2}{\partial\theta^2} f(x,y;\theta)\mu(dy)$$

(c) denote $h(x,y;\theta) = \log f(x,y;\theta)$, then for any $\theta \in \Theta$ there exists $\eta(\theta) > 0$ such that

$$E_\theta^Q \sup(|\frac{\partial^2}{\partial\theta^2} h(X_0,X_1;\theta')|;\ |\theta - \theta'| < \eta(\theta),\ \theta' \in \Theta) < \infty$$

(E_θ^Q denotes expectation under the stationary initial distribution Q and the true value θ of the parameter)

(d) for any $\theta \in \Theta$ and any $\varepsilon > 0$ holds

$$-\infty < E_\theta^Q \sup([h(X_0,X_1;\theta') - h(X_0,X_1;\theta)];\ |\theta' - \theta| > \varepsilon,\ \theta' \in \Theta) < 0$$

(e) $i(\theta) = - E_\theta^Q \frac{\partial^2}{\partial\theta^2} h(X_0,X_1;\theta)$ is finite, non-zero and continuous for all $\theta \in \Theta$.

Condition C: Let an a priori probability measure be given on $[\Theta,\mathfrak{C}]$ with the bounded density λ which is continuous and positive in a neighbourhood of the true value θ.

Condition K: Let θ be the true value of the parameter and K a non-negative measurable function such that there exists ε $(0 < \varepsilon < i(\theta))$ and

$$M(\theta) = \left(\frac{i(\theta)}{2\pi}\right)^{1/2} \int_{-\infty}^{+\infty} K(t) \exp\left(-\tfrac{1}{2}(i(\theta) - \varepsilon)t^2\right) dt$$

is finite; for any $h > 0$ and any $\delta > 0$ holds

$$\exp(-\delta n) \int_{|t|>h} K(\sqrt{n}\, t) \quad (\theta_n + t)\, dt \xrightarrow[n \to \infty]{} 0 \quad P_{\upsilon,\theta} \text{ - a.s.}$$

where θ_n is the ML-estimate and $P_{\upsilon,\theta}$ the probability measure of the Markov process under initial distribution υ and true value θ.

RESULTS

Let $f_n(\theta | x_0, \ldots, x_n)$ denote the a posteriori density and

$$g_n(t) = n^{-\frac{1}{2}} f_n(\theta_n + tn^{-\frac{1}{2}} | x_0, \ldots, x_n).$$

Theorem 1: If conditions $\bar{Z}$, B, C, K are satisfied then

$$\lim_{n \to \infty} \int K(t) \left| g_n(t) - \left(\frac{i(\theta)}{2\pi}\right)^{1/2} \exp\left(-\frac{t^2}{2} i(\theta)\right)\right| dt = 0$$

holds $P_{\upsilon,\theta}$ - a.s. under any initial distribution υ.

Theorem 2: If conditions $\bar{Z}$, B, C are satisfied and for some non-negative integer m

$$\int |\theta|^m \lambda(\theta)\, d\theta < \infty$$

then

$$\lim_{n \to \infty} \int |t|^m \left| g_n(t) - \left(\frac{i(\theta)}{2\pi}\right)^{1/2} \exp\left(-\frac{t^2}{2} i(\theta)\right)\right| dt = 0$$

holds $P_{\upsilon,\theta}$ - a.s. under any initial distribution υ.

The proofs proceed as is shown by Borwanker et al (1971) using some results proved by the author (see Gillert (1978)). As a corollary it follows (see Borwanker (1971)) that under any initial distribution the bayesian estimate T_n is strong consistent and $\sqrt{n}\,(T_n - \theta)$ has asymptotically normal distribution.

REFERENCES

Borwanker, J., Kallianpur, G., Prakasa Rao, B.L.S. (1971): The Bernstein-von Mises theorem for Markov processes. Ann. Math. Statist. 42, 1241-1253.

Gillert, H. (1978): Maximum-Likelihood-Schätzungen für Parameter in homogenen Markovschen Ketten, MOS, Series statistics 9, 217-226.

Technische Universität Dresden
Sektion Mathematik
DDR-8027 Dresden
Mommsenstraße 13

ON DYNAMIC MIN-MAX DECISION MODELS

Hans-Joachim Girlich and Heinz-Uwe Küenle
Leipzig

Key words: Markov games, inventory models

ABSTRACT

We consider two types of stochastic models: (i) Markov games with simultaneous choice of the actions, which have also been studied for the general state and action spaces by Maitra/Parthasarathy (1970), Idzik (1979), Couvenbergh (1980); (ii) games of perfect information. Although the case (i) is the natural model of a game against nature, the case (ii) makes it easier to compute optimal strategies and to obtain assertions about the structure of such strategies.

The aim of this paper is to show the equivalence of the two models with respect to optimal strategies under convexity conditions. Finally, we give an example of application to an inventory system.

THE BASIC MODEL

A dynamic min-max decision model is determined by a tuple $M = ((X,\mathcal{X}), (A,\mathfrak{A}), (B,\mathfrak{B}), q, \mathcal{A}, \mathcal{B}, \mathcal{C}, \mathcal{F}, x)$, where $(X,\mathcal{X})$, $(A,\mathfrak{A})$, and $(B,\mathfrak{B})$ are measurable spaces. X is called state space, A action space of the decision-maker, and B action space of the opponent. $\mathcal{A}$ and $\mathcal{B}$ are maps $\mathcal{A}: X \to \mathfrak{A}$, $\mathcal{B}: X \to \mathfrak{B}$, with the properties $\{(x,a) \in X \times A: a \in \mathcal{A}(x)\} \in \mathcal{X} \otimes \mathfrak{A}$ $\{(x,b) \in X \times B: b \in \mathcal{B}(x)\} \in \mathcal{X} \otimes \mathfrak{B}$, and there are measurable maps $e: X \to A$, $f: X \to B$ with $e(x) \in \mathcal{A}(x)$, $f(x) \in \mathcal{B}(x)$ for all $x \in X$. $\mathcal{A}(x)$ and $\mathcal{B}(x)$ are called admissible action sets at the state x. Here, q is a transition probability from $(X \times A \times B, \mathcal{X} \otimes \mathfrak{A} \otimes \mathfrak{B})$ to $(X,\mathcal{X})$. Let be $H_n := X \times A \times B \times \ldots \times X \times A \times B \times X$ $(3n+1$ f.), $\mathcal{H}_n := \mathcal{X} \otimes \mathfrak{A} \otimes \mathfrak{B} \otimes \ldots \otimes \mathcal{X} \otimes \mathfrak{A} \otimes \mathfrak{B} \otimes \mathcal{X}$ Transition probabilities π_n from $(H_n, \mathcal{H}_n)$ to $(A,\mathfrak{A})$, and ϱ_n from $(H_n \times A, \mathcal{H}_n \otimes \mathfrak{A})$ to $(B,\mathfrak{B})$ with $\pi_n(\mathcal{A}(x_n) | x_0, a_0, \ldots, x_n) = 1$

and $\rho_n(\mathcal{B}(x_n)|x_o,a_o,b_o,\ldots,x_n,a_n) = 1$ for all $h_n=(x_o,a_o,b_o,\ldots,x_n)\in H_n$, $a_n\in A$, $n\in\mathcal{N}$, are called decision rules. A sequence of decision rules is a strategy. E and F denote the sets of all possible strategies for the decision-maker and the opponent respectively. $\mathcal{E}$ is a subset of E and $\mathcal{F}$ of F. k is a real-valued bounded function on $X\times A\times B\times X$, the cost function. If $\mathcal{E}$ = E and $\mathcal{F}$ = F, we call M a game of perfect information.

In the usual way, a pair of strategies $\pi = (\pi_n)\in\mathcal{E}$, $\rho = (\rho_n)\in\mathcal{F}$ associates with each initial state $x_o\in X$ the total expected discounted cost for the decision-maker

$$V_{\pi\rho}(x_o) := \sum_{n=o}^{\infty}\alpha^n\int_A \pi_o(da_o|x_o)\int_B \rho_o(db_o|x_o,a_o)\int_X q(dx_1|x_o,a_o,b_o)\ldots$$

$$\int_X q(dx_n|x_{n-1},a_{n-1},b_{n-1})\int_A \pi_n(da_n|x_o,a_o,b_o,\ldots,x_n)$$

$$\int_B \rho_n(db_n|x_o,a_o,b_o,\ldots,x_n,a_n)\int_X q(dx_{n+1}|x_n,a_n,b_n)k(x_n,a_n,b_n,x_{n+1}) .$$

Here, we consider the game from the point of view of the first player only, therefore the maximal cost $V_\pi(x) := \sup_{\rho\in\mathcal{F}} V_{\pi\rho}(x)$ is our criterion of optimality. We call a strategy π^* ε-optimal (with respect to an $\varepsilon \geqq 0$), if $V_{\pi^*}(x) \leqq \overline{V}(x) + \varepsilon$ for all $x\in X$, where

$$\overline{V}(x) := \inf_{\pi\in\mathcal{E}}\sup_{\rho\in\mathcal{F}} V_{\pi\rho}(x),\quad \underline{V}(x) := \sup_{\rho\in\mathcal{F}}\inf_{\pi\in\mathcal{E}} V_{\pi\rho}(x)$$

for all $x\in X$. If $\underline{V} = \overline{V} =: V$, then V is called the value of the model M.

Now we can introduce an operator T :

$$Tu(x,a,b) := \int_X q(dy|x,a,b)(k(x,a,b,y)+\alpha u(y))$$

for every bounded $\mathcal{X}$-measurable u and all $x\in X$, $a\in A$, $b\in B$.
E^d denotes the set of all deterministic strategies of the decision-maker. We obtain the following sufficient optimality conditions.

Theorem 1. Let M be a game of perfect information.
If there exist a bounded $\mathcal{X}$-measurable function u' and a $\mathcal{X}$-$\mathcal{O}\!\mathcal{L}$-measurable function e with the property

$$\begin{aligned} u'(x) &= \inf_{a\in\mathcal{A}(x)}\sup_{b\in\mathcal{B}(x)} Tu'(x,a,b) \\ &\geqq \sup_{b\in\mathcal{B}(x)} Tu'(x,e(x),b) - (1-\alpha)\varepsilon \quad \text{for all } x\in X, \end{aligned} \tag{1}$$

then the strategy $\delta_e^{\infty} := (\delta_e,\delta_e,\ldots)\in E^d$ is ε-optimal in M.
For a proof see Küenle (1981).

F^S denotes the set of all strategies $\rho=(\rho_n)$, where $\rho_n(.|h,a)$ is independent of $a\in A$ for every $h\in H_n$, $n\in\mathcal{N}$. We call a min-max model with $\mathcal{E}$ = E and $\mathcal{F}$ = F^S a Markov game with simultaneous action choice and write M^S.

EQUIVALENCE ASSERTIONS

In our model, the usual Markov game M^s follows from a game of perfect information M by restriction the set of the opponent's strategies. Under some conditions we even obtain a kind of equivalence of both games.

Theorem 2. If there is in M^s for every $\varepsilon > 0$ an ε-optimal deterministic strategy, then (i) every ε-optimal strategy in M or M^s is also ε-optimal in M^s or M, respectively,
(ii) if M^s has a value, then M has the same value.

Proof. Let $\varrho' = (\varrho_n')$ be a strategy of the opponent, defined by

$$\varrho_n'(.|h,a) := \varrho_n(.|h,e_n(h)) \text{ for all } h\in H_n,\ a\in A,\ n\in\mathcal{N}, \text{ where}$$

$\delta := (\delta_{e_n}) \in E^d$ and $\varrho = (\varrho_n) \in F$ arbitrary. Then $\varrho' \in F^s$ and $V_{\delta\varrho} = V_{\delta\varrho'}$.

Hence, $\sup_{\varrho\in F} V_{\delta\varrho} \leq \sup_{\varrho'\in F^s} V_{\delta\varrho'}$, and it follows

$$\overline{V} := \inf_{\pi\in E} \sup_{\varrho\in F} V_{\pi\varrho} \leq \inf_{\pi\in E^d} \sup_{\varrho\in F} V_{\pi\varrho} \leq \inf_{\pi\in E^d} \sup_{\varrho\in F^s} V_{\pi\varrho}$$
$$= \inf_{\pi\in E} \sup_{\varrho\in F^s} V_{\pi\varrho} =: \overline{V}^s \leq \inf_{\pi\in E} \sup_{\varrho\in F} V_{\pi\varrho} = \overline{V}.$$

Thus, $\overline{V} = \overline{V}^s$. Since $\overline{V}$ is the upper value in M and $\overline{V}^s$ in M^s the part (i) is proved.

If M^s has a value, then, by $F^s \subseteq F$, we have

$$\overline{V}^s = \underline{V}^s := \sup_{\varrho\in F^s} \inf_{\pi\in E} V_{\pi\varrho} \leq \sup_{\varrho\in F} \inf_{\pi\in E} V_{\pi\varrho} =: \underline{V} \leq \overline{V},$$

consequently, $\overline{V}^s = \underline{V}^s = \underline{V} = \overline{V}$ and the value of M is also $\overline{V}^s = \overline{V}$.

Now, we give a useful sufficient condition for the existence of an optimal strategy.

Theorem 3. Let u' be a function with the following properties:
(i) property (1) from Theorem 1,
(ii) $Tu'(x,.,b)$ convex on $\mathcal{A}(x)$,
(iii) $Tu'(x,a,.)$ concave on $\mathcal{B}(x)$,
(iv) $Tu'(x,.,.)$ continuous on $\mathcal{A}(x)\times\mathcal{B}(x)$ for all $x\in X, a\in A, b\in B$,
where $\mathcal{A}$ and $\mathcal{B}$ are compact-convex valued and continuous.
Then there exists in M^s an optimal stationary deterministic strategy and u' is the value of the game M^s.

Proof. Under the assumptions the functions $\underline{v}$ and $\overline{v}$, defined by

$$\underline{v}(x,b) := \inf_{a\in\mathcal{A}(x)} Tu'(x,a,b)\ ,\ \overline{v}(x,a) := \sup_{b\in\mathcal{B}(x)} Tu'(x,a,b),$$

are upper semicontinuous and lower semicontinuous respectively.
From Himmelberg/Parthasarathy/van Vleck (1976) follows the existence of measurable selectors e and f with $\underline{v}(x,f(x)) = \sup_{b\in\mathcal{B}(x)} \underline{v}(x,b)$,

$\bar{v}(x,e(x)) = \inf_{a\in\mathcal{A}(x)} \bar{v}(x,a)$. The rest of the proof is similar to the proof of Theorem 5.1 by Iwamoto/Kai (1974).

AN INVENTORY MODEL

We consider the following special min-max decision model M' with the state space $X := [0,C]$, $\mathcal{A} := [0,C]$, $\mathcal{A}(x) := [x,C] \cap [0,C]$, $0 < m, C < \infty$, B is the set of all probability measures concentrated on $[0,m]$, with the $*$-σ-algebra $\mathcal{L}$, $\mathcal{B}(x) := B$ for all $x \in X$. The transition law is defined by $q(\mathfrak{X}|x,a,b) := b(\{\beta: a-\beta \in \mathfrak{X}\})$, $\mathfrak{X} \in \mathcal{X}$ and the cost function k according to $k(x,a,b,y) := (a-x)c + L(y)$, where $L \geqq 0$ is a continuous and convex function on X.

We can give the following interpretation of M': At a depot with the capacity C demand for a item occurs. The future demand in a period will be described by a random variable with the unknown distribution b. If the inventory on this item is equal to x, the decision-maker chooses an action a, i.e. he brings the stock to the level a. Then demand occurs and the system goes over to a new state y, given by the transition law q, which is determined by the distribution b. Moreover, at this step costs arise, namely, replenishing cost $(a-x)c$ with c being the cost factor per unit replenished, and the expected carrying and shortage costs $L(y)$. The system is controlled with an unbounded horizon so that an infinite number of replenishments will occur. Then we shall deal with the present worth of total costs. If the decision-maker adopts the replenishment policy $\pi \in E$ and the initial inventory is x, the maximal expected total discounted costs are $V_\pi(x)$.

Our aim is to find an optimal policy, its structure and value.

Theorem 4. Let (u_n) be a sequence of functions defined by

(2) $$u_o = 0, \quad u_{n+1} = \inf_{a\in\mathcal{A}(x)} \sup_{b\in B} Tu_n(x,a,b) .$$

In the case of M', this sequence exists and converges uniformly to a function u'.

u' is convex and has the properties (i) to (iv) described in Theorem 3.

Proof. The first hypothesis follows from the generating operator's property of being monotone and contractive. By a theorem found by Küenle (1982) u_n is integrable. If u is continuous and convex, then

$$Tu(x,a,b) := (a-x)c + \int_0^m (L(a-\beta) + \alpha u(a-\beta))b(d\beta)$$

is continuous and convex in x,a,b too.

Therefore, $v(a) := \max_{b\in B} Tu(x,a,b)$ is continuous and convex. Thus,

(4) $$u^*(x) := \min_{a \in A(x)} \max_{b \in B} Tu(x,a,b) = -cx + \begin{cases} v(S) , & x < S \\ v(x) , & x > S \end{cases}$$

is also continuous and convex. Here, S is defined by $v(S) = \min_{a \in A} v(a)$

Since u_o is continuous and convex, the same can be said for u_n and also for $u' := \lim_{n \to \infty} u_n$. Consequently, Tu' is continuous, Tu'(x,.,b) is convex, Tu'(x,a,.) is linear and therefore concave, q.e.d.

By Theorem 3 and 4 an optimal stationary deterministic strategy in M^s exists, which, by Theorem 2, is optimal in M' too. By Theorem 1 and Eq.(4) the stationary deterministic strategy δ_e^{∞} with

(5) $$e(x) := \begin{cases} S , & x < S \\ x , & x > S \end{cases}$$

is optimal, where S = S' for u = u' in Eq.(4).

In our model it is optimal to pursue the simple replenishment policy in every period to stock the inventory to the reorder level S.

The reader is referred to Girlich (1982) for a min-max inventory model, where an (s,S) policy is optimal.

REFERENCES

Couwenbergh H. (1980): Stochastic games with metric state space. Intern. Journal of Game Theory 9, 25-36.

Dietzsch V. (1977): Dynamische Minimax-Entscheidungsmodelle. Diss.(A), Karl-Marx-Universität Leipzig.

Girlich H.-J. (1982): Optimal decision rules. In: Mathematical Statistics, Warsaw 1981, Banach Center Publications.

Himmelberg C., T.Parthasarathy, F.van Vleck (1976): Optimal plans for dynamic programming problems. Math.Oper.Res. 1, No.4, 390-394.

Idzik A. (1979): Remarks on discounted stochastic games. In: Trans,of the Eighth Prague Conference, Prague 1978, Vol. C, Academia, Prague, 165-174.

Iwamoto S. and Yû Kai (1974): On deterministic stationary strategies for Markov games. Bull.Math.Stat.Res. 16, 71-82.

Küenle H.-U. (1981): Über die Optimalität von Strategien in stochastischen dynamischen Minimax-Entscheidungsmodellen I. Math.Operationsforsch.Stat.Optim.12, No.3,421-435.

Küenle H.-U. (1982): On ε-optimal strategies in discounted Markov games. In: Optimal Control Theory, Warsaw 1980, Banach Center Publications.

(1983): Über die Optimalität von Strategien in stochastischen dynamischen Minimax-Entscheidungsmodellen II. Math.Oper.Stat.Optim. 14, No.2.

Maitra A. and T. Parthasarathy (1970): On stochastic games. J. Optimization Theory Appl. 5, 289-300.

Sion M. (1958): On general minimax theorems. Pacific Journal of Mathematics 8, 171-176.

Karl-Marx-Universität Leipzig
Sektion Mathematik
DDR-7010 Leipzig
Karl-Marx-Platz 10
G.D.R.

ESTIMATION OF PRECIPITATION CHARACTERISTICS FROM TIME-INTEGRATED DATA

Jan Grandell
Stockholm

Key words: Robust estimation, Estimation of precipitation characteristics

---

ABSTRACT

The mean duration of dry periods is estimated from observations of the total amount of precipitation during fixed time intervals. Reasonable estimates are obtained when the probability of a dry period at an arbitrary time is known (or estimated by other data). We also consider estimation of the mean duration from an arbitrary time until precipitation occurs. This quantity, which is relevant in connection with precipitation scavenging studies, turns out to be easy to estimate.

1 INTRODUCTION

Sometimes there is a need for estimating precipitation characteristics like the mean duration of dry periods or the mean duration from an arbitrary time until precipitation occurs. That is the case, for example, when models for the residence time of pollutants in the atmosphere, see Grandell and Rodhe (1978), are fitted to data. In the presence of detailed (i.e. continuously recorded) precipitation data such quantities are simply and naturally estimated. However, most data do not contain such detailed information, but consist of precipitation amounts accumulated over 6, 12 or 24 hours. In such a case some model for the variation of the precipitation is needed.

Consider a climatologically fairly homogeneous time interval $[0,t_0]$ of the length of some months. During this interval the precipitation intensity is assumed to be described by a stationary and ergodic process $\{R(t);t\in R\}$. On $[0,t_0]$ we associate a sequence of random variables $T_{d,1},T_{p,1},\ldots,T_{d,N},T_{p,N}$ where $T_{d,k}$ $(T_{p,k})$ is the length (h) of the k:th dry (precipitation) period and N is the number of such periods in $[0,t_0]$. The chosen notation correspond to the case where 0 is in a dry period and t_0 in a precipitation period.

Let T_d (T_p) be the length of a "typical" dry (precipitation) period. Let F_d be the distribution function of T_d and put $\tau_d=E(T_d)$, $\sigma_d^2=\mathrm{Var}(T_d)$, $\tau_p=E(T_p)$ and $p_d=\Pr\{R(t) = 0\}= =\tau_d/(\tau_d+\tau_p)$. Let W be the duration from an arbitrary time point t until precipitation occurs, and note that W = 0 if R(t) > 0. Then, compare Matthes et al. (1978, p 343),

$$\Pr\{W > x\} = \frac{1}{\tau_d+\tau_p} \int_x^\infty (1-F_d(y))dy \tag{1}$$

$$E(W) = (\sigma_d^2 + \tau_d^2)/(2(\tau_d+\tau_p)). \tag{2}$$

Let $\{h_k;k=1,\ldots,n\}$, where $n=t_0/\Delta$, be the total amounts of precipitation during succesive time intervals of length Δ (h). The h_k:s will be regarded as our *observations*.

We shall consider estimation of τ_d, τ_p and E(W) based on observations of $h_1,\ldots,h_n$. Since there exists an enormous amount of measurements of accumulated amounts of precipitation it is not so important to have "effective" estimates. It seems much more important to have some knowledge of the "systematic error" and to find estimates which are "robust" with regard to the chosen precipitation model. Unfortunately we do not know any systematic method producing such estimates. Let $\tau^*_{d,n}(\Delta)$ be some estimate of τ_d based on $h_1,\ldots,h_n$ such that $\tau^*_{d,n}(\Delta) \to$ some fixed value $\tau_d(\Delta)$ as $n \to \infty$. With the *systematic error* we mean $\tau_d(\Delta) - \tau_d$. It seems further natural to require that the estimates for small Δ and fixed t_0 shall be close to the natural estimates based on detailed data.

For larger Δ it seems difficult to perform theoretical considerations, and therefore we shall illustrate the estimates on real precipitation data. In these comparisons we use the same data as Rodhe and Grandell (1972 and 1981), namely detailed data recorded in Stockholm during winter and summer 1966. Both periods consist of 182 days (i.e. t_0 = 4368 h) and N = 191 (112) for the winter (summer) period. For these data we use the natural estimates $\tau_d^*=(1/N)\Sigma T_{d,k}$, $\tau_p^*=(1/N)\Sigma T_{p,k}$ and $E(W)^*=(\Sigma T_{d,k}^2)/(2t_0)$.

Alexander (1981) has considered estimation of τ_d and τ_p when the fraction of time of precipitation, and not the total amount, is observed.

2 ESTIMATION OF τ_d AND τ_p

Consider an observation sequence $h_1,\ldots,h_n$. Since we make no assumptions on the amounts of precipitation we shall only use the fact if a h_k is 0 or not. Let D be the number of dry Δ-periods and K the number of sequences of dry Δ-periods, i.e.

$$D = \#\{h_k = 0\} \text{ and } K = \#\{h_k = 0,\ h_{k+1} > 0\}.$$

Since R(t) is assumed to be ergodic, it follows from (1) that

$$D/n \to \Pr\{h_k=0\} = \Pr\{W>\Delta\} = \frac{1}{\tau_d+\tau_p}\int_\Delta^\infty (1-F_d(y))dy \tag{3}$$

$$K/n \to \Pr\{h_k=0,\ h_{k+1}>0\} = \Pr\{\Delta<W\le 2\Delta\} = \frac{1}{\tau_d+\tau_p}\int_\Delta^{2\Delta} (1-F_d(y))dy \tag{4}$$

when t_0 (and thus n) tends to infinity. If T_d is exponentially distributed (3) and (4) are reduced to

(5) $$D/n \to p_d \exp(-\Delta/\tau_d) = (\tau_d/(\tau_d+\tau_p))\exp(-\Delta/\tau_d)$$
$$K/n \to (\tau_d/(\tau_d+\tau_p))\exp(-\Delta/\tau_d)(1 - \exp(-\Delta/\tau_d))$$

and we are led to the estimates

$$\tau_d^*(\Delta) = \Delta/\log(D/(D-K)) \text{ and } \tau_p^*(\Delta) = \tau_d^*(\Delta)\left(\frac{n-D}{D} - \frac{nK}{D^2} \right).$$

For small Δ we have $K \approx N$ and $\Delta D/K \approx \tau_d^*$ or $D \approx N\tau_d^*/\Delta$ and thus

(6) $$\tau_d^*(\Delta) \approx -\Delta/\log(1-(\Delta/\tau_d^*)) \approx \tau_d^* \text{ and } \tau_p^*(\Delta) \approx \tau_d^*((\tau_p^*/\tau_d^*) - 0) = \tau_p^*.$$

In the general case, i.e. when T_d has distribution function F_d, it follows from (3) and (4) that

(7) $$\tau_d^*(\Delta) \to \Delta/ \log \left(\int_\Delta^\infty (1-F_d(y))dy \Big/ \int_{2\Delta}^\infty (1-F_d(y))dy \right)$$

Assume now that $F_d(x) = xf_d(0) + O(x^2)$ for small positive x, where $f_d(x)$ is the density function of T_d and where thus $f_d(0)$ is a measure of the frequency of short dry periods. Then

(8) $$\tau_d^*(\Delta) \to \tau_d + \frac{3\Delta\tau_d}{2} \left(f_d(0) - \frac{1}{\tau_d} \right) + O(\Delta^2).$$

This indicates that τ_d is overestimated when short dry periods are common, which is reasonable to assume in the case of convective precipitation.

TABLE 1

	Winter ($\tau_d^* = 19.04$, $\tau_p^* = 3.73$)					Summer ($\tau_d^* = 37.03$, $\tau_p^* = 1.64$)				
Δ	$\tau_d^*(\Delta)$	$\tau_p^*(\Delta)$	n	D	K	$\tau_d^*(\Delta)$	$\tau_p^*(\Delta)$	n	D	K
2	24.85	4.95	2184	1681	130	48.40	2.29	2184	2001	81
6	33.14	9.07	728	477	79	57.64	3.65	728	617	61
12	45.13	18.78	364	197	46	64.28	6.06	364	276	47
24	46.59	19.20	182	77	31	80.20	13.09	182	116	30

From Table 1 it is seen that τ_d is overestimated and, roughly speaking, the estimates seem to be bad. Thus it is tempting to try to find some other estimate of τ_d.

Assume now that p_d is known. From (5) we are then led to the estimate

$$\hat{\tau}_d(\Delta) = \Delta/(\log(n/D) + \log(p_d)).$$

With similar arguments as in the derivation of (6) we get $\hat{\tau}_d(\Delta) \approx \Delta/(\log(p_d)-\log(p_d^*))$ for fixed t_0 and small Δ, where $p_d^* = \tau_d^*/(\tau_d^*+\tau_p^*)$. In practice p_d has to be estimated. If we, in addition to the h_k:s, also know if it is raining or not at the observation times, a natural estimate of p_d is $\#\{R(k\Delta) = 0\} / n$. Since $\#\{R(k\Delta) = 0\} \approx D+N$ for small Δ, we have in that case

$$\hat{\tau}_d(\Delta) \approx \Delta/(\log(n/D)+\log((D+N)/n)) = \Delta/\log(1+(N/D)) \approx \Delta/\log(1+(\Delta/\tau_d^*)) \approx \tau_d^*$$

for fixed t_0 and small Δ.

With similar arguments as in the derivation of (7) and (8) it follows, when t_0 tends to infinity, that

$$\hat{\tau}_d(\Delta) \to -\Delta/\log\left(\frac{1}{\tau_d}\int_\Delta^\infty (1-F_d(y))dy\right)$$

and thus for small Δ

$$\hat{\tau}_d(\Delta) \to \tau_d + \frac{\tau_d}{2}\left(f_d(0) - \frac{1}{\tau_d}\right) + O(\Delta^2). \tag{9}$$

Comparing (8) and (9) we realize that we, for small Δ, have two estimates with different systematic errors but with systematic errors of similar form. Thus it is natural to combine them in order to reduce the systematic error and we are led to the estimate

$$\hat{\tau}_d^*(\Delta) = 3\hat{\tau}_d(\Delta)/2 - \tau_d^*(\Delta)/2.$$

From (8) and (9) it follows that $\hat{\tau}_d^*(\Delta) \to \tau_d + O(\Delta^2)$ for small Δ.

TABLE 2

	Winter ($\tau_d^* = 19.04$)		Summer ($\tau_d^* = 37.03$)	
Δ	$\hat{\tau}_d(\Delta)$	$\hat{\tau}_d^*(\Delta)$	$\hat{\tau}_d(\Delta)$	$\hat{\tau}_d^*(\Delta)$
2	23.91	23.44	44.77	42.95
6	24.52	20.22	48.94	44.59
12	27.53	18.74	51.30	44.81
24	35.90	29.49	58.88	48.22

In Table 2 we have used p_d^* as the "known" p_d.

3 ESTIMATION OF E(W)

If T_d is exponentially distributed it follows from (2) that $E(W) = p_d\tau_d$ and thus we can no longer base the construction of the estimate on such an assumption. Therefore we shall adopt a more primitive approach. Define

$$W_k(\Delta) = j \text{ if } h_k = 0, h_{k+1} = 0, \ldots, h_{k+j-1} = 0, h_{k+j} > 0,$$

i.e. $W_k(\Delta)$ is the length of dry Δ-periods after time $\Delta(k-1)$. Define $T_{d,1}(\Delta),\ldots,T_{d,K}(\Delta)$ as the lengths of succesive sequences of dry Δ-periods. This means, for example, that if $h_1=0$, $h_2=0$, $h_3>0$, $h_4=0$, $h_5>0,\ldots$ then $T_{d,1}(\Delta)=2\Delta$, $T_{d,2}(\Delta)=\Delta,\ldots$ and $W_1(\Delta)=2\Delta$, $W_2(\Delta)=\Delta$, $W_3(\Delta)=0$, $W_4(\Delta)=\Delta$, $W_5(\Delta)=0,\ldots$. Put $\overline{W}(\Delta) = (1/n)\Sigma W_k(\Delta)$. It is easy to realize that

$$\overline{W}(\Delta) = \frac{1}{2n}\left(D\Delta + \frac{1}{\Delta}\sum_1^K T_{d,k}(\Delta)^2\right)$$

which simplifies the numerical calculations. Since $W_k(\Delta)$ has the same distribution as $\Delta[W/\Delta]$ where $[\cdot]$ means integer part it follows that $\overline{W}(\Delta) \to \Delta E([W/\Delta])$ as $t_0 \to \infty$. The difference $E(W) - \Delta E([W/\Delta])$ does of course depend on the actual distribution of W. If T_d is exponentially distributed we have

(10) $E(W) - \Delta E(W/\Delta) = (\tau_d/(\tau_d+\tau_p))\left(\tau_d - (\exp(-\Delta/\tau_d)/(1-\exp(-\Delta/\tau_d)))\right).$

If we estimate this difference by replacing τ_d and τ_p by $\tau_d^*(\Delta)$ and $\tau_p^*(\Delta)$ we are led to the estimate

$$E(W)^*(\Delta) = \frac{D\Delta}{2n} + \frac{1}{2t_0}\sum_1^K T_{d,k}(\Delta)^2 + \frac{D^2}{n}\left(\frac{\tau_d^*(\Delta)}{D-K} - \frac{\Delta}{K}\right).$$

For fixed t_0 and small Δ we have

$$E(W)^*(\Delta) \approx \frac{N\tau_d^*\Delta}{2t_0} + E(W)^* + \frac{(N\tau_d^*)^2}{\Delta t_0}\left(\frac{\tau_d^*\Delta}{N\tau_d^*} - \frac{\Delta}{N}\right) \approx E(W)^*$$

TABLE 3

	E(W)*	E(W)*(2)	E(W)*(6)	E(W)*(12)	E(W)*(24)
Winter	31.47	31.88	31.26	30.15	30.98
Summer	80.60	80.93	80.56	80.60	80.69

4 DISCUSSION AND RECOMMENDATIONS

The only estimate of τ_d which behaves reasonably is $\hat{\tau}_d^*(\Delta)$ where p_d is assumed to be known. The only recommendation we dare to give is to avoid estimation of τ_d if $h_1,\ldots,h_n$ is the sole information available. In connection with precipitation scavenging Rodhe and Grandell (1981) recomend the mean residence time of pollutants in the atmosphere to be computed according to the formula $E(W) + 1/(\alpha R_0)$ where $R_0 = E(R(t))$ and α a parameter depending on the pollutant. Since R_0 is naturally estimated by $(h_1+\ldots+h_n)/t_0$ the only estimation problem is E(W). The figures in Table 3 indicate that our proposed estimate behaves well.

A natural question is if τ_d really is very difficult to estimate from observations of $h_1,\ldots,h_n$, or if we just have chosen bad estimates. Of course we can not give a definite answer, but we believe that τ_d is a very "unstable" parameter. With "unstable" we mean that small changes in the precipitation, or in the definition of dry and precipitation periods, which influence the observation very little, change τ_d and τ_p rather much. An example of such a change is if two showers, separated by a short dry period, are considered as one shower. Assume that dry periods shorter that some δ are ignored. If $\delta < \Delta$ the observations are not influenced at all, at least if we only consider whether h_k is zero or not. Let $\tau_d(\delta)$ and $\tau_p(\delta)$ be the means of typical dry and precipitation periods in that case, and assume that the lengths of all periods are independent. If δ is small compared to τ_d but of similar order as τ_p, then, under the same assumptions as in the derivation of (8), we have

$$\tau_d(\delta) \approx \tau_d + \delta\tau_d f(0) \text{ and } \tau_p(\delta) \approx \tau_p + \delta(\tau_p + (\delta/2))f_d(0)$$

which shows that the means depend strongly on how common dry periods are. For the summer data we have $\tau_d(2) = 48.8$, $\tau_d(6) = 62.7$, $\tau_p(2) = 2.6$ and $\tau_p(6) = 4.5$. Since dry and precipitation periods are differently affected, and assumption of known p_d has a stabilizing effect. If we now consider E(W) the situation is quite different since there is only a small change that an arbitrary chosen time point happens to fall in a short dry period, and therefore E(W) is much more stable than τ_d. This indicates that E(W) really is simpler to estimate and that this fact does not depend on our choice of possible estimates.

ACKNOWLEDGEMENTS

I want to thank Gunnar Englund, Henning Rodhe and Torbjörn Thedéen for comments leading to improvments of this paper.

REFERENCES

Alexander K. (1981): Determination of rainfall duration statistics for rainout models from daily records. Water Resour. Res. 17, 521-528.

Grandell J. and Rodhe H. (1978): A mathematical model for the residence time of aerosol particles removed by precipitation scavenging. In: Trans. of the Eighth Prague Conference Vol. A, Academia, Prague, 247-261.

Matthes K., Kerstan J. and Mecke J. (1978): Infinitely divisible point processes. Wiley, New York.

Rodhe H. and Grandell J. (1972): On the removal time of aerosol particles from the atmosphere by precipitation scavenging. Tellus 24, 442-454.

(1981): Estimates of characteristic times for precipitation scavenging. J. Atmos. Sc. 38, 370-386.

Department of Mathematics
The Royal Institute of Technology
S-100 44 Stockholm, Sweden

THE CENTRAL LIMIT THEOREM FOR STATISTICS OF A SPECTRAL DENSITY WITH TIME SHIFT

A.Halilov, M.A.Mirzahmedov

Tashkent

Key words: statistics for spectral density, time shift, semi - invariant, spectral density of n - th order, strong mixing, central limit theorem, confidence interval.

ABSTRACT

The statistics of spectral density of the homogeneous in strong sense random field obtained by time shift are considered. The principal term of asymptotics of variance of statistics is found, the central limit theorem is proved and for statistics confidence intervals are constructed.

Let $\xi(t)$, $t\in Z^m$, be a real - valued homogeneous in strong sense random field (r.f.) with mean zero, Z^m is the m - dimensional integral lattice. If r.f. $\xi(t)$ have absolute moments up to n -th order inclusive, then there exist a mixed semi-unvariants $S_n(t_1,\cdots,t_n)$, defined by the relation

$$S_n(t_1,\ldots,t_n)=i^{-n}\frac{\partial^n}{\partial x_1\cdots\partial x_n}\ln E\left\{i\sum_{j=1}^{n}x_j\xi(t_j)\right\}\Big|_{x_1=\cdots=x_n=0}$$

The spectral density (s.d.) of n-th order, if it exist, is defined from the equality

$$S_n(t_1,\ldots,t_n)=\int_{\Pi^{m\times n}}f_n(\lambda_1,\ldots,\lambda_n)\exp\left\{i\sum_{j=1}^{n}(t_j,\lambda_j)\right\}\delta^*\left(\sum_{j=1}^{n}\lambda_j\right)d\lambda_1\cdots d\lambda_n$$

where

$$\delta^*_{\cdot}(x)=\sum_{t\in Z^m}\delta(x+2\pi t),$$

$\delta(x)$ - the multivariate delta-function of Dirac,

$$\Pi^{m\times n}=\underbrace{\Pi^m\times\cdots\times\Pi^m}_{n},\ \Pi^m=\underbrace{\Pi\times\cdots\times\Pi}_{m},\ \Pi=[-\pi,\pi]$$

S.d. $f_2(\lambda_1,\lambda_2)$ on the manifold $\lambda_1+\lambda_2=0$ is called s.d. of r.f. $\xi(t)$ and is denoted by $f(\lambda)$.

Put

$$\Pi(N)=\{t;\ t\in Z^m,\ 0\le t_j\le N_j,\ j=1,2,\ldots,m\},\ N=(N_1,\ldots,N_m).$$

The statistics $f_N(\lambda)$ of s.d. $f(\lambda)$ of the r.f. $\xi(t)$ based on sampll $\xi(t),\ t\in\Pi(N)$ we define in the following way:

$$f_N(\lambda)=\frac{1}{T_1\cdots T_m}\sum_{t\in\Pi(T)}\left|W_M^{L\cdot t}(\lambda)\right|^2,$$

where

$$L\cdot t=(L_1t_1,\ldots,L_mt_m),\ N_j=M_j+1+(T_j-1)L_j,\ j=1,\ldots,m$$

and L_j, M_j, T_j - are positive integral function of natural arguments N_j,

$$W_M^Q(\lambda)=\sum_{t\in Z^m}a_M(t-Q)\exp\{i(t,\lambda)\}\xi(t),$$

$a_M(t),\ t\in Z^m$ - some non-negative function, which is equal to zero outside of parallelepiped $\Pi(M)$ and such that for

$$\varphi_M(x)=\sum_{t\in Z^m}a_M(t)\exp\{i(t,x)\}$$

the equality

$$\int_{\Pi^m} |\varphi_M(x)|^2 dx = 1 \tag{1}$$

is fulfilled.

Assume that $L_j \le M_j < N_j$ and $L_j \to \infty, M_j \to \infty, T_j \to \infty$ at $N_j \to \infty$.

Let the s.d. $f(x)$ of the r.f. $\xi(t)$ integrable on Π^m, s.d. of fourth order $f_4(x_1, \dots, x_4)$ integrable on the hyperlanes $x_1 + \dots + x_4 = 2\pi t, t \in Z^m$ of cube $\Pi^{m \times 4}$. Under these conditions the exact relations for the bias and variance of the estimate $f_N(\lambda)$ are obtained.

Let for any x , certain given λ, $0 < \alpha \le 1$, $C > 0$ for the s.d. $f(x)$ the r.f. $\xi(t)$ the inequality

$$|f(x+\lambda) - f(\lambda)| \le C \sum_{j=1}^{n} |x_j|^\alpha \tag{2}$$

is fulfilled and the s.d. of fourth order is bounded:

$$|f(x_1, x_2, x_3, x_4)| \le C_1 .$$

The set of such r.f. $\xi(t)$ we denote by $X = X(\lambda, f, \alpha, C, C_1)$.

Let for sequence $|\varphi_M(x)|^2$ the following conditions are fulfilled:

1) for every N, $|\varphi_M(x)|^2$ is continuous, even, 2π - periodical on every arguments function and the equality (1) is fulfilled,

2) for any $\varepsilon > 0$ uniformly on $x \in \Pi^m \setminus [-\varepsilon, \varepsilon]^m$, $|\varphi_M(x)|^2 \to 0$, when $N_0 \to \infty$, where $N_0 = \min\{N_1, \dots, N_m\}$,

3) exists a sequence of vectors $B(N) = (B_1(N_1), \dots, B_m(N_m))$ such that $\min\{B_1(N_1), \dots, B_m(N_m)\} \to \infty$ at $N_0 \to \infty$ and

for all $y \in \prod_{j=1}^{m} [-B_j(N_j)/N_j ,\ B_j(N_j)/N_j]$

$$\int_{\Pi^m} |\varphi_M(x)|^2 |\varphi_M(x+y)|^2 dx = \int_{\Pi^m} |\varphi_M(x)|^4 dx + o\Big(\int_{\Pi^m} |\varphi_M(x)|^4 dx\Big),$$

4) for any $a>0$ and for every j, $j=1,2,\dots,m$

$$\sum_{1 \le k < L_j \pi/a} \sup_{x \in A_j(k,a)} |\varphi_M(x)|^2 = O(L_j),$$

where

$$A_j(k,a) = \{x=(x_1,\dots,x_m);\ x \in \Pi^m,\ ka/L_j \le |x_j| \le \pi\}$$

The set of such sequences of functions $|\varphi_M(x)|^2$ we denote by Y.

Theorem 1. If $\xi(t) \in X$, $|\varphi_M(x)|^2 \in Y$ and

$$(3)\quad T_j L_j \sim N_j,\ j=1,2,\dots,m,\ L_1 \dots L_m \ll \int_{\Pi^m} |\varphi_M(x)|^4 dx$$

at $N_0 \longrightarrow \infty$, then

$$D f_N(\lambda) \sim \frac{(2\pi)^m f^2(\lambda)}{N_1 \dots N_m} \int_{\Pi^m} |\varphi_M(x)|^4 dx (1+\gamma(\lambda)),$$

where

$$\gamma(\lambda) = \begin{cases} 1, & \text{if } \lambda = \pi t,\ t \in Z^m, \\ 0, & \text{if } \lambda \neq \pi t,\ t \in Z^m. \end{cases}$$

Here and further $a_N \ll b_N$, $a_N \sim b_N$ means respectively $a_N/b_N \longrightarrow 0$, $a_N/b_N \longrightarrow 1$ at $N_0 \longrightarrow \infty$.

Asymptotic normality of $f_N(\lambda)$ is proved by us in assumption that r.f. $\xi(t)$ satisfies defined condition of mixing. Introduce the coefficient of mixing (see, for example, R.L.Dobrushin (1970)) $\alpha(\tau)$ of the r.f. $\xi(t)$;

$$\alpha(\tau)=\sup\{\alpha(U',U'');|U'|+|U''|<\infty,\ \rho(U',U'')\geq\tau\},$$

where $\rho(U',U'')$ - is the euclidean distance between sets U' and U'', $|U|$ - is number of elements in finite set $U\subset Z^m$, and

$$\alpha(U',U'')=\sup\{|p(AB)-p(A)p(B)|: A\in\sigma(U'), B\in\sigma(U'')\},$$

where $\sigma(U)$ - is σ - algebra of ewents generated by random variables $\xi(t),\ t\in U,\ U\subset Z^m$.

Theorem 2. Let the coefficient of mixing $\alpha(\tau)$ r,f. $\xi(t)$ for some $\varepsilon>0$ satisfies the condition $\alpha(\tau)\leq C_2\tau^{-\varepsilon}$, s.d. $f_2(x_1,x_2),\dots,f_8(x_1,\dots,x_8)$ r.f. $\xi(t)$ are bounded, s.d. $f(x)$ satisfies the condition (2), the conditions (3) are fulfilled, $|\varphi_M(x)|^2\in Y$ and

$$M_j\sim C(j)N_j^{\beta_j},\ 0<\beta_j<1,\ N_1\sim N_j,\ j=1,\dots,m,$$

$$\max_x|\varphi_M(x)|^2\sim C_3\int_{\Pi^m}|\varphi_M(x)|^4dx$$

$$\frac{L_1^2\cdots L_m^2}{T_1\cdots T_m}\int_{\Pi^m}|\varphi_M(x)|^4dx=o(1).$$

Then

$$\lim_{N_0\to\infty}P\left\{\frac{f_N(\lambda)-Mf_N(\lambda)}{\sqrt{Df_N(\lambda)}}<x\right\}=\Phi(x),$$

where $\Phi(x)$ is the standard normal law.

Theorem 3. Let $X_0 \subset X_0(\lambda, f, \alpha, C, C_1)$ subset of r.f. $\xi(t)$, for which the conditions of theorem 2 are fulfilled. Then

$$\sup_{\xi(t)\in X_0} \lim_{N_0\to\infty} P\{f_N(\lambda) - f(\lambda) \geq h_N + \bar{\sigma}_N x_\beta\} =$$

$$= \sup_{\xi(t)\in X_0} \lim_{N_0\to\infty} P\{f_N(\lambda) - f(\lambda) \leq -h_N - \bar{\sigma}_N x_\beta\} = \beta,$$

where

$$\beta = 1 - \Phi(x_\beta),\ h_N = C\sum_{j=1}^{m} \int_{\Pi^m} |x_j|^\alpha |\varphi_M(x)|^2 dx,$$

$$\bar{\sigma}_N = f_N(\lambda) = \left(\frac{(2\pi)^m}{N_1\cdots N_m} \int_{\Pi^m} |\varphi_M(x)|^4 dx (1+\gamma(\lambda))\right)^{1/2}.$$

Note that in case $m=1$ similar results were obtained by I.G.Žurbenko (1980). Let $\tilde{\xi}(t),\ t\in R^m$ be a measurable real-valued homogeneous in strong sense r.f. with mean zero and continuous time.All results reduced above for r.f. with discrete time are correct in case of r.f. $\tilde{\xi}(t),\ t\in R^m$, with corresponding changes.

REFERENCES

Dobrushin R.L.(1970): Theory of probability and its applications, vol.15, No 3, 469-497 (in Russian).

Žurbenko I.G. (1980): Ukrainian mathematical journal, vol 32, No 4, 463-476 (in Russian).

Mathematical Faculty
University of Tashkent,
Tashkent, USSR

ON THE MARTINGALE CONVERGENCE THEOREM IN QUANTUM THEORY

Blahoslav Harman, Beloslav Riečan

Bratislava

Key words: Stochastic processes, quantum logic, martingales with discrete parameter

In the last time a mathematical model of quantum mechanical systems was intensively studied /for references see Gudder (1979) and Varadarajan (1968). In this model also some limit theorems were formulated and proved /see e.g. Dvurečenskij (1979), Dvurečenskij-Riečan (1980), Pulmannová (1982)/. The purpose of the paper is to prove a variant of the martingale convergence theorem in quantum theory.

1. NOTATIONS AND NOTIONS

In the quantum mechanical systems instead of a probability space (X,S,P) some more general mathematical objects are studied. Firstly, instead of a σ-algebra S a logic L is taken. Logics represent a type of lattices; a typical example of a logic is the space of all linear subspaces of a Hilbert space (ordered by the inclusion). More generally, a logic L is a σ-lattice with the first and the last elements 0 and 1, respectively, with an orthocomplementation $\perp : a \longrightarrow a^{\perp}$, $a, a^{\perp} \in L$, which satisfies /i/ $(a^{\perp})^{\perp} = a$ for all $a \in L$, /ii/ if $a \leq b$, then $b^{\perp} \leq a^{\perp}$, /iii/ $a \vee a^{\perp} = 1$ for all $a \in L$, and with the orthomodular law /if $a \leq b$, then $b = a \vee (b \wedge a^{\perp})$/.

Secondly, instead of a probability measure a state m on L is studied as a mapping $m: L \longrightarrow \langle 0,1 \rangle$ such that $m/1/ = 1$ and $m/\bigvee_i a_i/ = \sum_i m/a_i/$ for every sequence $(a_i)_i$ of pairwise orthogonal elements of L, i.e. such sequence that $a_i \leq a_j^{\perp}$ /$i \neq j$/.

Finally, instead of a classical random variable /i.e. a mapping $f: X \longrightarrow R$ inducing a mapping $f^{-1}: B/R/ \longrightarrow S$/ we study an observable, i.e. a σ-homomorphism from the Borel σ-algebra $B(R)$ to the logic L.

Two elements $a,b \in L$ are compatible ($a \leftrightarrow b$) if there are mutually orthogonal elements a_1, b_1, c in L such that $a = a_1 \vee c$, $b = b_1 \vee \vee c$. Two observables x and y are compatible (= simultaneously observable) ($x \leftrightarrow y$), if $x(E) \leftrightarrow y(E)$ for any $E,F \in B(R)$. For compatible observables a functional calculus can be introduced (see Varadarajan (1968), chapter 6). Especially, the product xy of two (compatible) observables x, y is the observable $z : B(R) \to L$ defined by $z = T \circ h^{-1}$, where $h : R \times R \to R$, $h(u,v) = uv$ and $T : B(R^2) \to L$ is such a σ-homomorphism that $T \circ p_1^{-1} = x$, $T \circ p_2^{-1} = y$, p_1, p_2 being the projections $p_1, p_2 : R^2 \to R$, $p_1(u,v) = u$, $p_2(u,v) = v$ for every $u,v \in R$.

2. INDEFINITE INTEGRAL

Let $x : B(R) \to L$ be an observable, $m : L \to R$ be a state. Denote by m_x the mapping $m_x : B(R) \to R$ defined by the equality $m_x(E) = m(x(E))$. Evidently, m_x is a probability measure on $B(R)$. An observable x is called integrable (in a state m), if there exists $\int_R t \, dm_x(t)$. We shall denote the integral by $\int x \, dm$.

Let $a \in L$. Then an observable $q_a : B(R) \to L$ can be defined by $q_a(\{1\}) = a$, $q_a(\{0\}) = a^{\perp}$. If S is a σ-algebra, $S \subset L$, $a \in S$ and $x(B(R)) \subset S$, then q_a, x are compatible (see Varadarajan (1968), lemma 6.7), hence we can construct the product $q_a x$ and define

$$\int_a x \, dm = \int q_a x \, dm \, .$$

Proposition 2.1. Let $S \subset L$ be a countable generated σ-algebra. Then there exists an observable $y : B(R) \to L$ such that $S = y(B(R))$.

Proof. Varadarajan (1968), theorem 1.6.

Proposition 2.2. Let x,y be observables, $x(B(R)) \subset y(B(R))$. Then there exists a Borel measurable function $f : R \to R$ such that $x = y \circ \circ f^{-1}$.

Proof. The assertion can be proved similarly as theorem 6.9 in Varadarajan (1968) or proposition 1 in Dvurečenskij-Riečan (1980).

Proposition 2.3. Let $x,y : B(R) \to L$ be observables, $f : R \to R$ be a Borel measurable function, $x = y \circ f^{-1}$. Let $a \in x(B(R))$ and $A \in B(R)$ be such that $a = y(A)$. Then

$$\int_a x \, dm = \int_A f \, dm_y \, .$$

Proof. Evidently $q_a = y \circ \chi_A^{-1}$. Now we prove that $y \circ (f \chi_A)^{-1} =$

$= (y \circ f^{-1})\ (y \circ \chi_A^{-1})$. Indeed,

$$y \circ (f\ \chi_A)^{-1} = y \circ (h(f,\ \chi_A))^{-1} =$$

$$= y \circ (h \circ U)^{-1} = y \circ U^{-1} \circ h^{-1},$$

where $h(u,v) = u.v$, $h : R \times R \longrightarrow R$, $U : R \longrightarrow R \times R$, $U(u) = (f(u),\ \chi_A(u))$.
Put $T = y \circ U^{-1}$. Then

$$y \circ (f \chi_A)^{-1} = T \circ h^{-1},$$

where $T : B(R^2) \longrightarrow L$ is such a σ-homomorphism that

$$T \circ p_1^{-1} = y \circ U^{-1} \circ p_1^{-1} = y \circ (p_1 \circ U)^{-1} = y \circ f^{-1}$$

$$T \circ p_2^{-1} = y \circ \chi_A^{-1},$$

hence $y \circ (f\ \chi_A)^{-1}$ is the product of the observables $y \circ f^{-1}$ and $q_a =$ $= y \circ \chi_A^{-1}$. Therefore

$$\int_a x\ dm = \int q_a\ x\ dm = \int (y \circ \chi_A^{-1})\ (y \circ f^{-1})\ dm =$$

$$= \int y \circ (f\ \chi_A)^{-1}\ dm = \int_R t\ dm_z(t),$$

where $z = y \circ (f \chi_A)^{-1}$. Put $g(t) = t$. Then by the integral tranformation theorem on the line

$$\int_R t\ dm_z(t) = \int_R g\ dm_z = \int g \circ (f\ \chi_A)\ dm_y = \int f \chi_A\ dm_y$$

since $m_z(E) = m(z(E)) = m(y((f\ \chi_A)^{-1}(E)) = m_y((f\ \chi_A)^{-1}(E))$.

Corollary 2.4. Let $S \subset L$ be a countably generated σ-algebra, x be an observable, $x(B(R)) \subset S$. Then there exists a Borel measurable function $f : R \longrightarrow R$ and an observable y such that for every $a \in S$

$$\int_a x\ dm = \int_A f\ dm_y$$

whenever $y(A) = a$.

Proposition 2.5. Let $\int_a x\ dm \geqq 0$ for every $a \in S$. Then $x \geqq 0$ m-almost everywhere, i.e. $m(x((-\infty,0))) = 0$.

Proof. By Proposition 2.2 $x = y \circ f^{-1}$, hence by Proposition 2.3

$$\int_A f\ dm_y = \int_a x\ dm \geqq 0$$

whenever $y(A) = a$. Therefore $f \geqq 0$ m_y - a.e. and hence $m(x((-\infty,0))) =$ $= m_y(f^{-1}((-\infty,0))) = 0$.

3. CONDITIONAL EXPECTATION

Proposition 3.1. Let $S \subset L$ be a countably generated σ-algebra, $x : B(R) \longrightarrow S$ be an integrable observable. Let $S_o \subset S$ be a σ-algebra. Then there exists an integrable observable u such that $u(B(R)) \subset S_o$ and

$$\int_a x \, dm = \int_a u \, dm$$

for every $a \in S_o$.

Proof. By Corollary 2.4 there exists an observable $y : B(R) \longrightarrow S$ and a Borel measurable function $f : R \longrightarrow R$ such that $x = y \circ f^{-1}$ and $\int_a x \, dm = \int_A f \, dm_y$ for every $a \in S$ and every $A \in B(R)$ such that $y(A) = a$. Put $T_o = y^{-1}(S_o)$. Then $T_o \subset B(R)$, T_o is a σ-algebra. Let $g : R \longrightarrow R$ be the conditional expectation $E(f \mid T_o)$, i.e. $g^{-1}(B((R)) \subset T_o$ and

$$\int_A g \, dm_y = \int_A f \, dm_y , \; A \in T_o .$$

Put $u = y \circ g^{-1}$. Then $u(B(R)) = y(g^{-1}(B(R))) \subset y(T_o) = S_o$. Let $a \in S_o$, $y(A) = a$. Then by Proposition 2.3

$$\int_a u \, dm = \int_A g \, dm_y = \int_A f \, dm_y = \int_a x \, dm .$$

Definition 3.2. Any observable u satisfying the conclusions of Proposition 3.1 will be called (a version) of the conditional expectation and will be denoted by $u = E(x \mid S_o)$.

4. MARTINGALE CONVERGENCE THEOREM

Definition 4.1. Let S be a countably generated Boolean σ-algebra, $S \subset L$. Let $(S_n)_n$ be a sequence of σ-algebras, $S_n \subset S_{n+1} \subset S$ $(n = 1,2,\dots)$ and $(x_n)_n$ be a sequence of observables. We shall say that $((x_n, S_n))_n$ is submartingale, if $x_n(B(R)) \subset S_n$ and

$$\int_a x_{n+1} \, dm \geq \int_a x_n \, dm$$

for every $a \in S_n$.

Proposition 4.2. Let $S \subset L$ be a countably generated Boolean σ-algebra, $(S_n)_n$ be a sequence of σ-algebras, $(x_n)_n$ be a sequence of observables. Then $((x_n, S_n))_n$ is a submartingale iff the following three conditions are satisfied:

(i) $S_n \subset S_{n+1}$ $(n = 1,2,\dots)$

(ii) $x_n(B(R)) \subset S_n$ $(n = 1,2,\dots)$

(iii) $E(x_{n+1} \mid S_n) \geqq x_n$ $(n = 1,2,\dots)$ m-almost everywhere

Proof. If $((x_n, S_n))_n$ is a submartingale, then for every $a \in S_n$

$$\int_a E(x_{n+1} \mid S_n)\, dm = \int_a x_{n+1}\, dm \geqq \int_a x_n\, dm,$$

hence $E(x_{n+1} \mid S_n) = x_n$ m-a.e. by Proposition 2.5. If the properties (i) - (iii) are satisfied, then $\int_a x_{n+1}\, dm = \int_a E(x_{n+1} \mid S_n)\, dm \geqq$ $\geqq \int_a x_n\, dm$ for every $a \in S_n$.

We shall say that a sequence $(x_n)_n$ of observables converges almost everywhere (a.e.) to an observable x, if for every $\varepsilon > 0$ $\lim_{n\to\infty} m(\bigvee_{k=n}^{\infty} (x_k - x)(R\setminus(-\varepsilon, \varepsilon))) = 0$.

Theorem 4.3. Let $((x_n, S_n))_n$ be a submartingale, $\sup_m \int x_n\, dm <$ $< \infty$, $S_n \subset S$, where S is a countably generated σ-algebra. Then there exists such an observable $x : B(R) \longrightarrow S_o = \sigma(\bigcup S_n)$ that $x_n \longrightarrow x$ a.e.

Proof. Let y be such an observable that $y(B(R)) = S$ and let $f_n : R \to R$ be Borel measurable functions such that $x_n = y \circ f_n^{-1}$ $(n = 1,2,\dots)$. Put $T_n = y^{-1}(S_n)$. Then $((f_n, T_n))_n$ is a submartingale and by Corollary 2.4 $\int f_n\, dm_y = \int x_n\, dm$, hence $\sup_n \int f_n\, dm_y < \infty$. Therefore by the classical martingale convergence theorem there exists a random variable f, measurable with respect to $\sigma(\bigcup_n T_n)$ and such that $f_n \longrightarrow f$ a.e. (with respect to m_y). Put $x = y \circ f^{-1}$. Then $f^{-1}(E) \in \sigma(\bigcup T_n)$ for every $E \in B(R)$, hence $x(E) = y(f^{-1}(E)) \subset$ $\subset y(\sigma(\bigcup T_n)) \subset \sigma(\bigcup S_n)$ and therefore x is $\sigma(\bigcup S_n)$ - measurable. Since $f_n \longrightarrow f$ a.e., we obtain for every $\varepsilon > 0$

$$0 = \lim_n m_y((f_n - f)^{-1}(R \setminus (-\varepsilon, \varepsilon))) =$$

$$= \lim_n m(y \circ (f_n - f)^{-1}(R \setminus (-\varepsilon, \varepsilon))) =$$

$$= \lim_n m((y \circ f_n^{-1} - y \circ f^{-1})(R \setminus (-\varepsilon, \varepsilon))) =$$

$$= \lim_n m((x_n - x)(R \setminus (-\varepsilon, \varepsilon))),$$

hence $x_n \longrightarrow x$ a.e. (with respect to m).

REFERENCES

Dvurečenskij A. (1979): Laws of large numbers and the central limit theorems on a logic. Math. Slovaca 29, 397 - - 410.

Dvurečenskuj A. - Riečan B. (1980): On the individual ergodic theorem on a logic. Comment. Mathem. Univ. Carol. 21, 2 , 385-291.

Gudder S.P. (1979): Stochastic methods in quantum mechanics. North Holland, New York.

Pulmannová S. (1982): Individual ergodic theorem in a logic. Math. Slovaca, to appear.

Vadarajan V.S. (1968): Geometry of quantum theory. Van Nostrand, New York.

Comenius University
Mlynská dolina
842 15 Bratislava
Czechoslovakia

SOME COMPLEXITY CONSIDERATIONS CONCERNING HYPOTHESES IN MULTIDIMENSIONAL CONTINGENCY TABLES

Tomáš Havránek
Prague

Key words: Contingency tables, log-linear hypotheses

ABSTRACT

A simple formal representation of hierarchical log-linear hypotheses (models) for multidimensional contingency tables is presented. This representation leads to a complexity hierarchy between these hypotheses.

INTRODUCTION

Consider multidimensional contingency tables under the multinomial independent sampling model. Then hypotheses to be tested are usually formulated in the language of hierarchical log-linear models or hypotheses (see e.g. Bishop, Fienberg and Holland, 1975). Instead of hierarchical log-linear we shall write further HLL only. HLL models are usually described by its minimal generating sets. If we apply a simple logical approach, we can view these sets as sentences of some formal language and analyze their syntactical and semantical properties. Such an analysis is indispensable for constructing model search procedures working on the space of HLL models in the sense of the GUHA methods (Hájek and Havránek, 1978). A procedure of this kind was suggested in (Havránek, 1982). Our present approach to contingency tables is inspired by works of Goodman (1970, 1971), Enke (1980) and Wermuth (1976, 1980) and has a close connection with the approach of Darroch and Speed (1979). We believe that a simple formal description with nice logical properties is the best way for easy interpretation of any statistical analysis if we consider sentences as formal codings of hypotheses in question in a statistical data analysis.

In the present paper we shall consider models (hypotheses) containing at least all main efects.

Let the dimension of the table be fixed, say n . Then each HLL sentence is of the form $\{A_1,\ldots,A_k\}$,where (a) $A_i \subseteq \{1,\ldots,n\}$ and (b) for any $i,j \in \{1,\ldots,n\}, i \neq j$, neither $A_i \subseteq A_j$ nor $A_j \subseteq A_i$ .In the following φ,λ,μ are HLL sentences.

Let & be the usual logical connective of conjunction.As usual $\varphi \& \lambda$ holds if φ and λ hold (i.e. HLL models described by φ,λ hold - the world model is used here not in the logical but statistical sense).

For each φ,λ there is a μ such that $\varphi \& \lambda$ is logically equivalent to μ (i.e. the set of HLL sentences is closed under &).The sentence μ can be obtained by the formal operation $\sqcap$ (meet) between HLL sentences.This operation is defined as follows:let $\varphi=\{A_1,\ldots,A_k\}$, $\lambda=\{B_1,\ldots,B_l\}$, then $\varphi \sqcap \lambda = \{A_1 \cap B_1, A_1 \cap B_2,\ldots,A_k \cap B_l\}$ (we suppose that from $\varphi \sqcap \lambda$ all redundant sets are removed).

The closedness under & is important from the statistical point of view: for two statistical models of some kind, their conjunction (intersection) is to be of the same kind.

Another operation is the operation $\sqcup$ (join): $\varphi \sqcup \lambda = \{A_1,\ldots,A_k, B_1,\ldots,B_l\}$ (again with removing redundant sets).If we denote Sent the set of all HLL sentences, then $\langle Sent; \sqcap, \sqcup, \{\{1,\ldots,n\}\}, \{\emptyset\}\rangle$ is a distributive lattice with unit and zero members. In the sequel we shall use the following fact (reduction rule):if $\varphi=\{A_1,\ldots,A_k\} \subseteq \lambda=\{B_1,\ldots,B_l\}$, then $\varphi \& \lambda$ is logically equivalent to φ.

CANONICAL REPPRESENTATION OF HYPOTHESES

It seems useful to use the formal means to find a representation of HLL sentences that enable to establish a complexity hierarchy of these sentences as well as to define some important subclasses of sentences. A HLL sentence containing only sets of cardinality n-1 will be called an <u>atomic</u> HLL sentence.

<u>Theorem</u>. Each HLL sentence is logically equivalent to a conjunction of atomic HLL sentences.

Proof. Let $\varphi=\{A_1,\ldots,A_k\}$.In the following steps the conjunction needed can be constructed:(i) transform φ into $\{A_1\} \sqcup \ldots \sqcup \{A_k\}$. Express each A_i as $\{A_{i1}\} \sqcap \ldots \sqcap \{A_{ik_i}\}$,where each A_{ij} is of cardinality n-1. (iii) apply the distributivity rule for $\sqcap$ and $\sqcup$.(iv) Transform expressions of the form $\{A_{ij}\} \sqcup \ldots \sqcup \{A_{rs}\}$ into $\{A_{ij},\ldots,A_{rs}\}$.

(v) Substitute $\sqcap$ by &. (vi) Remove all redundant atomic HLL sentences by the help of the reduction rule.In (ii) and (vi) we use substituting of logically equivalent expressions.In the other steps we do not add or remove any set.

The point (vi) is not necessary for obtaining a canonical representation (conjunction of atomic HLL sentences), but it makes the representation unique and not redundant. The described algorithm is not an optimal one,but the simpliest to be described.A better way is to use a 'halving' algorithm: to divide $\varphi=\{A_1,\ldots,A_k\}$in two parts and each part to transform separately etc.

Consider an example.Let n=4 and,as usual,we shall use letters A,B,C,D instead of 1,2,3,4 and write e.g. (AB,C,D) instead of {{A,B},{C},{D}}.We shall omit brackets where possible. Consider φ=(AB,AC,BD,CD). (i) (AB,AC,BD,CD) = AB $\sqcup$ AC $\sqcup$ BD $\sqcup$ CD. (ii) AB = (ABC $\sqcap$ ABD),AC=(ABC $\sqcap$ ACD),BD=(ABD $\sqcap$ BCD),CD=(ACD $\sqcap$ BCD). Hence (AB,AC,BD, CD)=(ABC $\sqcap$ ABD) $\sqcup$ (ABC $\sqcap$ ACD) $\sqcup$ (ABD $\sqcap$ BCD) $\sqcup$ (ACD $\sqcap$ BCD).Now by the halving algorithm (1) (ABC $\sqcap$ ABD) $\sqcup$ (ABC $\sqcap$ ACD) = ABC $\sqcap$ (ABC,ACD) $\sqcap$ (ABC,ABD) $\sqcap$ (ABD,ACD). By the reduction rule we have (1) = ABC $\sqcap$ (ABD,ACD).Similarly, (2) (ABD $\sqcap$ BCD) $\sqcup$ (ACD $\sqcap$ BCD) = (ABD,ACD) $\sqcap$ BCD.Now (1) $\sqcup$ (2) = (ABC $\sqcap$ (ABD,ACD)) $\sqcup$ ((ABD,ACD) $\sqcap$ BCD) = (ABC,ABD,ACD) $\sqcap$ (ABC,BCD) (ABD,ACD) $\sqcap$ (ABD,ACD,BCD) = (ABC,BCD)&(ABD,ACD) using the reduction rule and substituting by &.

THE COMPLEXITY HIERARCHY

For a sentence $\varphi=\{A_1,\ldots,A_k\}$ call k the size of the sentence. We can see that the complexity of sentences can be measured by the maximal size of an atomic HLL sentence contained in the minimal canonical representation.By the length of a canonical representation we mean the number of & in it.Generally we have size $\leq$ n and length $\leq$ $\binom{n}{2}$.

Consider now an example for n=4. In the following table we have represented types of HLL sentences (other sentences can be obtained by permuting the letters). Numbers of types reffer to (Enke,1980) and they correspond in the inverse order to the usual understanding of complexity of HLL models.

The hierarchy ruled by the maximal size of the atomic HLL sentence contained in the canonical representation can be here described as follows:

H2 = {4,7,8,13,14,15,18,19,21,24},

H3 = {2,3,5,6,10,11,12,16}, H4 = {1}.

TABLE

no.of type	sentence	canonical representation	no.of sentences
1.	(ABC,ABD,ACD,BCD)	(ABC,ABD,ACD,BCD)	1
2.	(ABC,ABD,ACD)	(ABC,ABD,ACD)	4
3.	(ABC,ABD,CD)	(ABC,ABD,ACD)&(ABC,ABD,BCD)	6
4.*	(ABC,ABD)	(ABC,ABD)	6
5.	(ABC,AD,BD,CD)	(ABC,ABD,ACD)&(ABC,ABD,BCD)& (ABC,ACD,BCD)	4
6.	(ABC,AD,BD)	(ABC,ABD)&(ABC,ACD,BCD)	12
7.*	(ABC,AD)	(ABC,ABD)&(ABC,ACD)	12
8.*	(ABC,D)	(ABC,ABD)&(ABC,ACD)&(ABC,BCD)	4
10.	(AB,AC,AD,BC,BD,CD)	(ABC,ABD,ACD)&(ABD,ACD,BCD)& (ABC,ABD,BCD)&(ABC,ACD,BCD)	1
11.	(AB,AC,AD,BC,BD)	(ABC,ABD)&(ABC,ACD,BCD)&(ABD,ACD,BCD)	6
12.	(AB,AC,AD,BC)	(ABC,ACD)&(ABC,ABD)&(ABD,ACD,BCD)	12
13.	(AB,AC,BD,CD)	(ABC,BCD)&(ABD,ACD)	3
14.*	(AB,AC,AD)	(ABC,ABD)&(ABC,ACD)&(ABD,ACD)	4
15.*	(AB,AC,BD)	(ABC,ABD)&(ABC,BCD)&(ABD,ACD)	12
16.	(AB,AC,BC,D)	(ABC,BCD)&(ABC,ABD)&(ABC,ACD)& (ABD,ACD,BCD)	4
18.*	(AB,CD)	(ABC,ACD)&(ABC,BCD)&(ABD,ACD)& (ABD,BCD)	3
19.*	(AB,AC,D)	(ABD,ACD)&(ABC,ABD)&(ABC,ACD)& (ABC,BCD)	12
21.*	(AB,C,D)	(ABC,ABD)&(ABC,ACD)&(ABC,BCD)& (ABD,ACD)	6
24.*	(A,B,C,D)	(ABC,ABD)&(ABC,ACD)&(ABC,BCD)& (ABD,ACD)&(ABD,BCD)&(ACD,BCD)	1

(* decomposable (multiplicative) hypotheses - see Bishop,Fienberg and Holland,1975)

If we denote by $\leq_s$ the ordering (for a given dimension n) given by the maximal size of an atomic HLL in the canonical representation, we can see that $\varphi\&\lambda \leq_s \varphi,\lambda$, i.e. this complexity ordering is in accordance with conjunction of sentences. The simpliest class in this hierarchy is the class of HLL sentences with the maximal size 2 in the canonical representation. This class coincides with zero-partial-association hypotheses (models) or graphical models ((Wermuth, 1976,1980) and (Darroch,Lauritzen and Speed, 1980))and it is the least class containing decomposable hypotheses and closed with respect to conjunction.

For a table of dimension n, there are n-2 complexity classes between the class H2 and Hn. If we define $CH_i = \cup_{j=2}^{i} H_j$,then CH_n is the set of all HLL sentences and each CH_i is closed under conjunction. So each of these sets has this property of the set of graphical models.

More discutable is the ordering of sentences within each class H_i by the lenght of the representation. For example, we obtain for n=4 and H2 the following ordering:

4 < 7,13 < 8, 14, 15 < 18, 19 < 21 < 24 .

Clearly, 4. (ABC,ABD) seems to be a simple hypothesis from the interpretational point of view. Some hesitations concern the position of 13. (AB,AC,BD,CD) which is not decomposable (but decomposable hypotheses are usualy considered to be the simpliest).On the other hand (AB,AC,BD,CD) is a conjunction of two very simple hypotheses (ABC,BCD) and (ABD,ACD).

We can conclude that the inner structure of the classes H_i should be investigated further.

REFERENCES

Bishop Y.M.M., Fienberg S.E., Holland P.W. (1975): Discrete multivariate analysis:theory and practice. MIT Press, Cambridge (Mass.).

Darroch J.N., Lauritzen S.L., Speed T.P. (1980): Markov fields and log-linear models for contingency tables. Annals of Statistics 8, 522-539.

Darroch J.N., Speed T.P. (1979): Multiplicative and additive models and interactions. Res.rep.49., Department of Theoretic Statistics, University of Aarhus.

Enke H. (1980): To some reasonable test procedures in multiple contingency tables to investigate certain epidemiological or medico-sociological relationships. Biometrical Journal 22, 779-793.

Goodman L.A. (1970): The multivariate analysis of qualitative data: interactions among multiple classifications. J. Amer.Statistical Assoc. 65, 226-256.

(1971): Partitioning of chi-square, anylysis of contingency

tables,and estimation of expected frequencies in multidimensional contingency tables. J.Amer.Statist. Assoc. 66, 339-344.

Havránek T. (1982): O analýze mnohorozměrných kontingenčních tabulek (On analysis of multidimensional contingency tables). In: Robust 82, JČMF, Prague, 11-18.

Hájek P., Havránek T. (1978): Mechanizing hypothesis formation - mathematical foundations for a general theory. Springer-Verlag, Berlin-Heidelberg-New York.

Wermuth N. (1976): Model search among multiplicative models. Biometrics 32, 253-263.

(1980): Linear recursive equations, covariance selection and path analysis. J.Amer.Statist.Assoc. 75, 963-972.

Czechoslovak Academy of Sciences
Center of Biomathematics,Institute of Physiology
Vídeňská 1083
142 20 Praha 4
Czechoslovakia

THEOREMS ON SELECTORS IN TOPOLOGICAL SPACES I

Adam Idzik

Warsaw

Key words : measurable multifunctions, Čech-complete spaces, the selection theorem of Kuratowski and Ryll-Nardzewski.

ABSTRACT

Theorems similar to the theorem of Kuratowski and Ryll-Nardzewski on the existence of measurable selectors are presented. Various concepts of completness of topological spaces and various concepts of measurability of multifunctions are examined.

INTRODUCTION

The problem of the existence of measurable selectors for measurable multifunctions has been investigated in recent years by many authors (for the literature see Wagner (1977), (1980)). Almost all papers consider measurable multifunctions as "set valued functions which assign to each element t of a measurable space T a subset of a topological space X in a manner satisfying any one of several possible definitions of measurability" (see Himmelberg (1975)). Different but equivalent definitions of the measurability of multifunctions were studied in Idzik (1978). In Idzik (1981) we studied the problem of the existence of selectors for multifunctions which are defined on arbitrary sets T and X with special measurable structures. In this paper we develop the theory of selectors in topological spaces.

PRELIMINARY DEFINITIONS

Let T be an arbitrary set and $\mathfrak{T}$ be a field of its subsets, i.e. $T_1, T_2 \in \mathfrak{T}$, then $T_1 \cap T_2$ and $T - T_1 \in \mathfrak{T}$. And let X be a topological space and $\mathfrak{F}$, $\mathfrak{G}$ denote families of all closed, open subsets of X respectively.

By $P(X)$ we denote the family of all nonempty subsets of X. A function $\varphi: T \to P(X)$ is called a multifunction from T into X. A function $f:T \to X$ is called a selector of a multifunction $\varphi: T \to P(X)$ if $f(t) \in \varphi(t)$ for every $t \in T$. For a family $\mathcal{S}$ of subsets of T by $\mathcal{S}_a$ ($\mathcal{S}_m$) we denote a countably additive (countably multiplicative) family generated by $\mathcal{S}$; for example by $\mathcal{S}_{am}$ we understand a countably multiplicative family generated by $\mathcal{S}_a$. Finally we denote $N = \{1, 2, 3, \ldots\}$.

(1.1) Definition. A multifunction $\varphi: T \to P(X)$ is measurable (a-measurable, m-measurable, am-measurable, etc.) if for every $F \in \mathcal{F}$ $\varphi^{-1}(F) = \{t \mid \varphi(t) \cap F \neq \emptyset\} \in \mathcal{T}$ ($\mathcal{T}_a$, $\mathcal{T}_m$, $\mathcal{T}_{am}$, etc.).

Analogously, if we exchange in Definition 1.1 "$F \in \mathcal{F}$" on "$G \in \mathcal{G}$", we define a weakly measurable multifunction (weakly a-measurable, weakly m-measurable, etc.).

Let us recall that the diameter of a set $A \subset X$ is less than a cover $\mathcal{A} = \{A_s\}_{s \in S}$ of the set X, and we shall write $\delta(A) < \mathcal{A}$, provided that there exists an $s \in S$ such that $A \subset A_s$. We say that a cover $\mathcal{B} = \{B_p\}_{p \in P}$ is a refinement of another cover $\mathcal{A} = \{A_s\}_{s \in S}$ of the same set X if for every $p \in P$ there exists an $s(p) \in S$ such that $B_p \subset A_{s(p)}$. A family $\{A_s\}_{s \in S}$ of subsets of a set X is star-finite if for every $s_o \in S$ the set $\{s \in S \mid A_s \cap A_{s_o} \neq \emptyset\}$ is finite.

(1.2) Definition. Let $\mathcal{A} = \{\mathcal{A}_n\}_{n \in N}$, where $\mathcal{A}_n \subset P(X)$ is a cover of X. If $Z \in \mathcal{F}$ and for any countable family $\mathcal{D} \subset \mathcal{F}$ such that $D \subset Z$ for $D \in \mathcal{D}$, $\mathcal{D}$ has the finite intersection property and for each $n \in N$ contains sets of diameter less than $\mathcal{A}_n$, we have $\bigcap \{D \mid D \in \mathcal{D}\} \neq \emptyset$, then Z is called complete with respect to $\mathcal{A}$.

We say that a set $Z \in \mathcal{F}$ is $*$ - complete with respect to a family $\mathcal{A} = \{\mathcal{A}_n\}_{n \in N}$, where $\mathcal{A}_n \subset P(X)$, if $\mathcal{A}_n^* = \bigcup_{m \in N} \mathcal{A}_{n+m-1}$ ($n \in N$) is a cover of X and Z is complete to the family $\mathcal{A}' = \{\mathcal{A}_n^*\}_{n \in N}$.

(1.3) Definition. A family $\{\mathcal{B}_n\}_{n \in N}$ of covers of the space X, separates points if for every $x \in X$ and for every sequence $\{B_n\}_{n \in N}$ such that $x \in B_n \in \mathcal{B}_n$ we have $\bigcap_{n \in N} B_n = \{x\}$.

And we say that a family $\mathcal{B} = \{\mathcal{B}_n\}_{n \in N}$, where $\mathcal{B}_n \subset P(X)$, $*$-separates points if $\mathcal{B}_n^* = \bigcup_{m \in N} \mathcal{B}_{m+n-1}$ ($n \in N$) is a cover of X and the family $\{\mathcal{B}_n^*\}_{n \in N}$ separates points.

Of course if a closed set $Z \subset X$ is $*$ - complete with respect to a family $\mathcal{A} = \{\mathcal{A}_n\}_{n \in N}$ and $\mathcal{A}_n$ is a cover of the space X ($n \in N$), then it is also complete and, if a family $\mathcal{B} = \{\mathcal{B}_n\}_{n \in N}$ $*$-separates points and $\mathcal{B}_n$ is a cover of the space X ($n \in N$), then it separates points.

THEOREMS ON SELECTORS

From now on, the terminology and notations are taken from Engelking (1977) and Kuratowski (1966). Theorems on selectors which we present in this section are similar to the theorem of Kuratowski and Ryll-Nardzewski (1965) on the existence of measurable selectors.

To emphasize relations between the definitions of the completness introduced in the previous section and the definition of the Čech-completeness we recall the following

(2.1) Lemma [see Theorem 3.9.2 in Engelking (1977)]. A Tychonoff space X is Čech-complete if and only if there exists a countable family $\{\mathcal{A}_n\}_{n\in N}$ of open covers of the space X with the property that any family of closed subsets of X, which has the finite intersection property and contains sets of diameter less than $\mathcal{A}_n$ for $n\in N$, has nonempty intersection.

(2.2) Definition. In a topological space X by a Čech family we understand a countable family of open covers of the space X with the property described in Lemma 2.1.

Now, we reformulate several theorems on selectors which have been proved in Idzik (1981). We state them without proofs.

(2.3) Theorem [cf. Theorem 1 and Theorem 4 in Idzik (1981)]. Let $\varphi: T\to P(X)$ be a measurable multifunction which has values complete with respect to a family $\mathcal{A}=\{\mathcal{A}_n\}_{n\in N}$ of covers of X $(\mathcal{A}_n\subset P(X)\ ;\ n\in N)$. If there exists a family $\mathcal{B}=\{\mathcal{B}_n\}_{n\in N}$ such that $\mathcal{B}_n$ is a countable closed refinement of $\mathcal{A}_n$ $(n\in N)$ and $\mathcal{B}$ separates points, then there exists an am-measurable selector for φ.

(2.4) Theorem. Let X be a Lindelöf space and let $\varphi: T\to P(X)$ be a measurable multifunction which has values complete with respect to a Čech family $\mathcal{A}=\{\mathcal{A}_n\}_{n\in N}$. If $\mathcal{A}$ separates points, then there exists an am-measurable selector for φ.

Proof. For every $n\in N$ there exists a countable subcover $\mathcal{A}'_n=\{A_{nm}\}_{m\in N}$ of $\mathcal{A}_n$. Because X is normal and countably paracompact, then there exists a closed cover $\mathcal{B}_n=\{B_{nm}\}_{m\in N}$ of X such that $B_{nm}\subset A_{nm}$ for $m\in N$ [see Theorem 5.2.3 in Engelking (1977)]. Now, if $\mathcal{A}$ separates points then $\mathcal{B}=\{\mathcal{B}_n\}_{n\in N}$ also separates points and by Theorem 2.3 there exists an am-measurable selector for φ.

(2.5) Theorem [cf. Theorem 2 and Theorem 5 in Idzik (1981)]. Let $\varphi: T\to P(X)$ be a measurable multifunction which has values $\varkappa$-complete with respect to a family $\mathcal{A}=\{\mathcal{A}_n\}_{n\in N}$ $(\mathcal{A}_n\subset P(X);\ n\in N)$. If there exists a family $\mathcal{B}=\{\mathcal{B}_n\}_{n\in N}$ such that $\mathcal{B}_n$ is a countable star-finite closed refinement of $\mathcal{A}_n$ ($n\in N$) and $\mathcal{B}$ *-separate points, then there exists an m-measurable selector for φ.

As a consequence of Theorem 2.5 we have

(2.6) Theorem. Let X be a normal and countably paracompact space and let $\varphi: T \to P(X)$ be a measurable multifunction which has values *-complete with respect to a Čech family $\mathcal{A} = \{\mathcal{A}_n\}_{n\in N}$ in X. If $\mathcal{A}_n$ has a countable open refinement $\mathcal{B}_n$ ($n\in N$) and the family $\mathcal{B} = \{\mathcal{B}_n\}_{n\in N}$ *-separate points, then there exists an m-measurable selector for φ.

Proof. By Theorem 5.2.6 in Engelking (1977) the countable open cover $\mathcal{B}_n^* = \bigcup_{m\in N} \mathcal{B}_{n+m-1}$ of X has a countable star-finite open refinement $\mathcal{B}_n^o = \{B_{nm}\}_{m\in N}$ ($n\in N$) which is a cover of X. Furthemore, there exists a closed cover $\mathcal{D}_n = \{D_{nm}\}_{m\in N}$ of X such that $D_{nm} \subset B_{nm}$ for $m,n\in N$ [see Theorem 5.2.3 in Engelking (1977)]. The closed cover $\mathcal{D}_n$ is a fortiori a star-finite refinement of $\mathcal{B}_n^*$ (and also a star-finite refinement of $\mathcal{A}_n^*$). Observe that $\mathcal{B}_n^{**} = \bigcup_{r\in N} \mathcal{B}_{n+r-1}^* = \bigcup_{m,r\in N} \mathcal{B}_{n+m+r-2} = \mathcal{B}_n^*$, and the family $\mathcal{B}^* = \{\mathcal{B}_n^*\}_{n\in N}$ *-separate points, because by our assumption $\mathcal{B}^*$ separates points. Thus the family $\mathcal{D} = \{\mathcal{D}_n\}_{n\in N}$ also *-separates points. Analogously we check that values of φ are *-complete with respect to the family $\mathcal{A}^* = \{\mathcal{A}_n^*\}_{n\in N}$. Applying Theorem 2.5 for the families $\mathcal{A}^*$ and $\mathcal{D}$ we conclude that there exists an m-measurable selector for φ.

WEAKLY ČECH-COMPLETE SPACES

In all theorems on selectors in the previous section we considered a topological space X in which there existed a countable family $\mathcal{B} = \{\mathcal{B}_n\}_{n\in N}$, where $\mathcal{B}_n$ was a countable closed cover of X and $\mathcal{B}$ separated points. The family $\mathcal{B}$ with such properties we shall call a separating family. A topological space in which there exists a separating family is T_1-space. In this section we examine relations of such families to Čech-complete spaces.

We recall that a Tychonoff space X is Čech-complete if for every compactification cX of the space X, c(X) is a G_δ-set in cX.

(3.1) Definition. We say that a Tychonoff space X is weakly Čech-complete if there exists a compactification cX of the space X such that c(X) is an $F_{\sigma\delta}$-set in cX.

(3.2) Proposition. If there exists a perfectly normal compactification of a Tychonoff space X and X is Čech-complete, then it is weakly Čech-complete.

(3.3) Proposition. Every Tychonoff σ-compact space X (i.e., $X = \bigcup_{n\in N} X_n$, where each X_n is compact) is weakly Čech-complete.

Analogously to Lemma 2.1 we have the following theorem for weakly Čech-complete spaces.

(3.4) Theorem. If a Tychonoff space X is weakly Čech-complete, then there exists a countable family $\mathcal{B}=\{\mathcal{B}_n\}_{n\in N}$ of countable closed covers of the space X such that X is complete with respect to $\mathcal{B}$.

Proof [it is a modification of the proof of Theorem 3.9.2 in Engelking (1977)]. Let $X=\bigcap_{n\in N}\bigcup_{m\in N}F_{nm}$, where F_{nm} is a closed set in a compactification cX of the space X and let $\mathcal{B}_n=\{X\cap F_{nm}\}_{m\in N}$ ($n\in N$). We shall show that X is complete with respect to the countable family $\{\mathcal{B}_n\}_{n\in N}$ of countable closed covers of the space X. Indeed, let $\{D_r\}_{r\in N}$ be a family of closed subsets of X, which has the finite intersection property and contains sets of diameter less than $\mathcal{B}_n$ for every $n\in N$. The family $\{\overline{D}_r\}_{r\in N}$, where $\overline{D}$ is the closure of D in cX, consists of closed subsets of cX and has the finite intersection property. Thus there exists a point $x\in\bigcap_{r\in N}\overline{D}_r$. Now, let us observe that to prove that $x\in\bigcap_{r\in N}D_r$ it suffices to show that $x\in X$. For any $n\in N$ choose an $r_n\in N$ such that $\delta(D_{r_n})<\mathcal{B}_n$ and an $m_n\in N$ such that $D_{r_n}\subset X\cap F_{nm_n}$. We have $x\in\overline{D}_{r_n}\subset$ $\subset\overline{X\cap F_{nm_n}}\subset\overline{F}_{nm_n}\subset\bigcup_{m\in N}F_{nm}$ for every $n\in N$, and $x\in\bigcap_{n\in N}\bigcup_{m\in N}F_{nm}=X$.

Putting in Theorem 2.3 $\mathcal{A}=\mathcal{B}$ we have

(3.5) Theorem. Let X be a weakly Čech-complete space and $\varphi:T\to P(X)$ be a measurable multifunction with closed values. If a countable family $\mathcal{B}=\{\mathcal{B}_n\}_{n\in N}$ of countable closed covers of the space X, such that X is complete with respect to $\mathcal{B}$ (the existence of such a family follows from Theorem 3.4), separates points i.e., $\mathcal{B}$ is a separating family, then there exists an am-measurable selector for φ.

(3.6) Remark. Theorem 3.5 is analogous to the theorem of Kuratowski and Ryll-Nardzewski (1965) on the existence of measurable selectors.

The following question arises naturally in relation to Lemma 2.1.

(3.7) Question. Is the converse theorem to Theorem 3.4 true ?

Acknowledgement. I am very grateful to J. Kaniewski and C. Ryll-Nardzewski for many valuable advices and suggestions during the preparation of this paper.

REFERENCES

Engelking R. (1977): General Topology, PWN - Polish Scientific Publishers, Warszawa.

Himmelberg C.J. (1975): Measurable relations, Fund. Math. 87, 53 - 72.

Idzik A. (1978) : On equivalent definitions of the measurability of multifunctions and Filippov's lemma (in Polish), Roczniki PTM, Seria III: Matematyka Stosowana 14,65-72.

(1981) : Theorems on selectors of measurable multifunctions, Bull. Acad. Polon. Sci. Sér. Sci. Math.,No.11-12,597-603.

Kuratowski K. (1966): Topology, Vol. I, Academic Press, New York-London; PWN, Warszawa.

Kuratowski K. and C. Ryll-Nardzewski (1965): A general theorem on selectors, Bull. Acad. Polon. Sci. Sér. Sci. Math. Astronom. Phys. 13, 397-403.

Wagner D.H. (1977): Survey of measurable selection theorems, SIAM J. Control and Optimization 15, 859-903.

(1980): Survey of measurable selection theorems; an update, Lectures Notes in Math., Vol.794, Springer, 176-219.

Polish Academy of Sciences
Institute of Computer Science
P.O.Box 22
00-901 Warsaw PKiN
Poland

A GAME-THEORETIC ARROW-DEBREU MODEL

Adam Idzik
Warsaw

and

Per Barndorff Simonsen
Copenhagen

Key words: A noncooperative game with constraints and an Arrow-Debreu model with countably many agents.

1.0. ABSTRACT

Arrow and Debreu (1) were the first to consider N-person games with constraints and showed the existence of equilibrium points in some special cases. Furtermore, using game-theoretic methods, they proved the existence of an equilibrium for a competitive economy. Makarov and Rubinov (7) developed a game-theoretic Arrow-Debreu model (see chapters 5 and 6). But their proof of the existence of an equilibrium in such a model contains a few errors. Namely, the constraints for players are not well-defined (see 7 p. 284).

In our paper we generalize their model. We define a non-cooperative game with constraints and prove the existence of an equilibrium for it. We extend the Arrow-Debreu model to countably many agents. Under some assumptions we prove the existence of an equilibrium in such a generalized Arrow-Debreu model.

2.0. A NONCOOPERATIVE GAME WITH CONSTRAINTS.

Let I be a countable, finite or infinite, set of players. For a closed subset S of a metric space E we denote by 2^S a space of all nonempty closed subsets of S. 2^S is the metric space defined by the Hausdorff metric.

Let X,Y be two subsets of metric spaces, Y a compact set and $\varphi : X \to 2^Y$. We say that φ is upper semi-continuous (u.s.c.) if $x^n \to x^o$, $y^n \to y^o$, $y^n \in \varphi(x^n)$ implies $y^o \in \varphi(x^o)$. And we say that φ is lower semi-continuous (l.s.c.) if from the conditions: $x^n \to x^o$, $y^o \in \varphi(x^o)$ follows the existence of a sequence $\{y^n\}$ such that $y^n \to y^o$ and $y^n \in \varphi(x^n)$. φ is continuous iff it is both u.s.c. and l.s.c. (see (5) p. 173 and (6) §43,II). The graph of φ is defined as the set $\Gamma(\varphi) = \{(x,y) | y \in \varphi(x)\} \subset X \times Y$.

Immediately from the definition follows: (2.0.1.) Lemma. Let X and Y be compact subsets of metric spaces and $\varphi: X \to 2^Y$. Then φ is u.s.c. iff the graph $\Gamma(\varphi)$ of φ is closed in X x Y.

(2.0.2.) Theorem. Let $\{E_i \mid i \in I\}$ be a family of euclidean spaces and for each $i \in I$ let K_i be a nonempty compact convex subset of E_i. Let $K = \prod\{K_i \mid i \in I\}$. If $\{\varphi_i \mid i \in I\}$ is a family of u.s.c. functions, $\varphi_i : K \to 2^{K_i}$ such that $\varphi_i(x)$ is convex for all $x \in K$, then there exists an $\bar{x} \in K$ such that $\bar{x}_i \in \varphi_i(\bar{x})$, where $\bar{x}_i$ is the projection of $\bar{x}$ on K_i $(i \in I)$.

Proof. Define $\varphi: K \to 2^K$ by $\varphi(x) = \prod\{\varphi_i(x) \mid i \in I\}$. It is easy to chech that φ is a well-defined u.s.c. function (using Lemma 2.0.1.) and that $\varphi(x)$ is convex for all $x \in K$. Thus, by a theorem of Fan (4)p.122 there is an $\bar{x} \in K$ such that $\bar{x} \in \varphi(\bar{x})$ i.e. $\bar{x}_i \subset \varphi_i(\bar{x})$ for each $i \in I$.

(2.0.3.) Lemma. (cf. Berge (2)p.116). Let X and Y be subsets of euclidean spaces and Y be compact. Let $\varphi: X \to 2^Y$ be a continuous function. Then for every continuous function $f : X \times Y \to R$ (R - the real line) the function $\varphi_f : X \to 2^Y$ defined by

$$\varphi_f(x) = \{y \mid y \in \varphi(x),\ f(x,y) = \max_{\bar{y} \in \varphi(x)} f(x,\bar{y})\}$$

is u.s.c.

From Theorem 2.0.2. and Lemma 2.0.3. follows: (2.0.4.) Theorem. Let $\{E_i \mid i \in I\}$ be a family of euclidean spaces. For each $i \in I$ let K_i be a nonempty compact, convex subset of E_i and $K = \prod\{K_i \mid i \in I\}$. Let $\varphi_i: K \to 2^{K_i}$ be a continuous function and $f_i: K \to R$ be a continuous real function $(i \in I)$. If the set

$$\tilde{\varphi}_i(x) = \{y \mid y \in \varphi_i(x),\ f_i(y,x_i') = \max_{\tilde{x}_i \in \varphi_i(x)} f_i(\tilde{x}_i,x_i')$$

is convex for each $x \in K$ and $i \in I$, where x_i' is the projection of x on $K_i' = \prod\{K_j \mid j \in I,\ j \neq i\}$, then there is an $\bar{x} \in K$ such that

$$f_i(\bar{x}) = \max_{\tilde{x}_i \in \varphi_i(\bar{x})} f_i(\tilde{x}_i, \bar{x}_i') \quad \text{and} \quad \bar{x} \in \varphi_i(\bar{x})$$

for each $i \in I$.

For an indexed family of sets $\{Z_i \mid i \in I\}$ we denote: $Z = \prod\{Z_i \mid i \in I\}$, $Z_i' = \prod\{Z_j \mid j \in I,\ j \neq i\}$ and $z_i \in Z_i$, $z \in Z$, $z_i' \in Z_i'$ for $i \in I$.

By a noncooperative game with constraints we understand an indexed family of triples $\{(Z_i, f_i, \varphi_i) \mid i \in I\}$. I is the set of players; Z_i is a closed subset of a metric space E_i, the set of strategies for player i; f_i is a real function defined on Z, a payoff function for player i; and $\varphi_i: Z \to 2^{Z_i}$ is a constraint for player i $(i \in I)$.

The game with constraints works as follows: each of the players $i \in I$ chooses the strategy $z_i \in Z_i$ (independently of the other players). In this way the players choose the point $z \in Z$ which defines the usual noncooperative game without constraints of the same players with the payoff functions f_i but with strategies restricted to $\varphi_i(z)$.

A point $\bar{z} \in Z$ we call an equilibrium point of the noncooperative game with constraints, if it is an equilibrium point of the game associated with it, in the above sense.

This means that

(2.0.5.) $z_i \in \varphi_i(z)$

and

(2.0.6.) $f_i(z) = \max_{z_i \in \varphi_i(z)} f_i(z_i, \bar{z}_i')$

for each $i \in I$.

From Theorem 2.0.4. it is easy to derive

(2.0.7.)Theorem. (cf. (7), Theorem 17.1) If a noncooperative game with constraints $\{(Z_i, f_i, \varphi_i) \mid i \in I\}$ has the following properties:

(2.0.8.) Z_i is a nonempty compact, convex subset of a euclidean space E_i

(2.0.9.) f_i is a continuous function defined on Z and concave with respect to the variable z_i,

(2.0.10.)for each $z \in Z$ the set $\varphi_i(z)$ is convex,

(2.0.11.)φ_i is continuous,

for each $i \in I$, then the game has an equilibrium point.

3.0. An Arrow-Debreu model with countably many agents.

Let us consider a closed economic system i.e. a system upon which there are no external influences. The system consists of producers, consumers, a pricing agency, a market and a bank. Furthermore, the system takes into consideration only a finite number of commodities, say l.

3.1. Basic definitions.

Let M and N denote countable, finite or infinite, sets of producers and consumers respectively. Each of the producers $i \in M$ has a production set $X_i \subset E$, where E is a fixed l - dimensional euclidean space, i.e. his production possibilities. A vector $x_i = (x_i^1, \ldots, x_i^l) \in X_i$ is a production bundle for producer i ($i \in M$) and $x_i^k \geqslant 0$ ($x_i^k \leqslant 0$) is production (consumption) of commodity k ($k = 1, \ldots, l$).

The consumers are described by consumption sets $Y_j \subset E$, and utility functions $u_j : X \times Y \times P_l \rightarrow R$ ($j \in N$), where $X = \prod\{X_i \mid i \in M\}$, $Y = \prod\{Y_j \mid j \in N\}$ and $P_l = \{p \in E \mid \sum_{k \in \{1,\ldots,l\}} p^k = 1;\ p^k \geqslant 0\}$.

Elements of X,Y,Z we denote by x,y,z respectively, where $Z = X \times Y \times P_l$.

Furthermore, each of the producers $i \in M$ has a value-production function $v_i: Z \rightarrow R$ and some function $\varphi_i: Z \rightarrow 2^E$ ($\varphi(z) \subset X_i$ for $z \in Z$), which restricts the production. We call φ_i a productionconstraining function for producer i.

The pricing agency establishes prices on all commodities, i.e. it chooses $p \in P_l$.

The number $u_{j_0}(x,y,p)$ is a measure of the utility of the consumption bundle y_{j_0} for consumer j_0 when the production x, the consumptions y_j ($j \in N$; $j \neq j_0$) of the rest of consumers and the prices are fixed.

And analogously, the number $v_{i_o}(x,y,p)$ is a measure of the value of the production bundle x_{i_o} for producer x_{i_o}, when the production $x_i(i\in M, i\neq i_o)$ of the rest of the producers, the consumption y and the prices p are fixed.

The production-constraining function φ_i restricts the production set X_i of producer i to the set $\varphi_i(x,y,p)$, when planned production is x, planned consumption is y and planned prices are p.

Commercial connections among producers and consumers are directed by a market of the economic system. The producer $i\in M$ can supply on the market only some part (none-zero) of his production bundle $x_i\in X_i$; namely $\mu_i^k x_i^k$ of commodity $k(\mu_i^k > 0,\ k\in\{1,\dots,l\})$.
Analogously each of the consumers can buy only some part (non-zero) of his consumption bundle; namely $\nu_i^k y_i^k$ of commodity $k(\nu_j^k > 0, k\in\{1,\dots,l\})$ $(j\in N)$.

Sequences of vectors $\mu=\{\mu_i \mid i\in M\}$ and $\nu = \{\nu_j \mid j\in N\}$ we call market-restrictions of producers and consumers respectively. They are fixed in our model.

Financial connections are directed by a bank of the economic system. The consumer j receives from the bank (or pays to the bank) an ammount $\alpha_j(z)$, where α_j: $Z\to R$, if $\alpha_j(z)>0$ (or $\alpha_j(z)<0$), planned production of producer i is x_i, consumer j_o wants to buy the consumption bundle y_{j_o} and the prices are $p(i\in M;\ j,j_o\in N)$.

The function α_j we call an allocation function of consumer $j(j\in N)$ or simply an allocation function.

Let us introduce the following notations: for two vectors $\lambda = (\lambda^1,\dots,\lambda^l)\in E$ and $m = (m^1,\dots,m^l)\in E$ by $\lambda\bullet m$ we understand a vector $(\lambda^1 m^1,\dots,\lambda^l m^l)\in E$.

In this section we assume that

(3.1.1.) $\sum_{i\in M} \|\mu_i\bullet x_i\| < \infty$ for $x_i\in X_i$

(3.1.2.) $\sum_{j\in N} \|\nu_j\bullet y_j\| < \infty$ for $y_j\in Y_j$

(3.1.3.) $\sum_{j\in N} |\alpha_j(z)| < \infty$ for $z\in Z$

where $\|m\|$ is the norm of $m = (m^1,\dots,m^l)\in E$ defined by $\|m\| = \max_{k\in\{1,\dots,l\}} |m^k|$

In this section we also assume that the following natural condition is satisfied:

(3.1.4.) $\sum_{j\in N}\alpha_j(z) \leqslant \sum_{i\in M} p(\mu_i\bullet x_i)$ for $z\in Z$

The number $\sum_{i\in M} p(\mu_i\bullet x_i)$ we call an income of the economic system. The condition (3.1.4.) means that at most all income of the economic system is allocated by the bank to the consumers.

In our economic system producers, consumers and the pricing agency compose parts of the system, which exist independently of each other Furthermore, each part tends to maximize some functions: producers tend to maximize their value-production functions, consumers their utility functions and the pricing agency an income function of its activity.

By an equilibrium in our model we understand a point $\bar{z} = (\bar{x},\bar{y},\bar{p})$ of the space Z which satisfies the following conditions:

(3.1.5.) $$\sum_{i\in M} \mu_i \circ \bar{x}_i \geq \sum_{j\in N} \nu_j \circ \bar{y}_j$$

(3.1.6.) $$\max_{x_i \in \varphi_i(\bar{z})} v_i(x_i,\bar{x}_i',\bar{y},\bar{p}) = v_i(\bar{z}) \quad \text{and} \quad \bar{x}_i \in \varphi_i(\bar{z}) \quad (i\in M)$$

(3.1.7.) $$\max u_j(\bar{x},y_j,\bar{y}_j',\bar{p}) = u_j(\bar{z}) \quad \text{and} \quad \bar{y}_j \in \Psi_j(\bar{z}),$$

where $\Psi_j: Z\rightarrow 2^{Y_j}$ is defined by

(3.1.8.) $$\Psi_j(z) = \{m \mid m\in Y_j,\ p(\nu_j \circ m) \leq \alpha_j(z)$$

Inequality (3.1.5.) is the material balance between demand and supply. $\sum_{i\in M} \mu_i \circ \bar{x}_i$ is the restricted production presented by producers at the market in equilibrium and $\sum_{j\in N} \nu_j \circ \bar{y}_j$ is the restricted consumption in equilibrium.

Equality (3.1.6.) says that in equilibrium the value-production functions for i-th producer obtains a maximum at $\bar{x}_i$ which belongs to the set constrained by the equilibrium $\bar{z}$.

Similarly equality (3.1.7.) says that in equilibrium, the consumption y_j of the j-th consumer is a maximum of the utility function u_j under the balance condition described by the function Ψ_j $(j\in N)$.

3.2. Proof of the existence of an equilibrium.

Now, let us assume that

(3.2.1.) X_i is a compact, convex subset of E $(i\in M)$,

(3.2.2.) Y_j is a compact, convex subset of E $(j\in N)$,

(3.2.3.) φ_i is a continuous function and $\varphi_i(z)$ is convex for $z\in Z$, $(i\in M)$,

(3.2.4.) α_j is a continuous function $(j\in N)$,

(3.2.5.) v_i and u_j are continuous functions and concave with respect to the variables x_i and y_j respectively $(i\in M,\ j\in N)$,

(3.2.6.) $\Psi_j(z) \neq \emptyset$ for each $z\in Z$ $(j\in N)$.

Before a formulation of the main result of this section we prove two lemmas.

(3.2.7.) Lemma. If α_j is a continuous function, then a function $\widetilde{\Psi}_j : Z\rightarrow 2^F$ defined as $\widetilde{\Psi}_j(z) = \{m \mid m\in E,\ p(\nu_j \circ m) \leq \alpha_j(z)\}$ is l.s.c. $(j\in N)$.

Proof. Let us fix j and let $z^n \rightarrow z^o$, $m^o \in \Psi_j(z^o)$.

If $p^o(\nu_j \circ m^o) < \alpha_j(z^o)$, then for sufficiently large n $m^o \in \widetilde{\Psi}_j(z^n)$.

If $p^o(\nu_j \circ m^o) = \alpha_j(z^o)$, then $m^n = m^o + \frac{1}{p^n \nu_j}\left[\alpha_j(z^n) - p^n(\nu_j \circ m^o)\right] e \to m^o$ and $m^n \in \tilde{\Psi}_j(z^n)$, where $e = (1,1,\dots,1) \in E$, because
$p^n(\nu_j \circ m^n) = \alpha_j(z^n)$ and

$$\| m^n - m^o \| = \frac{1}{p^n \nu_j} |\alpha_j(z^n) - p^n(\nu_j \circ m^o)|$$

$$\leq \frac{1}{\nu_j^{min}} |\alpha_j(z^n) - \alpha_j(z^o)| + \frac{1}{\nu_j^{min}} |p^n(\nu_j \circ m^o) - p^o(\nu_j \circ m^o)|$$

where $\nu_j^{min} = \min_{k \in \{1,\dots,l\}} \nu_j^k > 0$

(3.2.8.) Lemma. If condition (3.2.2.), (3.2.4.) and (3.2.6.) are fulfilled, then the function Ψ_j is continuous $(j \in N)$.

Proof. The u.s.c. of Ψ_j follows from the definition. It is sufficient to prove the l.s.c.

Let $z^n \to z^o$ and $m^o \in \Psi_j(z^o)$. If $\Psi_j(z^o) = \{m^o\}$ then by (3.2.6.) there exists $m^n \in \Psi_j(z^n)$. We claim that $m^n \to m^o$. If not, there exists a subsequence $\{m^{n_k}\}$ and $m^{n_k} \to \bar{m}^o \neq m^o$, because Y_j is a compact set. But Ψ_j is u.s.c. and thus $\bar{m}^o \in \Psi(z^o) = \{m^o\}$, which is impossible.

If there exists $m^1 \in \Psi_j(z^o)$ and $m^1 \neq m^o$, then because of (3.2.2.) $\bar{m}^r = \frac{1}{r} m^1 + (1 - \frac{1}{r}) m^o \in \operatorname{Int} Y_j$ $(r = 1,2,\dots)$. It is easy to check that $\bar{m}^r \in \Psi_j(z^o)$ and $\bar{m}^r \to m^o$.

Let us assume that the function Ψ_j is not l.s.c. Then there exist an $\bar{\varepsilon} > 0$ and a sequence $\{m^{n_k}\}$ such that $\|m^{n_k} - m^o\| \geq \bar{\varepsilon}$, where m^{n_k} is defined by $\|m^{n_k} - m^o\| = \min \{\|m - m^o\| \mid m \in \Psi_j(z^{n_k})\}$.
It follows from above that there exists r_o such that
$\bar{m}^{r_o} \in \operatorname{Int} Y_j$, $\bar{m}^{r_o} \in \Psi_j(z^o)$ and $\|\bar{m}^{r_o} - m^o\| < \frac{\bar{\varepsilon}}{2}$.
Because $\bar{m}^{r_o} \in \operatorname{Int} Y_j$ and Lemma 3.2.7. there exists n_o such that
$\min \{\|\bar{m}^{r_o} - m\| \mid m \in \Psi_j(z^n)\} \leq \frac{\bar{\varepsilon}}{2}$ for $n \geq n_o$. Thus
$\min \{\|m - m_o\| \mid m \in \Psi_j(z^n)\} \leq \|\bar{m}^{r_o} - m^o\| + \min \{\|m - \bar{m}^r\| \mid m \in \Psi_j(z^n)\} < \bar{\varepsilon}$
for $n \geq n_o$. But this contradicts the definition of the sequence $\{m^{n_k}\}$.

(3.2.9.) Theorem. If the conditions (3.2.1.) - (3.2.6.) are fulfilled, then the generalized Arrow-Debreu model has an equilibrium.

Proof. With the generalized Arrow-Debreu model we associate a noncooperative game with constraints as follows: the players are the producers, the consumers and the special player: the agency setting prices.

Sets of strategies of players are $X_i (i \in M)$, $Y_j (j \in N)$ and P_l respectively.

The constraints are defined by functions: $\varphi_i (i \in M)$, $\Psi_j (j \in N)$ and π, where $\pi : Z \to 2^{P_l}$ is defined by $\pi(z) = P_l$ for $z \in Z$; and payoff functions by $v_j (i \in M)$, $u_j (j \in N)$ and h where $h: Z \to R$ is defined by
$h(z) = p\left(\sum_{j \in N} \nu_j \circ y_j - \sum_{i \in M} \mu_i \circ x_i\right)$ for $z \in Z$.

By Lemma 3.2.8. $\Psi_j (j \in N)$ are continuous functions and thus by our assumptions from Theorem 2.0.7. follows the existence of an equilibrium point in a noncooperative game with constraints so defined.

Let $z = (\bar{x}, \bar{y}, \bar{p})$ denote this equilibrium point . We will show that it defines an equilibrium in the generalized Arrow-Debreu model. By our definitions conditions (3.1.6.) and (3.1.7.) are satisfied. Also (3.1.5) is satisfied. Indeed, suppose that it is not fulfilled. Then there is a nonempty set C of commodities such that

(3.2.10.) $$\sum_{j\in N} \nu_j^k \bar{y}_j^k - \sum_{i\in M} \mu_i^k \bar{x}_i^k > 0 \quad \text{for } k \in C$$

and

(3.2.11.) $$\sum_{j\in N} \nu_j^k \bar{y}_j^k - \sum_{i\in M} \mu_i^k \bar{x}_i^k \leqslant 0 \quad \text{for } \{1,\dots,l\} \setminus C$$

But in equilibrium the special player maximizes the function $h(\bar{x},\bar{y},p)$ with respect to p, and we have $\sum_{k\in C} \bar{p}^k = 1$. Thus from (3.2.10.) we obtain

(3.2.12.) $$\sum_{j\in N} \bar{p}\,(\nu_j \circ \bar{y}_j) > \sum_{i\in M} \bar{p}\,(\mu_i \circ \bar{x}_i$$

and from (3.1.8.)

(3.2.13.) $$p\,(\nu_j \bullet y_j) \leqslant \alpha_j(\bar{z})$$

Summing the inequalities in (3.2.13.) we have

$$\sum_{j\in N} \bar{p}\,(\nu_j \bullet \bar{y}_j) \leqslant \sum_{j\in N} \alpha_j(\bar{z}) \leqslant \sum_{i\in M} \bar{p}\,(\mu_i \bullet \bar{x}_i),$$

by assumption (3.1.4.) But this contradicts (3.2.12.).

In this way (3.1.5.) is satisfied too.

REFERENCES

(1) Arrow, K.J. and G. Debreu, 1954, Existence of an equilibrium for a competitive economy, Econometrica 22, 265-290.

(2) Berge, C.,1963, Topological spaces (Oliver and Boyd Ltd, Edinburg-London).

(3) Debreu, G.,1952, A social equilibrium existence theorem, Proc. Nat. Acad. Sci. USA 38, 886-893.

(4) Fan, K.,1952, Fixed-point and minimax theorems in locally convex topologocal linear spaces, Proc. nat. Acad. Sci. USA 38, 121-126.

(5) Kuratowski, K.,1966, Topology, Vol.1 (Academic Press,N.Y.-London).

(6) Kuratowski, K.,1968, Topology, Vol.2 (Academic Press,N.Y.-London).

(7) Makarov, V.L. and A.M. Rubinov,1973, Mathematical theory of economic dynamics and equilibrium (in Russian) (Nauka, Moskva).

Adam Idzik
Polish Academy of Sciences
Institute of Computer Science
P. O. Box 22
00-901 Warsaw PKiN
Poland

Per Barndorff Simonsen
Frederiksborgvej 11 Port
2400 Copenhagen NV.
Denmark.

ASYMPTOTIC RESULTS FOR AN EPIDEMIC PROCESS ON RANDOM GRAPHS

Zvetan Ignatov
Sofia

Key words: Epidemic Process, Random Graph, Poisson-Dirichlet Measure, Limit Distribution.

ABSTRACT

We consider the uniform distribution on the set of all mappings of the finite set $M_n=\{1,2,\dots,n\}$ into itself. Each mapping is a digragh and may consist of disjoined components. We investigate the asymptotic behaviour of the components of a random mapping and give some applications to one epidemic process.

ASYMPTOTIC BEHAVIOUR OF THE COMPONENTS OF RANDOM MAPPINGS

Let $\mathcal{M}_n$ be the set of all mappings of the finite set M_n, $M_n:=\{1,2,\dots,n\}$ into itself and P_n be the uniform distribution on $\mathcal{M}_n$, i.e. $P_n(h)=n^{-n}$ for every $h\in\mathcal{M}_n$. The random mapping h is a digraph G_h, whose points belong to the set M_n and the points x and y are joined by an oriented arc goes from x to y iff $y=h(x)$. Each digraph may consist of disjoined components and each component includes only one cycle.

Imagine that an arc connecting two arbitrary vertices x and y, $x,y\in M_n$, carries infection in two directions: from x to y if x had been infected first, and conversely. If one bacterium is placed at the element $x\in M_n$ then the infected area in the mapping h is the component of G_h which contains the element x. This scheme of an epidemic process, so called " two-sided epidemic process ", is introduced by Gertsbakh (1977).

The main purpose of this paper is to investigate the limit distribution (joint) of the components of the random mappings.

Let the mapping $h \in \mathcal{M}_n$ have j components $A_1, A_2, \dots, A_j$ and $z_i = \min\{x : x \in A_i\}$. We arrange $z_1, z_2, \dots, z_j$ in increasing order $z_{i_1}, z_{i_2}, \dots z_{i_j}$ and denote by

(1) $$x_k(h) := \begin{cases} |A_{i_k}|/n & , 1 \le k \le j \\ 0 & j < k \end{cases}$$

where $|A_{i_k}|$ is the number of the elements of the component A_{i_k} containing the element z_{i_k}. For example, if $n=5$, $h(1)=5$, $h(2)=3$, $h(3)=2$, $h(4)=2$, $h(5)=5$ then $j=2$, $A_{i_1}=\{1,5\}$ $A_{i_2}=\{2,3,4\}$, $x_1(h)=2/5$, $x_2(h)=3/5$, $x_3(h)=x_4(h)=\dots=0$

Let Σ be the simplex $\Sigma := \{(y_1, y_2, \dots) : y_i \ge 0, i=1,2,\dots, \sum_1^\infty y_i = 1\}$ and τ be the topology of the pointwise convergence. The mapping

$$\varphi_n : \mathcal{M}_n \to \Sigma, \quad \varphi_n(h) := (x_1(h), x_2(h), \dots)$$

generates in (Σ, τ) a measure $\varphi_n(P_n)$ which we denote by η_n

We shall introduce a family of measures $\mathcal{H}_\alpha$, $\alpha > 0$ on Σ which has been studied by Ignatov (1982).

The measure $\mathcal{H}_\alpha$ is correct defined by the projections $\mathcal{H}_\alpha^m$ of $\mathcal{H}_\alpha$ on $\Sigma^m := \{(y_1, \dots, y_m) : y_i \ge 0, i=1,\dots,m, \sum_1^m y_i \le 1\}$ which have a joint density

$$\frac{d\mathcal{H}_\alpha^m}{dy_1 \cdots dy_m} = \alpha^m \left(\prod_1^{m-1} \left(1 - \sum_1^k y_i\right)\right)^{-1} (1 - y_1 - \dots - y_m)^{\alpha - 1}$$

<u>Theorem 1</u>. The measure η_n tends weakly to $\mathcal{H}_{1/2}$

Proof. Let η_n^m be the projection of η_n on Σ^m. It is enough to prove the weak convergence of η_n^m to $\mathcal{H}_{1/2}^m$, $m=1,2,\dots$

If a_n is the number of mappings $h \in \mathcal{M}_n$ having only one component, then

(2) $$a_n = (n-1) \sum_{k=0}^{n-1} \frac{n^k}{k!}$$

The formula (2) is obtained by Katz (1955).

Let $n > m$ and r_i, $i=1,\dots,m$ be natural integers for which $\sum_{i=1}^m r_i \le n$. Using (2) we get

$$\eta_n^m \left(\frac{r_1}{n}, \dots, \frac{r_m}{n} \right) = P_n \{ h : h \in \mathcal{M}_n, \, n x_i(h) = r_i, \, i = 1, \dots, m \} =$$

(3)

$$\frac{\binom{n-1}{r_1 - 1} a_{r_1} \binom{n - r_1 - 1}{r_2 - 1} a_{r_2} \cdots \binom{n - r_1 - \dots - r_{m-1} - 1}{r_m - 1} a_{r_m} \left(n - \sum_{i=1}^m r_i\right)^{n - \sum_{i=1}^m r_i}}{n^n} =$$

$$\frac{(m-1)!\,(n-z_1-\dots-z_m)^{n-z_1-\dots-z_m}\prod_{i=1}^{m}\left(\sum_{k=0}^{z_i-1}\frac{z_i^k}{k!}\right)}{(n-z_1-\dots-z_m)!\;n^n\prod_{i=1}^{m-1}(n-z_1-\dots-z_i)}$$

If d_i, b_i, $i=1,\dots,m$ are fixed real numbers, such that $0<d_i<b_i$, $i=1,\dots,m$; $\sum_1^m b_i<1$ then from the Stirling's formula and from

(4) $$\sum_{k=0}^{n-1}\frac{n^k}{k!}=\left(\frac{1}{2}+O\left(\frac{1}{\sqrt{n}}\right)\right).\exp(n)$$

we receive from (3) uniformly with respect to $\{z_i : d_i n\le z_i\le b_i n; i=1,m\}$

(5) $$\eta_n^m\left(\frac{z_1}{n},\dots,\frac{z_m}{n}\right)=\left[n^m.2^m\sqrt{1-\frac{z_1}{n}-\dots-\frac{z_m}{n}}\prod_{i=1}^{m-1}\left(1-\frac{z_1}{n}-\dots-\frac{z_i}{n}\right)\right]^{-1}(1+o(1))$$

The formula (4) is given in the paper of Kolcin (1976).

Thus we have from (5) :

(6) $$\lim_{n\to\infty}\sum_{\substack{z_1,\dots,z_m\\ nd_i\le z_i\le nb_i\\ i=1,2,\dots,m}}\eta_n^m\left(\frac{z_1}{n},\dots,\frac{z_m}{n}\right)=\int_{d_1}^{b_1}\dots\int_{d_m}^{b_m}\frac{dy_1\dots dy_m}{2^m\sqrt{1-y_1-\dots-y_m}\prod_{i=1}^{m-1}(1-y_1-\dots-y_i)}$$

Using the theorem 2.2. in the book of Billingsley (1968) and the fact $\mathcal{H}_{1/2}^m(\Sigma^m)=1$ it follows from (6) that η_n^m, $n=1,2,\dots$ tends weakly to $\mathcal{H}_{1/2}^m$.

Let $\tilde{\Sigma}\subset\Sigma$ be the set of all decreasing sequences $(y_1,y_2,\dots)$ i.e. $\tilde{\Sigma}:=\{(y_1,y_2,\dots): y_1\ge y_2\ge\dots; y_1+y_2+\dots=1\}$ and let W map Σ onto $\tilde{\Sigma}$ arranging the elements of each sequence $(y_1,y_2,\dots)\in\Sigma$ in decreasing order.If $\zeta_n=W(\eta_n)$ is a measure generated by W on $\tilde{\Sigma}$ then we have:

<u>Theorem 2</u>. The sequence ζ_n, $n=1,2,\dots$ tends weakly to the Poisson-Dirichlet measure ν_α with parameter $\alpha=1/2$.

Proof. The Poisson-Dirichlet measure ν_α is introduced by Kingman (1975) and studied by Watterson(1976).

Let us define the function

$$f_\alpha(t)=\frac{1}{\Gamma(\alpha).\exp(\alpha C)}\left\{t^{\alpha-1}+\sum_{k=1}^{\infty}\frac{(-\alpha)^k}{k!}\int\dots\int_{B_k(t)}\frac{(t-u_1-\dots-u_k)^{\alpha-1}du_1\dots du_k}{u_1\dots u_k}\right.$$

where $B_k(t):=\{(u_1,\dots,u_k): u_1\ge1,\dots,u_k\ge1,\ u_1+\dots+u_k\le t\}$;

C is the Euler's constant and $\Gamma(\cdot)$ is the gamma function.

Put $p_\alpha(t)=\Gamma(1+\alpha).t^{\alpha-2}.f_\alpha((1-t)/t).\exp(\alpha.C)$ Then the projection ν_α^m of ν_α on $\tilde{\Sigma}^m:=\{(y_1,y_2,\dots): y_1\ge y_2\ge\dots\ge y_m\ge0, \sum_{i=1}^m y_i\le1\}$ has a density

$$\frac{d\nu_\alpha^m}{dy_1\dots dy_m}=\alpha^{m-1}\frac{(1-y_1-\dots-y_{m-1})^{\alpha-2}}{y_1 y_2\dots y_{m-1}}.p_\alpha\left(\frac{y_m}{1-y_1-\dots-y_{m-1}}\right)$$

Ignatov (1982) received that $W(\mathcal{H}_\alpha)=\nu_\alpha$. It is easy to prove that W is a continuous mapping on Σ with regard to the topology τ , which is enough for the proof of theorem2.

As a consequence of theorem 2, after some calculations, we get for $0<t<1$

$$\lim_{n\to\infty} \mathfrak{Z}_n(y_m \le t) = \nu_{1/2}(y_m \le t) = \underset{\substack{1\ge y_1\ge y_2\ge\dots\ge y_m\\ 0\le y_m\le t}}{\iint\cdots\int} \frac{d\nu^m_{1/2}(y_1,\dots,y_m)}{dy_1\dots dy_m} =$$

$$\sum_{l=0}^{m-1}\sum_{\frac{1}{t}>s\ge l} \frac{(-1)^{s-l}\binom{s}{l}}{2^{s-l}(s-l)!} \underset{\substack{u_1\ge t,\dots,u_s\ge t\\ u_1+\dots+u_s\le 1}}{\iint\cdots\int} \frac{du_1\,du_2\dots du_s}{u_1\dots u_s\sqrt{1-u_1-\dots-u_s}}$$

This result coincids with theorem 9 in the paper of Aolcin (1976).

APPLICATIONS

We shall give some results without proofs or with some hints.

Let m bacteria be placed at random at the vertices of G_h, $h\in\mathcal{M}_n$, and let $P_n(g_1,\dots,g_m)$ be the probability that: g_1 components of the random mapping h are infected only with one bacterium, g_2 components are infected with two bacteria, ... , g_m components are infected with m bacteria, where $g_1\ge 0,\dots,g_m\ge 0$, $\sum_1^m i g_i = m$.

We have

$$(7)\qquad \lim_{n\to\infty} P_n(g_1,\dots,g_m) = \int_\Sigma \Big(m!\prod_{j=1}^m (j!)^{-g_j} \sum y_1^{l_1}\cdot y_2^{l_2}\cdots\Big)d\nu_{1/2}$$

where the sum extends over all sequences $l_1, l_2, \dots$ of integers $0\le l_i\le m$ such that, for $1\le j\le m$, exactly g_j of the l_i are equal to j . From the paper of Watterson (1976) follows for the right expression in (7)

$$\lim_{n\to\infty} P_n(g_1,\dots,g_m) = \frac{m!\,2^{g_1+\dots+g_m-m}}{1.3.5\dots(2m-1)}\prod_{j=1}^m \left(j^{g_j}\cdot g_j!\right)^{-1}$$

If $V_n(h,m)$ is the number of all infected elements in $h\in\mathcal{M}_n$ when m bacteria are placed at random then

$$(8)\qquad \lim_{n\to\infty} P_n\Big\{h: \frac{V_n(h,m)}{n} > u\Big\} = \sum_{\kappa_1,\kappa_2,\dots}\ \int_{D(u;\kappa_1,\kappa_2,\dots)} m!\, y_1^{\kappa_1}\cdot y_2^{\kappa_2}\cdots(\kappa_1!\,\kappa_2!\dots)^{-1}d\nu_{1/2}$$

where the sum is taken over all nonnegative integers $\kappa_1,\kappa_2,\dots$ with $\kappa_1+\kappa_2+\dots = m$ and $D(u;\kappa_1,\kappa_2,\dots) := \{(y_1,y_2,\dots): \sum_1^\infty \delta(\kappa_i)y_i > u\}$, $\delta(\kappa_i):=1$ if $\kappa_i>0$ and $\delta(\kappa_i):=0$ if $\kappa_i=0$. In the case $m=1$ using the Watterson's (1976) identity:

$$\frac{1}{2}\cdot\frac{1}{x\sqrt{1-x}} = \sum_{\kappa=0}^{\infty}\int_{H(\kappa,x)} d\nu_{1/2}$$

where $0 < x < 1$, $H(k,x) := \{(y_1, y_2, \ldots) : y_1 \geq y_2 \geq \quad , y_{k+1} = x\}$ we get from (8)

$$\lim_{n\to\infty} P_n \left\{ h : \frac{V_n(h,m)}{n} > u \right\} = \int_u^1 \frac{dx}{2\sqrt{1-x}} = \sqrt{1-u} \quad , 0<u<1 \tag{9}$$

The identity (9) coincides with one result of Mutafciev (1982).

Let $x_{(1)}(h) \geq x_{(2)}(h) \geq \ldots \geq x_{(n)}(h)$ be the order statistics of the random variables $x_1(h), x_2(h), \ldots, x_n(h)$ from (1) defined on $(\mathcal{M}_n, P_n)$ and let $L_n(h) := n \cdot x_{(1)}(h)$ i.e. $L_n(h)$ is the number of the elements of the largest component in h. As a consequence of theorem 2, after some calculations, we receive one curious identity

$$\lim_{n\to\infty} n^{1/2 - n} \sum_{h \in \mathcal{M}_n} \left(L_n(h)\right)^{-\frac{1}{2}} = \frac{3}{2}\sqrt{\pi \exp C}$$

where C is the Euler's constant.

REFERENCES

Billingsley P. (1968) : Random discrete distributioms. Wiley. New York-London-Sydney-Toronto.

Gertsbakh I. (1977) : Epidemic process on a random graph; some preliminary results. Journal Appl. Probability, 14, 427-438.

Ignatov Z. (1982) : On a constant arizing in the asymptotic theory of symmetric groups and on the Poisson-Dirichlet measurs. Theor. Probability Apll. 27, No 1, 129-141

Katz L. (1955) : Probability of indecomposability of a random mapping function. Ann. Math. Statist., 26, No. 3, 512-517.

Kingman J. (1975) : Random discrete distributions. J. R. Statist. Soc. B, 37, 1-22.

Kolcin V. (1976) : On disposal of particles in cells and random mappings of finite set. Theor. Probability Appl. 21, No. 1, 48-62.

Mutafciev L. (1982) : Some limit distributions for a random graph. / to appear /

Watterson G. (1976) : The stationary distribution of the infinitely-many neutral alleles diffusion model. Journal Appl. Probability, 13, 639-651.

Bulgarian Academy of Sciences
Institute of Mathematics
1090 Sofia, P.O.Box 373
BULGARIA

MAXIMUM ENTROPY SPECTRAL ANALYSIS AND ARMA PROCESSES II

Shunsuke Ihara

Matsuyama

Key words: Maximum entropy spectral analysis, ARMA process.

ABSTRACT

In the previous paper (1982), the author generalized a result due to Burg (1967) and showed that under a prior knowledge the stationary process having the maximum entropy is a Gaussian ARMA process. In this paper we propose a practical procedure to find the spectral density function of the ARMA process with the maximum entropy.

§ 1. INTRODUCTION

We consider the practical application of the result obtained by the author (1982). We use the same notation as in the previous paper, unless otherwise stated.

Let $X = \{X_n\}_{n \in \mathbb{Z}}$ ($\mathbb{Z} = \{0, \pm 1, \pm 2, \cdots\}$) be a real valued purely nondeterministic stationary (in the wide sense) process with zero mean. Denote by $\rho_X(n)$, $n \in \mathbb{Z}$, the covariance function and by $f(\lambda)$, $-\pi \leq \lambda \leq \pi$, the spectral density function (s.d.f.) of X:

$$E[X_{n+k}X_k] = \rho_X(n) = \int_{-\pi}^{\pi} e^{in\lambda} f(\lambda)\, d\lambda, \quad n,k \in \mathbb{Z}.$$

We denote by $\bar{h}(X)$ the entropy rate of X (if exsists).

The maximum entropy (ME) method is as follows. Suppose that we are given only some partial knowledge, denote it by (A), about the stationary process. Denote by $\mathcal{P}(A)$ the class of all stationary processes which agree with the prior knowledge (A). The ME method is to take a process $X^* = \{X_n^*\}$ having the maximum entropy rate in

$\mathcal{P}(A)$ as the fitting model to the original process. The process X^* is said to be the ME (maximum entropy) process for the condition (A).

In order to state our previous result, we define a stationary process $Z(\beta) = \{Z_n(\beta)\}_{n\in\mathbb{Z}}$ by

$$(1) \qquad Z_n(\beta) = \sum_{k=0}^{\infty} \beta^{-k} X_{n-k}, \quad n \in \mathbb{Z}$$

where β is a constant such that $|\beta| > 1$. We note that the variance of $Z(\beta)$ is

$$(2) \qquad \rho_{Z(\beta)}(0) = \sum_{k,\ell=0}^{\infty} \beta^{-(k+\ell)} \rho_X(\ell-k).$$

<u>THEOREM 1</u> (Ihara (1982)). Let β_j, $j = 1,\cdots,J$, be constants such that $|\beta_j| > 1$. Assume that $\rho_X(k)$, $k = 0,\cdots,K$, and the variances $\rho_{Z(\beta_j)}(0)$ of $Z(\beta_j)$, $j = 1,\cdots,J$, are known, namely

$$(3) \qquad \rho_X(k) = u_k, \quad k = 0,\cdots,K,$$

$$(4) \qquad \rho_{Z(\beta_j)}(0) \equiv \sum_{k,\ell=0}^{\infty} \beta_j^{-(k+\ell)} \rho_X(\ell-k) = v_j, \quad j = 1,\cdots,J.$$

Here u_k and v_j are given constants such that there exists a stationary process satisfying (3) and (4). Then the ME process for the conditions (3) and (4) is a Gaussian ARMA(J+K,J) process with s.d.f. $f^*(\lambda) \equiv f^*(\lambda;\beta_1,\cdots,\beta_J)$ of the form

$$(5) \qquad f^*(\lambda) = a \prod_{j=1}^{J} |\beta_j^{-1} - e^{i\lambda}|^2 \times \prod_{k=1}^{J+K} |\alpha_k^{-1} - e^{i\lambda}|^{-2},$$

where $a > 0$ and α_k, with $|\alpha_k| > 1$, are constants uniquely determined by the conditions (3) and (4).

In case where the prior knowledge is (3) and no knowledge of type (4) is given, the result has been shown by Burg (1967).

We note here the following. Define stationary processes $Y^{(j)} = \{Y_n^{(j)}\}$, $j = 0,\cdots,J$, and $Y = \{Y_n\}$ by

$$(6) \qquad Y_n^{(0)} = X_n, \quad Y_n^{(j)} = \sum_{k=0}^{\infty} \beta_j^{-k} Y_{n-k}^{(j-1)}, \quad 1 \leq j \leq J,$$

$$\text{and} \quad Y_n = Y_n^{(J)} \quad (n \in \mathbb{Z}).$$

Then it can be shown that $\rho_Y(n)$, $n = 0,\cdots,J+K$, are derived linearly from $\rho_X(k)$, $k = 0,\cdots,K$, and $\rho_{Z(\beta_j)}(0)$, $j = 1,\cdots,J$. In other words, there exist constants $b_{n,m}$, $n,m = 0,\cdots,J+K$, (depend only on $\beta_1,\cdots,\beta_J$) such that

(7) $$\rho_Y(n) = \sum_{k=0}^{K} b_{n,k}\rho_X(k) + \sum_{j=1}^{J} b_{n,K+j}\rho_{Z(\beta_j)}(0).$$

The purpose of this paper is to give a method for the practical application of Theorem 1. In § 2, we shall give an algorithm to find the s.d.f. $f^*(\lambda)$ of (5) from the real data. As pointed out in Ihara (1982) the most important point is how to choose the constants β_j's, which appear in (4) and (5), appropriately. In § 3, we shall show that our estimate is consistent in some sense.

§ 2. DERIVATION OF THE MAXIMUM ENTROPY PROCESS

Let us assume that we have N observations $x_1,\cdots,x_N$ of the stationary process $X = \{X_n\}$. We shall propose an algorithm to get $f^*(\lambda)$ of (5) from the sample $\{x_1,\cdots,x_N\}$.

At first we define the constants $\beta_1,\cdots,\beta_J$ in the following manner. Choose constants $\beta_0 > 1$, $\delta > 0$ and an integer $J = 2J'$. $\beta_1,\cdots,\beta_J$ are given by

(8) $$\beta_{2j} = (1+\delta)^{j-1}\beta_0, \quad \beta_{2j-1} = -\beta_{2j}, \quad j = 1,\cdots,J'.$$

If the size N of the sample is not large, β_0 should be chosen apart from 1.

The estimates $u_{N,k}$ $(k = 0,\cdots,K)$ of the k-th lag covariance $\rho_X(k)$ and $v_N(\beta_j)$ $(j = 1,\cdots,J)$ of the variance $\rho_{Z(\beta_j)}(0)$ of the process $Z(\beta_j)$ are given by

(9) $$u_{N,k} = \frac{1}{N}\sum_{n=1}^{N-k} x_{n+k}x_n,$$

and

(10) $$v_N(\beta_j) = \frac{1}{N-N_0}\sum_{n=N_0+1}^{N} |\tilde{z}_n(\beta_j)|^2,$$

where

$$\tilde{z}_n(\beta_j) = \sum_{k=0}^{n-1} \beta_j^{-k}x_{n-k}$$

and $0 < N_0 < N$ is an integer. Using the ergodic property, we can prove for almost all sample $\{x_n\}$ that $v_N(\beta_j)$ converges to $\rho_{Z(\beta_j)}(0)$ as N tends to infinity. Thus $u_{N,k}$ and $v_N(\beta_j)$ are consistent estimates of $\rho_X(k)$ and $\rho_{Z(\beta_j)}(0)$, respectively.

The ME process for the conditions given by replacing u_k and $v(\beta_j)$ by $u_{N,k}$ and $v_N(\beta_j)$ in (3) and (4), respectively, is the Gaussian ARMA process with s.d.f. $f_N^*(\lambda) \equiv f_N^*(\lambda;\beta_1,\cdots,\beta_J)$ of the form

(11) $$f_N^*(\lambda) = a_N \prod_{j=1}^{J} |\beta_j^{-1} - e^{i\lambda}|^2 \times \prod_{k=1}^{J+K} |\alpha_{N,k}^{-1} - e^{i\lambda}|^{-2}.$$

Practically, $f_N^*(\lambda)$ is derived in the following way. By (7), the estimates of the first J+K+1 lags of the covariance of the process $Y = \{Y_n\}$ of (6) are given by

(12) $$\rho_Y(n) = \sum_{k=0}^{K} b_{n,k}u_{N,k} + \sum_{j=1}^{J} b_{n,K+j}v_N(\beta_j), \quad n = 0,\cdots,J+K.$$

It is well known (Burg (1967), (1975), Ulrych and Bishop (1975)) that the s.d.f. $g_N^*(\lambda)$,

$$g_N^*(\lambda) = a_N \prod_{k=1}^{J+K} |\alpha_{N,k}^{-1} - e^{i\lambda}|^{-2},$$

of the AR(J+K) process satisfying (12) is derived by solving a Yule-Walker equation. Then the s.d.f. $f_N^*(\lambda)$ is given by

$$f_N^*(\lambda) = g_N^*(\lambda) \prod_{j=1}^{J} |\beta_j^{-1} - e^{i\lambda}|^2.$$

Hannan (1980) has shown the consistency of certain estimates of the order of an ARMA process. Let $(\hat{L},\hat{J})$ be such an estimate of the order. Then we can take $\hat{L} - \hat{J}$ as K in (3).

§ 3. THEORETICAL CONSIDERATION OF THE ESTIMATION

We assume that the original unknown process $X = \{X_n\}$ is a Gaussian ARMA(J_0+K_0,J_0) process having

(13) $$f(\lambda) = a \prod_{j=1}^{J_0} |\gamma_j^{-1} - e^{i\lambda}|^2 \times \prod_{k=1}^{J_0+K_0} |\alpha_k^{-1} - e^{i\lambda}|^{-2},$$

as the s.d.f.. Then we can show the following

<u>THEOREM 2.</u> Assume that $K \geq K_0$, $J \geq J_0$ and that $\beta_1,\cdots,\beta_J$ are taken such as $\beta_j = \gamma_j$, $j = 1,\cdots,J_0$. Let $f_N^*(\lambda) = f_N^*(\lambda;\beta_1,\cdots,\beta_J)$ be the s.d.f. of (11). Then for almost all sample $\{x_n\}_{n\in\mathbb{Z}}$ of the process $X = \{X_n\}$, it holds that

(14) $$\lim_{N\to\infty} f_N^*(\lambda) = f(\lambda).$$

In other words, rearranging the order of $\alpha_{N,k}$, it holds that

$$\lim_{N\to\infty} \alpha_{N,k} = \begin{cases} \alpha_k, & 1 \leq k \leq K_0+J_0, \\ \infty, & K_0+J_0 < k \leq K+J_0, \\ \beta_{k-K}, & K+J_0 < k \leq K+J. \end{cases}$$

Proof. We note that $\rho_X(k)$, $k = 0,\cdots,K_0$, and $\rho_{Z(\beta_j)}(0)$, $j = 1,\cdots,J_0$, specify the s.d.f. $f(\lambda)$ of (13), and hence determine $\rho_X(k)$, $k > K_0$, and $\rho_{Z(\beta_j)}(0)$, $J_0 < j \leq J$. Since the operation to compute $f_N^*(\lambda)$ from $\{x_1,\cdots,x_N\}$ is continuous, and $u_{N,k}$ and $v_N(\beta_j)$ tend to $\rho_X(k)$ and $\rho_{Z(\beta_j)}(0)$, respectively, as $N \to \infty$, we get (14).

This theorem implies that, if the original process is an ARMA process and $\beta_1,\cdots,\beta_J$ are chosen such suitably as $\{\beta_1,\cdots,\beta_J\}$ contains $\{\gamma_1,\cdots,\gamma_{J_0}\}$ as a subset, we obtain a s.d.f. close to the true s.d.f. by use of the ME method. However, in practice, γ_j's are unknown and we can not choose $\beta_1,\cdots,\beta_J$ in such a way stated above. On the basis of this consideration, we have proposed a method to choose $\beta_1,\cdots,\beta_J$ as mentioned in § 2. For our method of the choice of $\beta_1,\cdots,\beta_J$, we can prove the following property.

THEOREM 3. Let X be the Gaussian ARMA(J_0+K_0,J_0) process with s.d.f. $f(\lambda)$ of (13) and $\{x_n\}$ be a sample of X. Then for each $\varepsilon > 0$ there exist $\delta > 0$ and N_1 such that the s.d.f. $f_N^*(\lambda)$ of (11) determined from $\{x_n\}$ as in § 2 satisfies the relation

$$(15)\qquad \left| \frac{f_N^*(\lambda) - f(\lambda)}{f(\lambda)} \right| < \varepsilon, \quad -\pi \leq \lambda \leq \pi, \quad N \geq N_1.$$

Proof. Taking δ in (8) sufficiently small and rearranging the order of $\beta_1,\cdots,\beta_J$, β_j $(1 \leq j \leq J_0)$ can be chosen sufficiently close to β_j. Define constants $\tilde{\beta}_1,\cdots,\tilde{\beta}_J$ by

$$\tilde{\beta}_j = \begin{cases} \gamma_j, & 1 \leq j \leq J_0, \\ \beta_j, & J_0 < j \leq J, \end{cases}$$

and denote by $\tilde{f}_N^*(\lambda) = f_N^*(\lambda;\tilde{\beta}_1,\cdots,\tilde{\beta}_J)$ the s.d.f. of the ME process for $(\tilde{\beta}_1,\cdots,\tilde{\beta}_J)$. Then, if δ is sufficiently small and N is sufficiently large, we have

$$(16)\qquad \sqrt{1 - \varepsilon} < \left| \frac{f_N^*(\lambda)}{\tilde{f}_N^*(\lambda)} \right| < \sqrt{1 + \varepsilon},$$

since $\tilde{\beta}_j$ and $v_N(\tilde{\beta}_j)$ are close to β_j and $v_N(\beta_j)$, respectively. On the other hand, it follows from Theorem 2 that

$$(17)\qquad \sqrt{1 - \varepsilon} < \left| \frac{\tilde{f}_N^*(\lambda)}{f(\lambda)} \right| < \sqrt{1 + \varepsilon}.$$

Thus we get (15) from (16) and (17).

When δ is small in (8), J is large. Therefore, seemingly, $f_N^*(\lambda)$ is a s.d.f. of a high order ARMA process and the order (J_0+K_0,J_0) is overestimated. However, Theorem 2, 3 imply that, if β_ℓ is apart from all γ_j, $1 \leq j \leq J_0$, then there is an integer k_ℓ such that α_{N,k_ℓ} is close to β_ℓ. Then, for such β_ℓ's, cancelling $|\beta_\ell^{-1} - e^{i\lambda}|^2$ and $|\alpha_{N,k_\ell}^{-1} - e^{i\lambda}|^{-2}$ in the right hand side of (11), we are given a lower order rational s.d.f. as an estimate of the true s.d.f..

REFERENCES

Burg J.P. (1967): Maximum entropy spectral analysis. Presented at the 37th Annu. Int. Mtg., Soc. of Explor. Geophys., Oklahoma City, OK.

(1975): Maximum entropy spectral analysis. Ph.D. Thesis, Dept. of Geophys., Stanford Univ..

Hannan E.J. (1980): The estimation of the order of an ARMA process. Ann. of Statist., 8, 1071-1081.

Ihara S. (1982): Maximum entropy spectral analysis and ARMA processes. submitted.

Ulrych T.J. and T.N. Bishop (1975): Maximum entropy spectral analysis and autoregressive decomposition. Reviews of Geophys. and Space Phys., 13, 183-200.

Department of Mathematics
Ehime University
Matsuyama
Japan

DISCRETE FINITE STATE RANDOM FIELDS AND THEIR REDUCED VERSIONS AS INFORMATION SOURCES

Martin Janžura
Prague

Key words: Random field, reduced random field, entropy rate, asymptotic rate, scanning.

ABSTRACT

The multidimensional version of McMillan's theorem (Föllmer (1973)) is used to establish the entropy rate and the asymptotic rate of the information source represented by a stationary random field. Provided we are given only a reduced version of the random field (the reduced parameter set being rectangular) similar results are obtained by means of scanning. The relations between the entropy rate and the asymptotic rate, resp., of the reduced and original field are stated.

INTRODUCTION

We consider a (discrete finite state) stationary random field to act as an information source. The main results concerning the entropy rate and the asymptotic rate of such sources are derived in Section 1.

The idea of Section 2 may be illustrated as follows.

Let us imagine the first component of the parameter vector to be discrete time and the others to be some space coordinates. As, mostly, the observation region is limited because of a priori restriction of the observation capabilities only a reduced version of the random field actually must be considered. Two special cases are treated in Section 2: either all the values from the "observation rectangle" may be obtained at given instant or just one of them.

1. ENTROPY RATE AND ASYMPTOTIC RATE OF A STATIONARY RANDOM FIELD

Let $\{X_t\}_{t\in T}$ be a stochastic process which assumes values in a finite set S whose parameter set T is the d-dimensional lattice Z^d. For every $V \subset T$ we denote by $\mathcal{F}_V$ the σ-algebra of subsets of $\Omega = S^T$ generated by the projection $p_V: \omega \to \omega_V = \{\omega_t\}_{t\in V}$. By random field we mean a distribution of the stochastic process $\{X_t\}_{t\in T}$, i. e. a probability measure on the space $(\Omega, \mathcal{F}_T)$.

For every point $t\in T$ let θ_t be the corresponding shift, i. e. the transformation on Ω defined through: $[\theta_t(\omega)]_{t_1} = \omega_{t+t_1}$.
Then we set $\mathcal{S} = \{E\in \mathcal{F}_T;\ \theta_t^{-1}E = E$ for every $t\in T\}$.

In the sequel we shall suppose that μ is a stationary random field, i. e. invariant with respect to all the shifts:

$$\mu\theta_t^{-1} = \mu \quad \text{for every } t\in T.$$

For $a^1, a^2 \in T$ we write $a^1 \le a^2$ iff $a_i^1 \le a_i^2$ for every $i=1,\ldots,d$; $V(a^1,a^2) = \{t\in T;\ a^1 \le t < a^2\}$ (for the sake of simplicity $V(a) = V(0,a)$). Furthermore we introduce the lexicographical ordering "$\lessdot$" on T and put $T^- = \{t\in T; t\lessdot 0\}$.

For $V\subset T$ we denote $|V| = \text{card } V$.

We shall write $a\to\infty$ iff $\min_{i=1,\ldots,d} a_i \to \infty$.

By "log" we mean the natural logarithm.

<u>Theorem 1.1.</u> If μ is a stationary random field then it holds

$$\lim_{a\to\infty} -\frac{1}{|V(a)|}\log\mu(\omega_{V(a)}) = h_\mu \quad \text{in the mean } [\mu]$$

where $h_\mu = E_\mu[-\log[\sum_{s\in S}\chi_{\{\omega_0=s\}}\cdot E_\mu[\chi_{\{\omega_0=s\}}|\mathcal{F}_{T^-}]|\mathcal{S}]$ a.s. $[\mu]$ is non-negative invariant essentially bounded function.

<u>Proof.</u> As to the convergence cf. Föllmer (1973), relation 2.8. The properties of the function h_μ are evident except the essential boundedness which follows from the ergodic representation of invariant measures, cf. Farrell (1962).

<u>Theorem 1.2.</u> Let $H(\mu_{V(a)}) = E_\mu[-\log\mu(\omega_{V(a)})]$ be the entropy of the measure $\mu_{V(a)} = {}^{\mu}/F_{V(a)}$. Then $\lim_{a\to\infty} \frac{1}{|V(a)|} H(\mu_{V(a)}) = E_\mu[h_\mu]$ where $\frac{1}{|V(a)|} H(\mu_{V(a)})$ is a nonincreasing function of "a" with respect to the ordering "≤".

<u>Proof.</u> The convergence follows immediately from Theorem 1.1. To prove the monotonicity let us take $a = (a_1,\ldots,a_d)$, $b = (a_1,\ldots,a_d-1)$. We denote $W_k = \{t\in V(a);\ 0 \le t_i < a_i,\ i=1,\ldots,d-1,\ t_d = k\}$; it means $V(a) = \bigcup_{k=0}^{a_d-1} W_k$, $V(b) = \bigcup_{k=0}^{a_d-2} W_k$. Then $\{Y_k = \{X_t\}_{t\in W_k}\}_{k=-\infty}^{\infty}$ is a stationary random process (which assumes values in S^{W_0}) and according to well-known theorem it holds $\frac{1}{a_d} H(\mu_{V(a)}) = \frac{1}{a_d-1} H(\mu_{V(b)})$ which is sufficient to finish the proof.

<u>Remark.</u> Having already proved the monotonicity we actually do not need McMillan's theorem to prove the convergence.

<u>Definitions.</u> The quantities $H(\mu) = E_\mu[h_\mu]$ and $\mathcal{H}(\mu) = \text{ess}_\mu \sup h_\mu$ (both of them being finite owing to Theorem 1.1.) will be called the "entropy rate" or the "asymptotic rate", respectively. Furthermore in accordance with Winkelbauer (1964) for every $\varepsilon > 0$, $a > 0$, we define the quantity $L_a(\varepsilon,\mu) = \min\{\text{card } E;\ E\subset S^{V(a)}, \mu(E)>1-\varepsilon\}$ called the "ε-length".

<u>Theorem 1.3.</u> If $\mathcal{H}(\mu)$ is the asymptotic rate of the stationary random field μ then it holds

1. $\limsup_{a\to\infty} \frac{1}{|V(a)|} \log L_a(\varepsilon,\mu) \le \mathcal{H}(\mu)$ for any $0<\varepsilon<1$,
2. $\lim_{\varepsilon\to 0} \liminf_{a\to\infty} \frac{1}{|V(a)|} \log L_a(\varepsilon,\mu) \ge \mathcal{H}(\mu)$.

<u>Proof.</u> The first statement may be easily concluded from Theorem 1.1. To prove the second statement let us take $\lambda > 0$. If $h_\mu \ge \mathcal{H}(\mu) - \lambda$ a.s. $[\mu]$ then for any $\varepsilon\in(0,1)$ it may be shown $L(\varepsilon) = \liminf_{a\to\infty} \frac{1}{|V(a)|} \log L_a(\varepsilon,\mu) \ge \mathcal{H}(\mu) - \lambda$. If not we put $\mu^1(.) = \frac{\mu(.\cap Z_\lambda)}{\mu(Z_\lambda)}$ where $Z_\lambda = \{\omega: h_\mu(\omega) \ge \mathcal{H}(\mu) - \lambda\}$.
Evidently $\mu(Z_\lambda) > 0$ and moreover the set Z_λ may be taken to be from F_{T^-}. Thus it must hold $h_\mu = h_{\mu^1}$ a. s. $[\mu^1]$ and therefore $h_{\mu^1} \ge \mathcal{H}(\mu) - \lambda$ a. s. $[\mu^1]$. Since the definition of the ε-length implies $L_a(\frac{\varepsilon}{\mu(Z_\lambda)},\mu^1) \le L_a(\varepsilon,\mu)$ the inequality $L(\varepsilon) \ge \mathcal{H}(\mu) - \lambda$ holds for $\varepsilon \le \mu(Z_\lambda)$. Obviously, $L(\varepsilon)$ is nondecreasing for $\varepsilon \to 0$.

2. ENTROPY RATE AND ASYMPTOTIC RATE OF REDUCED RANDOM FIELDS

For $0 < a^o = (a_2^o,\dots,a_d^o) \in Z^{d-1}$ let $g: V(a^o) \to \{0,\dots,|V(a^o)|-1\}$ be a one-to-one mapping (way of scanning).

If we denote $b_n^1 = \left(\left\lfloor \frac{n}{|V(a^o)|}\right\rfloor, g^{-1}(n-|V(a^o)|\left\lfloor\frac{n}{|V(a^o)|}\right\rfloor)\right)$, $b_n^2 = (n, g^{-1}(n-|V(a^o)|\left\lfloor\frac{n}{|V(a^o)|}\right\rfloor))$ (by the notation $\lfloor x\rfloor$ we mean the largest integer less than or equal to x) then it holds

1. $b_n^2 = b_{k_n}^1$, $k_n = n + |V(a^o)|\,(n-\left\lfloor\frac{n}{|V(a^o)|}\right\rfloor)$;
2. $\{b_n^1\}_{n=-\infty}^{\infty} = T_1 = \{a\in T;\ 0 \le a_i < a_i^o,\ i=2,\dots,d\}$,

   $\{b_n^2\}_{n=-\infty}^{\infty} = T_2 = \{a\in T_1;\ a_1 = g(a_2,\dots,a_d) + k|V(a^o)|,\ k\in Z\}$;
3. $b^1_{n+|V(a^o)|} = b_n^1 + (1,0,\dots,0)$, $b^2_{n+|V(a^o)|} = b_n^2 + (|V(a^o)|,0,\dots,0)$.

Thus for $j \in \{1,2\}$ the stochastic process $\{Y_n^j = X_{b_n^j}\}_{n=-\infty}^{\infty}$ with the distribution $\mu^j = \mu/F_{T_j}$ is periodic with the period $|V(a^o)|$.

We denote $V_j^n = \{b_k^j;\ k=0,\dots,n-1\}$.

<u>Theorem 2.1.</u> For $j \in \{1,2\}$ there exists a nonnegative essentially bounded F_{T_j} - measurable invariant (under the corresponding shift) function h^j such that $\lim_{n\to\infty} -\frac{1}{n}\log \mu(\omega_{V_j^n}) = h^j$ in the mean $[\mu]$.

<u>Proof.</u> The statement follows from Theorem 1.1. in connection with Theorem 2.3. ofJacobs (1959).

<u>Remark.</u> To prove the preceding theorem it is sufficient to suppose that the measure μ is invariant under the shift $\theta_{(1,0,\dots,0)}$ only. As the measure μ^j and the function h^j depend on the choice of "a^o" we should more exactly write μ^{j,a^o} or $h^j_{a^o}$, respectively. Besides, for fixed a^o the function $h^2_{a^o}$ apparently depends on the way of scanning while the function $h^1_{a^o}$ does not.

For $j \in \{1,2\}$ we shall denote by $H(\mu^{j,a^o}) = E_\mu[h^j_{a^o}]$ and $\mathcal{H}(\mu^{j,a^o}) = \operatorname{ess\,sup}_\mu h^j_{a^o}$ the corresponding entropy rate or asymptotic rate, respectively.

<u>Theorem 2.2.</u> If $0 < k \in Z$, $a^k = (k,a_2^o,\dots,a_d^o)$ then $H(\mu^{1,a^o}) = \lim_{k\to\infty}\frac{1}{|V(a^k)|}H(\mu_{V(a^k)})$ is nonincreasing function of "a^o" with respect to the ordering "$\le$". Moreover $H(\mu) = \lim_{a^o\to\infty} H(\mu^{1,a^o})$.

Proof. The statement follows from Theorem 2.1 and Theorem 1.2.

Theorem 2.3. If $0 < k_i \in Z$, $a_i^1 = k_i . a_i^o$ for $i=2,\dots,d$ then it holds

$$H(\mu) \le H(\mu^{1,a^1}) \le H(\mu^{1,a^o}).$$

Proof. Lemma 1.1. of Wilkelbauer (1964) implies

$$\frac{1}{|V(a)|} \log L_a(\varepsilon,\mu) \le \frac{1}{k.|V(a^1)|} \log L_{k.|V(a^1)|}(\varepsilon_1,\mu^{1,a^1}) \le$$

$$\le \frac{1}{k.|V(a^o)|} \log L_{k.|V(a^o)|}(\varepsilon_2,\mu^{1,a^o})$$

where $a = (k,k_1 a^1)$, $\varepsilon_1 = \frac{\varepsilon}{k_1^{d-1}}$, $\varepsilon_2 = \frac{\varepsilon_1}{\prod_{i=2}^{d} k_i}$.

The proof of the first inequality hence follows from Theorem 1.3 since its first statement remains valid for a periodic process as well.

If we define $\nu = \sum_{k=0}^{|V(a^1)|-1} \mu^{1,a^1} \tau^{-k}$ (τ being the corresponding shift) then it holds $H(\mu^{1,a^1}) \le H(\nu)$ (which follows from Theorem 2.3 of Jakobs (1959)) and $\log L_n(\varepsilon,\nu) \le \log L_n(\frac{\varepsilon}{2},\mu^{1,a^1}) + \log |V(a^1)| + |V(a^1)| . \log|S|$ (which follows from Lemma 1.3 of Winkelbauer (1964)).

The validity of the second statement of Theorem 1.3 for the stationary process ν is sufficient to finish the proof.

Theorem 2.4. Let $0 < a^o \in Z^{d-1}$ and the way of scanning g on $V(a^o)$ be fixed then $H(\mu^{1,a^o}) \le H(\mu^{2,a^o})$, $H(\mu^{1,a^o}) \le H(\mu^{2,a^o})$ hold.

Proof. If we define the transformation $T_n : (a_1,\dots,a_d) \mapsto$

$\mapsto (a_1 + 1 [\mathrm{mod}\ n], a_2,\dots,a_d)$ on $V_1^{n.|V(a^o)|}$ then $V_1^{n.|V(a^o)|} =$

$$= \bigcup_{i=0}^{|V(a^o)|-1} T_n^i(V_2^n).$$

The relation above yields the inequality

$$H(\mu^{1,a^o}_{n.|V(a^o)|}) \le |V(a^o)|\ H(\mu_n^{2,a^o}) + \frac{(|V(a^o)|-1)\ |V(a^o)|}{2} \log|S|,$$

and similarly (in connection with Lemma 1.1 of Winkelbauer (1964))

$$\log L_{n.|V(a^o)|}(\varepsilon,\mu^{1,a^o}) \le |V(a^o)| \log L_n(\frac{\varepsilon}{|V(a^o)|^2},\mu^{2,a^o}) + \frac{(|V(a^o)|-1)|V(a^o)|}{2} \log|S|$$

follows.

Then the rest of the proof is a consequence of Theorem 1.2 and Theorem 1.3.

Acknowledgement: The author wishes to express his appreciation to Dr. A. Perez for the most helpful discussions and supervising the preparation of this paper.

REFERENCES

Farrell R. H. (1962): Representation of Invariant Measures. Illinois J. Math. 6, 447-467.

Föllmer H. (1973): On Entropy and Information Gain in Random Fields. Z. Wahrscheinlichkeitstheorie verw. Gebiete 26, 207-217.

Jacobs K. (1959): Die Übertragung diskreter Informationen durch periodische und pastperiodische Kanäle. Math. Annalen Bd 137, 125-135.

Winkelbauer K. (1964): On Discrete Information Sources. Trans. of the Third Prague Conference, Prague 1962, Academia, Prague, 765-830.

Czechoslovak Academy of Sciences
Institute of Information Theory
and Automation
Pod vodárenskou věží 4
182 08 Praha 8 - Libeň
Czechoslovakia

STRATEGICAL TEST

- A GENERALIZATION OF THE WALD´S SEQUENTIAL TEST

Radim Jiroušek

Prague

Key words: Sequential tests, decision-making.

ABSTRACT

It is well known that the Wald´s sequential test terminates with probability one (and simultaneously is in some sense optimal) if a sampled sequence of random variables is independent and identically distributed (i.i.d.). In this paper we shall deal with the Wald´s sequential test constructed for a special sequence of dependent and unidentically distributed random variables. The speciality of the studied sequence of random variables lies in the fact that a controlled mixture of two i.i.d. sequences of random variables is considered.

INTRODUCTION

The studied model is fitted for the following example. An investigator examines a defective population of items. Each item may be put to one of two different examinations but never to both (e.g.the examinations are destructive). Using Wald's sequential probability ratio test he should decide in which way the population is injured. The questions arise, in which way he should control a choice of the examination at every step, and moreover, whether he can influence by the means of this control the probability of terminating the Wald's sequential test.

STRATEGICAL TEST

Let X denote sample space of all random variables which will be dealt with. Let N be the set of all positive integers $\{1,2,3,\ldots\}$.

<u>Definition:</u> A function

$$\mu : N \times X^{\infty} \to \{1,2\}$$

will be called a strategical function if it is *unpredicting*; i. e. if for all $i \in N$ and $x_k' \neq x_k$ $k \geq i$ it holds:

$$\mu(i,x_1,x_2,\ldots,x_{i-1},x_i,x_{i+1},\ldots) = \mu(i,x_1,x_2,\ldots,x_{i-1},x_i',x_{i+1}',\ldots).$$

Regarding this, a shortened notation $\mu(i,x_1,\ldots,x_{i-1})$ will be used throughout the paper.

Let H_o and H_1 be two alternative hypotheses concerning the probability distributions of a sequence

(1) $$(\xi_1^1,\xi_1^2),\ (\xi_2^1,\xi_2^2),\ (\xi_3^1,\xi_3^2),\ (\xi_4^1,\xi_4^2),\ \ldots$$

of pairs of random variables. (Each pair corresponds with the pair of two alternative examinations from the example mentioned above.) Using some fixed strategical function μ one can transform each sequence

$$(x_1^1,x_1^2),\ (x_2^1,x_2^2),\ (x_3^1,x_3^2),\ (x_4^1,x_4^2),\ \ldots$$

-which is a sequence of realizations of random variables (1) - into a sequence

$$x_1,x_2,x_3,x_4,\ \ldots \in X^{\infty}$$

in the following recursive way:

(2) $$x_i = x_i^{\mu(i,x_1,\ldots,x_{i-1})}$$

This transformation corresponds to the way the sampled random variables (ξ_i^1 or ξ_i^2) are being chosen at all the steps. During the first step, x_1 is observed as a realization of the random variable $\xi_1^{\mu(1)}$. When the outcomes $x_1,\ldots,x_{i-1}$ of the first i-1 steps are known, the random variable $\xi_i^{\mu(i,x_1,\ldots,x_{i-1})}$ will be sampled at the i-th step. Thus, in the described way, the outcomes $x_1,x_2,\ldots$ of the mixed sequence of random variables are obtained.

Now, let us state, what is understood under the term *strategical sequential test*.

Let A,B be constants ($0<B<1<A<\infty$) and μ be a strategical function. Let (ξ_1^1,ξ_1^2), (ξ_2^1,ξ_2^2),... be an infinite sequence of pairs of random variables and (x_1^1,x_1^2), (x_2^1,x_2^2),... a sequence of their realizations. Let $x_1,x_2,\ldots$ be the mixed sequence obtained according to the procedure (2). The Wald's sequential probability ratio test with boundaries (A,B) applied to the sequence $x_1,x_2,\ldots$ will be understood to be a strategical sequential test applied to the sequence (x_1^1,x_1^2), (x_2^1,x_2^2),..., or, sometimes as well, strategical sequential test applied to the sequence (ξ_1^1,ξ_1^2), (ξ_2^1,ξ_2^2),... This strategical test is defined by boundaries (A,B) and strategical function μ, or, as it will be used, by the triplet (A,B,μ).

Expressed in a different way, the strategical sequential test is performed according to the following simple scheme .

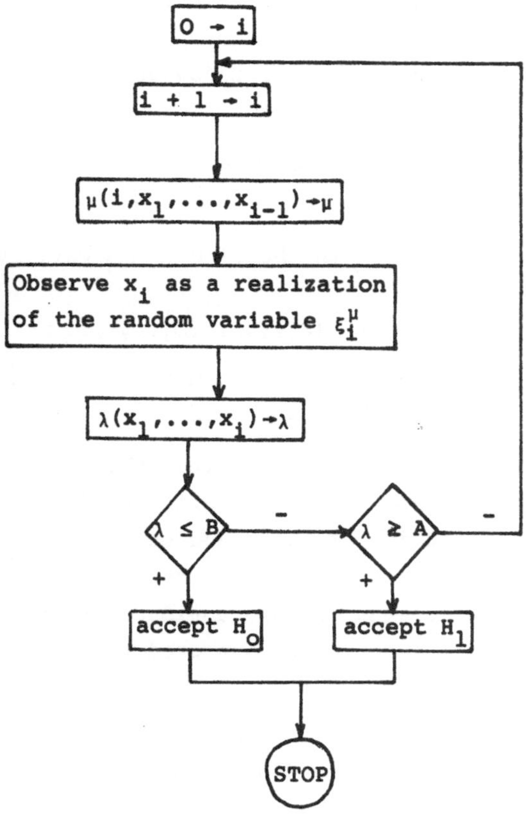

<u>Definition:</u> The strategical sequential test defined by the triplet (A,B,μ) applied to the sequence (ξ_1^1,ξ_1^2), (ξ_2^1,ξ_2^2),... is convergent if it terminates with probability one under both distinguished hypotheses.

INDEPENDENT REPETITION OF THE PAIR (ξ^1,ξ^2)

Let ξ^1,ξ^2 be two different random variables and H_o, H_1 be two alternative hypotheses concerning the probability distributions of both the random variables. By $p_j(x/H_k)$ a probability density of ξ^j under H_k $(k=0,1;j=1,2)$ will be denoted.

Furtner, for both $j=1,2$ denote

(3)
$$z_j(x) = \log \frac{p_j(x/H_1)}{p_j(x/H_o)}$$

and for $j=1,2$ and $k=0,1$ denote

$$E(|z_j|/H_k) = \int |z_j(x)|p_j(x/H_k)\, dx$$

the expectation of an absolute value of the random variable z_j under the hypothesis H_k.

Throughout the paper, the existence of all four average values $E(|z_j|/H_k)$ is supposed and moreover

(4)
$$0 < E(|z_j|/H_k) < +\infty$$

for all of them.

In correspondence with the heading of the paragraph, the considered sequence

$$(\xi_1^1,\xi_1^2),\ (\xi_2^1,\xi_2^2),\ (\xi_3^1,\xi_3^2),\ldots$$

is supposed to be a sequence of independent repetitions of the pair (ξ^1,ξ^2).

Considering that the aim is to study the convergence of the strategical sequential test, the attention will now be paid to a probability distribution of the mixed sequences $x_1,x_2,\ldots$ obtained according to the procedure (2). Note that this probability distribution naturally depends on the considered strategical function. Since the strategical function is supposed to be fixed it will not be incorporated into a notation.

The probability density concerning the first i variables under the hypothesis H_k will be denoted by $p(x_1,\ldots,x_i/H_k)$.

Similarly, $p(x_i/x_1,\dots,x_{i-1},H_k)$ will denote the probability density of the i-th variable over the space X under the hypothesis H_k along with the fact the first i-1 variables have been $x_1,\dots,x_{i-1}$.

As it has been told above, the random variable $\xi^{\mu(1)}$ is observed at the first step. Thus

$$p(x_1/H_k) = p_{\mu(1)}(x_1/H_k).$$

Similarly, when the outcomes $x_1,\dots,x_{i-1}$ of the first i-1 steps are known the random variable $\xi^{\mu(i,x_1,\dots,x_{i-1})}$ is observed at the i-th step, and therefore

$$p(x_i/x_1,\dots,x_{i-1},H_k) = p_{\mu(i,x_1,\dots,x_{i-1})}(x_i/H_k). \tag{5}$$

For the simplicity sake, let us understand $p(x_i/x_1,\dots,x_{i-1},H_k)$ to be $p(x_1/H_k)$ for i=1, and similarly (for i=1)

$$\mu(1) = \mu(i,x_1,\dots,x_{i-1}).$$

Under this agreement the equality (5) holds for all i=1,2,3,...

Wald's sequential test is based on computing a likelihood ratio at every step. If $x_1,\dots,x_n$ have been observed at the first n steps then the likelihood ratio λ is at the n-th step equal to the following expression:

$$\lambda(x_1,\dots,x_n) = \frac{p(x_1,\dots,x_n/H_1)}{p(x_1,\dots,x_n/H_o)} = \prod_{i=1}^{n} \frac{p(x_i/x_1,\dots,x_{i-1},H_1)}{p(x_i/x_1,\dots,x_{i-1},H_o)} .$$

According to the equalities (5) and (3) it can be modified in the following way:

$$\log \lambda(x_1,\dots,x_n) = \log \prod_{i=1}^{n} \frac{p_{\mu(i,x_1,\dots,x_{i-1})}(x_i/H_1)}{p_{\mu(i,x_1,\dots,x_{i-1})}(x_i/H_o)} =$$

$$= \sum_{i=1}^{n} z_{\mu(i,x_1,\dots,x_{i-1})}(x_i).$$

CONVERGENCE OF THE STRATEGICAL TEST

An importance of the following theorem is in the fact that there are several assertions holding for Wald's sequential tests terminating with probability one.

An idea of the proof presented is due to A. Perez.

<u>Theorem:</u> A strategical sequential test (A,B,μ) applied to a sequence of i.i.d pairs of random variables is convergent if inequalities (4) hold.

<u>Proof:</u> Convergence of the strategical test means that

$$\Pr(\text{test does not terminate during the first } n \text{ steps}/H_k) \to 0$$

for increasing n and both k=0,1.

According to the definition of the stopping rule of the Wald's sequential test (and thus of the strategical test, too)

$$\begin{aligned}&\Pr(\text{test does not terminate during the first } n \text{ steps}/H_k) =\\ &= \Pr((\forall\ i=1,\dots,n\ \lambda(x_1,\dots,x_i) \in (B,A))/H_k) \le\\ &\le \Pr((\lambda(x_1,\dots,x_n) \in (B,A))/H_k).\end{aligned}$$

Using the introduced probability density, the last expression can be expressed in the following form

$$\Pr((\lambda(x_1,\dots,x_n) \in (B,A))/H_k) = \int_{X^n(A,B)} p(x_1,\dots,x_n/H_k)\, dx_1\dots x_n \tag{6}$$

where $X^n(A,B) = \{(x_1,\dots,x_n):\lambda(x_1,\dots,x_n) \in (B,A)\}$. Let us estimate this expression for the hypothesis H_o. For the hypothesis H_1 it can be estimated analogically. Over the set $X^n(A,B)$ the following inequality holds

$$\lambda(x_1,\dots,x_n) = \frac{p(x_1,\dots,x_n/H_1)}{p(x_1,\dots,x_n/H_o)} > B\ ,$$

and therefore also

$$p(x_1,\dots,x_n/H_o) < B^{-1}\ p(x_1,\dots,x_n/H_1)\ .$$

That is why for all $0<t<1$ the expression (6) can be estimated

$$\begin{aligned}&\Pr((\lambda(x_1,\dots,x_n) \in (B,A))/H_o) <\\ &< \int_{X^n(A,B)} p(x_1,\dots,x_n/H_o)^t\ B^{t-1}\ p(x_1,\dots,x_n/H_1)^{1-t}\ dx_1 \dots x_n \le\\ &\le B^{t-1} \int p(x_1,\dots,x_n/H_o)^t\ p(x_1,\dots,x_n/H_1)^{1-t}\ dx_1 \dots x_n \qquad (7)\end{aligned}$$

The expression (7) will be estimated by induction. But first, let us note that the expression

$$\int p_j(x/H_o)^t\ p_j(x/H_1)^{1-t} dx$$

is a widely known generalized entropy which is according to the assumption (4) less than 1 for all $0<t<1$ (c.f. e.g. Vajda (1970)). Let us denote

$$H_{MAX}(t) = \max_{j=1,2} \left(\int p_j(x/H_o)^t\, p_j(x/H_1)^{1-t} dx\right) < 1.$$

Now, we can easily estimate the expression (7) for n = 1. According to the equality (5)

$$\int p(x_1/H_o)^t p(x_1/H_1)^{1-t} dx_1 = \int p_{\mu(1)}(x/H_o)^t\, p_{\mu(1)}(x/H_1)^{1-t} dx \le H_{MAX}(t).$$

Let us, now, consider an arbitrary $n > 1$ and suppose that

$$\int p(x_1,\dots,x_i/H_o)^t\, p(x_1,\dots,x_i/H_1)^{1-t} dx_1\dots x_i \le (H_{MAX}(t))^i$$

holds for all $0<i<n$.

$$\int p(x_1,\dots,x_n/H_o)^t\, p(x_1,\dots,x_n/H_1)^{1-t}\, dx_1\dots x_n =$$

$$\int p(x_1,\dots,x_{n-1}/H_o)^t p(x_1,\dots,x_{n-1}/H_1)^{1-t} p(x_n/x_1,\dots,x_{n-1},H_o)^t$$
$$p(x_n/x_1,\dots,x_{n-1},H_1)^{1-t} dx_1\dots x_n =$$

$$= \int p(x_1,\dots,x_{n-1}/H_o)^t\, p(x_1,\dots,x_{n-1}/H_1)^{1-t}$$
$$\left[\int p_{\mu(n,x_1,\dots,x_{n-1})}(x/H_o)^t\, p_{\mu(n,x_1,\dots,x_{n-1})}(x/H_1)^{1-t} dx\right] dx_1\dots x_{n-1} \le$$

$$\le H_{MAX}(t) \int p(x_1,\dots,x_{n-1}/H_o)^t\, p(x_1,\dots,x_{n-1}/H_1)^{1-t}\, dx_1\dots x_{n-1} \le$$

$$\le (H_{MAX}(t))^n \to 0.$$

Q.E.D.

APPENDIX

The paper is finished by a comment concerning the number of alternative random variables. For a simplicity sake, through the paper two alternative variable ξ^1 and ξ^2 have been considered. For that reason, all considered strategical functions have been supposed to have values from the set {1,2}. Naturally, generalization to more than two variables can be studied. One can consider an arbitrary number(finite or infinite) of alternative random variables. When infinite number of variables $(\xi^j)_{j\in N}$ is studied one must ensure validity of the following expression:

$$(\exists t)\ (0<t<1)\left(\sup_{j\in N}\left(\int p_j(x/H_o)^t\, p_j(x/H_1)^{1-t} dx\right) < 1\right)$$

to keep the theorem true.

REFERENCES

Wald Abraham (1947): Sequential Analysis. John Wiley.

Rao C. R. (1965): Linear Statistical Inference. John Wiley.

Vajda Igor (1970): On the Amount of Information Contained in a Sequence of Independent Observations. Kybernetika 6, No. 2, pp. 306-324.

Perez Albert (1972): Generalization of Chernoff's result on the asymptotic discernibility of two random processes. In: Colloquia math. societatis János Bolyai; 9. European meeting of statisticians, Budapest.

Czechoslovak Academy of Sciences
Institute of Information Theory
and Automation
Pod vodárenskou věží 4
182 08 Praha 8 - Libeň
Czechoslovakia

SEQUENCES OF STOCHASTIC PROGRAMMING PROBLEMS WITH INCOMPLETE INFORMATION

Vlasta Kaňková
Prague

Key words: Stochastic optimization problem, stochastic programming problem, time interval, time repeated problems, mathematical expectation, optimal value, optimal solution, estimates, stochastic decision problem, sample distribution function, conditional probability, convergence rate

ABSTRACT

Solving stochastic optimization problems with respect to a time interval we get often rather better results then solving the corresponding separated problems. This is caused by a stochastic dependency of the random elements in problems repeated in a time.

In this paper we shall deal with the former approach. Moreover, we shall assume that the optimum is sought with respect to the mathematical expectation. Then if probability laws are known the problem leads to several deterministic parametric optimization problems [Lemma 1]. However, it happens sometimes that the probability laws are unknown. Then it arises a stochastic decision problem how to find out estimates of the optimal value and the optimal solution. The aim of this paper is to present some of these estimates. They will be based on N realizations of some random vectors. We shall study their convergence rates too.

To get our results we use the papers Dupačová (1976), Kaňková (1974), Kaňková (1978). More precisely, we generalize the results for problems investigated without the time dependency to some problems with the dependency.

INTRODUCTION

Let (Ω,S,P) be a probability space where P is a complete probability measure,

$\xi_i(\omega) = \xi_i$, i=1,2,... be s-dimensional random vectors defined on (Ω,S,P),

$g_1(z,x)$ be a real valued, continuous bounded function defined on $Z \times K$, where $K \subset E_r$ is a non-empty, compact, convex set and $Z \subset E_s$ satisfies the condition $P\{\omega: \xi_i(\omega) \in Z\} = 1$, $i=1,2,\dots$,

$g_i(z_1,z_2,x)$, $i=2,3,\dots,N$ be real valued, continuous, bounded functions defined on $(Z \times Z \times K)$,

$K_i(z_{i-1}) = K_i$, $i=2,\dots,N$ be mappings of Z into the space of non-empty, compact, convex subsets of K

$K_1 \subset K$ be a non-empty, compact, convex set

(E_i denotes the i-dimensional Euclidean space).

We shall introduce the stochastic optimization problem (w.r.t. the discrete time interval $1 \div N$) as a problem of finding $(x_1,x_2,\dots,x_N)$, $x_1 \in K_1$, $x_i = X_i(\xi_{i-1}) \in K_i(\xi_{i-1})$, $i=2,\dots,N$ for which

$$E\{g_1(\xi_1,x_1) + \sum_{i=2}^{N} g_i(\xi_{i-1},\xi_i,x_i)\} \tag{1}$$

is maximal.

The aim of this paper is to suggest some estimates of the optimal value and the optimal solutions of (1) under the condition that the probability laws of the random sequence $\{\xi_i\}_{i=1}^{\infty}$ are unknown. These estimates will be obtained on the realization bases of the first n terms of a suitable random sequence.

SOME AUXILARY ASSERTIONS

Lemma 1. If for every $i \in \{2,\dots,N\}$

(i) $\bar{x}_i(\xi_{i-1}) = \bar{x}_i$ is a solution of the problem to find out

$$\max_{x_i \in K_i(\xi_{i-1})} E_{\xi_i|\xi_{i-1}} g_i(\xi_{i-1},\xi_i,x_i),$$

(ii) $E_{\xi_i|\xi_{i-1}} g_i(\xi_{i-1},\xi_i,\bar{x}_i(\xi_{i-1}))$ is a measurable function w.r.t. the σ-algebra given by $\xi_1,\dots,\xi_{i-1}$,

and if $\bar{x}_1$ is a solution of the problem to find out

$$\max_{x_1 \in K_1} Eg_1(\xi_1,x_1)$$

then $(\bar{x}_1,\bar{x}_2(\xi_1),\dots,\bar{x}_N(\xi_{N-1}))$ is a solution of the problem given by (1).

($E_{\cdot|\cdot}$ denotes the conditional expectation).

Proof. From the assumptions we get for $i=2,\dots,N$, $x_i \in K_i(\xi_{i-1})$, $i=2,\dots,N$

$$Eg_i(\xi_{i-1},\xi_i,x_i) \le E \max_{x_i \in K_i(\xi_{i-1})} E_{\xi_i|\xi_{i-1}} g_i(\xi_{i-1},\xi_i,x_i).$$

But as

$$E\sum_{i=2}^{N} g_i(\xi_{i-1},\xi_i,\bar{x}_i(\xi_{i-1})) = \sum_{i=2}^{N} E \max_{x_i \in K_i(\xi_{i-1})} g_i(\xi_{i-1},\xi_i,x_i)$$

too, it is easy to see that the assertion of Lemma is valid.

Let, further $F_i(.)$, $i=1,2,\ldots,N$ be s-dimensional distribution functions,

$f(z_1,z_2)$ be a real valued, measurable function defined on $E_s \times E_s$.

We shall assume

a) $F_1(z)$ is the distribution function of the $\xi_1(\omega)$,

b) $P_{\xi_i|\xi_{i-1}}\{\omega:\xi_i(\omega)<z_2|\xi_{i-1}(\omega)=z_1\} = F_i(f(z_1,z_2))$, $i=2,\ldots,N$

($P_{\cdot|\cdot}$ denotes the conditional probability measure).

Remark. To get an example of ξ_i, $i=1,2,\ldots,N$ fulfilling a),b) we put $\xi_i = \xi_{i-1} + \eta_i$, $i=2,\ldots,N$, $\xi_1 = \eta_1$, where η_i, $i=1,2,\ldots,N$ are independent random vectors with the distribution functions equal $F_i(.)$. Then $f(z_1,z_2) = z_2 - z_1$.

THE MAIN RESULTS

Let $\eta_i^j(\omega) = \eta_i^j = [\eta_{i1}^j,\ldots,\eta_{is}^j]$, $j=1,2,\ldots$ be for every $i=1,2,\ldots,N$ a sequence of independent s-dimensional random vectors defined on (Ω,S,P). We shall assume that for every $i=1,2,\ldots,N$ the distribution function of η_i^j, $j=1,2,\ldots$ is equal to $F_i(z)$.

We shall define the functions $U_i^j(z,\omega) = U_i^j(z)$, $F_n^i(z,\omega) = F_n^i(z)$ for $z = [z_1,\ldots,z_s] \in E_s$, $\omega \in \Omega$, $i=1,2,\ldots,n$, $n=1,2,\ldots$ by

$$U_i^j(z,\omega) = 1 \Leftrightarrow \eta_{i\ell}^j < z_\ell \text{ for all } \ell=1,2,\ldots,s,$$

$$U_i^j(z,\omega) = 0 \Leftrightarrow \eta_{i\ell}^j \geq z_\ell \text{ at least for one } \ell \in \{1,2,\ldots,s\},$$

$$F_n^i(z,\omega) = \frac{1}{n}\sum_{j=1}^{n} U_i^j(z,\omega).$$

Lemma 2. Let the assumptions (ii),a),b) be fulfilled. If we define

$$I_n^i(x_i,z_{i-1},\omega) = \int g_i(z_{i-1},z_i,x_i)dF_n^i(f(z_{i-1},z_i),\omega) \qquad i=2,\ldots,N,$$

$$I_n^1(x_1,\omega) = \int g_1(z\ ,x_1)dF_n^1(z,\omega)$$

then

1) $P\{\omega: |\max_{x_1 \in K_1} I_n^1(x_1,\omega) - \max_{x_1 \in K_1} Eg_1(\xi_1,x_1)| \xrightarrow[(n\to\infty)]{} 0\} = 1$

and

2) $P\{\omega: |\max_{x_i \in K_i(\xi_{i-1})} I_n^i(x_i, \xi_{i-1}, \omega) -$

$- \max_{x_i \in K_i(\xi_{i-1})} E_{\xi_i|\xi_{i-1}} g_i(\xi_{i-1}, \xi_i, x_i)| \xrightarrow[(n\to\infty)]{} 0\} = 1, \; i=2,\dots,N$

Proof. According to Theorem 1 Kaňková (1974) it is

$$P\{\omega: \max_{x_i \in K_i(\xi_{i-1})} I_n^i(x_i, \xi_{i-1}, \omega) - \max_{x_i \in K_i(\xi_{i-1})} E_{\xi_i|\xi_{i-1}} g_i(\xi_{i-1}, \xi_i, x_i)| \xrightarrow[(n\to\infty)]{} 0 \mid \xi_{i-1} = z_1\} = 1$$

for every $z_1 \in Z$, $i=2,\dots,N$.

But from this and from elementary properties of conditional probabilities follows the assertion 2 of Lemma 2.

The assertion 1 follows directly from Theorem 1 Kaňková (1974).

From Lemma 2 we can get immediately the following Theorem .

Theorem 1. If the assumptions of Lemma 2are fulfilled then

$$P\{\omega: |\max_{x_1 \in K_1} I_n^1(x_1, \omega) + \sum_{i=2}^{N} \max_{x_i \in K_i(\xi_{i-1})} I_n^i(x_i, \xi_{i-1}, \omega) - \max_{x_1 \in K_1} Eg_1(\xi_1, x_1) - \sum_{i=2}^{N} \max_{x_i \in K_i(\xi_{i-1})} E_{\xi_i|\xi_{i-1}} g_i(\xi_{i-1}, \xi_i, x_i)| \xrightarrow[(n\to\infty)]{} 0\} = 1$$

Corollary 1. If the assumptions of Lemma 2 are fulfilled then

$$E\{\max_{x_1 \in K_1} I_n(x_1, \omega) + \sum_{i=2}^{N} \max_{x_i \in K_i(\xi_{i-1})} I_n^i(x_i, \xi_{i-1}, \omega)\} \xrightarrow[(n\to\infty)]{} E\{\max_{x_1 \in K_1} Eg_1(\xi_1, x_1) + \sum_{i=2}^{N} \max_{x_i \in K_i(\xi_{i-1})} E_{\xi_i|\xi_{i-1}} g_i(\xi_{i-1}, \xi_i, x_i)\}.$$

Proof. The assertion of Corollary follows from Theorem 1 and the Lebesgue's dominated convergence theorem.

Theorem 1 presentssome estimates of the optimal value. Now we shall study their convergence rate.

Theorem 2. Let the assumptions of Theorem 1 be fulfilled. If $g_1(z_1, x_1)$, $g_i(z_{i-1}, z_i, x_i)$, $i=2,\dots,N$ are for every z_{i-1}, $z_i \in Z$ Lipschitz functions of x_i with Lipschitz constant L_i, $i=1,2,\dots,N$ not depending on $z_i \in Z$, $i=1,2,\dots,N$, then for every $y \in E_1, y > 0$, it holds

$$P\{\omega: |\max_{x_1 \in K_1} I_n^1(x_1,\omega) + \sum_{i=2}^{N} \max_{x_i \in K_i(\xi_{i-1})} I_n^i(x_i,\xi_{i-1},\omega) - \max_{x_1 \in K_1} Eg_1(\xi_1,x_1) -$$
$$- \sum_{i=2}^{N} \max_{x_i \in K_i(\xi_{i-1})} E_{\xi_i|\xi_{i-1}} g_i(\xi_{i-1},\xi_i,x_i)| > y\} \le K_1 \cdot \exp\{-\frac{ny^2}{K_2}\}$$

where K_1, K_2 are real valued constants.

Proof. From Theorem 2 Kaňková (1978) we get for arbitrary $y \in E, y>0$

$$P\{\omega: |\max_{x_1 \in K_1} I_n(x_1,\omega) - \max_{x_1 \in K_1} Eg_1(\xi_1,x_1)| \le K_1^1 \exp\{-\frac{ny^2}{K_2^1}\},$$

$$P\{\omega: |\max_{x_i \in K_i(\xi_{i-1})} I_n^i(x_i,\xi_{i-1},\omega) -$$
$$- \max_{x_i \in K_i(\xi_{i-1})} E_{\xi_i|\xi_{i-1}} g_i(\xi_{i-1},\xi_i,x_i)| > y | \xi_{i-1} = z_1\} \le K_1^i \exp\{-\frac{ny^2}{K_2^i}\},$$
$$i=2,\dots,N, \quad z_1 \in Z$$

where K_1^i, K_2^i are real valued constants.

From this, from elementary properties of the conditional probability and from triangular inequality we get

$$P\{\omega: |\max_{x_1 \in K_1} I_n(x_1,\omega) + \sum_{i=2}^{N} \max_{x_i \in K_i(\xi_{i-1})} I_n(x_i,\xi_{i-1},\omega) -$$
$$- \max_{x_1 \in K_1} Eg_1(\xi_1,x_1) - \sum_{i=2}^{N} \max_{x_i \in K_i(\xi_{i-1})} E_{\xi_i|\xi_{i-1}} g_i(\xi_{i-1},\xi_i,x_i)| > y\} \le$$
$$\le P\{\omega: |\max_{x_1 \in K_1} I_n(x_1,\omega) - \max_{x_1 \in K_1} Eg_1(\xi_1,x_1)| > \frac{y}{N}\} +$$
$$+ \sum_{i=2}^{N} P\{\omega: \max_{x_i \in K_i(\xi_{i-1})} I_n^i(x_i,\xi_{i-1},\omega) - \max_{x_i \in K_i(\xi_{i-1})} E_{\xi_i|\xi_{i-1}} g_i(\xi_{i-1},\xi_i,x_i)| > \frac{y}{N}\} \le$$
$$\le N \cdot \max_{i \in \{1,2,\dots N\}} K_1^i \exp\{-\frac{ny^2}{N^2 \max_{i \in \{1,2,\dots,N\}} K_2^i}\}.$$

If we set $K_1 = N \cdot \max_{i \in \{1,2,\dots,N\}} K_1^i$, $K_2 = N^2 \cdot \max_{i \in \{1,2,\dots,N\}} K_2^i$

we see that the Theorem is true.

In the end let us present the following two remarks

1) Let a random sequence $\{\zeta_i\}_{i=1}^{\infty}$ has the same probability laws as $\{\xi_i\}_{i=1}^{\infty}$. Let further, $\{\eta_i\}_{i=1}^{\infty}$ be a sequence of independent, identically distributed random vectors. Then if, for example, it happens one of the following cases: a) $\zeta_i = A.\ \eta_i$, $i=1,2,\dots,N$, where A is a constant, b) $\zeta_i = \zeta_{i-1} + \eta_i$, $i=2,\dots,N$, $\zeta_1 = \eta_1$, c) $\zeta_i = \zeta_{i-1} \cdot \eta_i$, $i=2,\dots,N$, $\zeta_1 = \eta_1$, we can (under the assumptions

of this paper) to put $\eta_i^j = \eta_j$, i=1,2,...N j=1,2,... So we can find out the optimal value estimates on the base of the first n terms of the random sequence $\{\zeta_i\}_{i=1}^{\infty}$. Further, it can also be seen (Lemma 2) that the results of this paper can be used for the learning models as well.

2) We have presented some estimates of the optimal value. Of course, these estimates define some estimates of the optimal solution. Under rather general conditions they converge with probability one to the theoretical optimal solution. Using the inequalities for the functions strongly concave in the sense of (20) in Тарасенко (1980), we can study the rate converge of them too.

REFERENCES

Dupačová J. (1976): Experience in Stochastic Programming Models. In: IX International Symposium on Mathematical Programming, Budapest 1976.

Kaňková V. (1974): Optimal Solution of a Stochastic Optimization Problem with Unknown Parameters. In: Trans. of the Seventh Prague Conference, Prague 1974, Academia, Prague 1978.

Kaňková V. (1978): On Approximative Solution of Stochastic Optimization Problem. In: Trans. of the Eighth Prague Conference, Prague 1978. Academia, Prague 1978.

Loév M. (1959): Probability Theory. Second edition. D. von Nostrand Company. New York 1959.

Тарасенко Р. С. (1980): Об оценке скорости сходимости адаптивного метода случайного поиска. Проблемы случайного поиска 1980.8.

Czechoslovak Academy of Sciences
Institute of Information Theory
and Automation
Pod vodárenskou věží 4
182 08 Praha 8 - Libeň
Czechoslovakia